Wildland Fire in Ecosy...

Effects of Fire on Soil and Water

I0035016

Editors

Daniel G. Neary, Project Leader, Forestry Sciences Laboratory, Rocky Mountain Research Station, U.S. Department of Agriculture, Forest Service, Flagstaff, AZ 86001

Kevin C. Ryan, Project Leader, Fire Sciences Laboratory, Rocky Mountain Research Station, U.S. Department of Agriculture, Forest Service, Missoula, MT 59807

Leonard F. DeBano, Adjunct Professor, School of Renewable Natural Resources, University of Arizona, Tucson, AZ 85721

Authors

Jan L. Beyers, Research Ecologist, Pacific Southwest Research Station, U.S. Department of Agriculture, Forest Service, Riverside, CA 92507

James K. Brown, Research Forester, Systems for Environmental Management, Missoula, MT 59802 (formerly with Fire Sciences Laboratory, Rocky Mountain Research Station, U.S. Department of Agriculture, Forest Service).

Matt D. Busse, Research Microbiologist, Pacific Southwest Research Station, U.S. Department of Agriculture, Forest Service, Redding, CA 96001

Leonard F. DeBano, Adjunct Professor, School of Renewable Natural Resources, University of Arizona, Tucson, AZ 85721

William J. Elliot, Project Leader, Rocky Mountain Research Station, U.S. Department of Agriculture, Forest Service, Moscow, ID 83843

Peter F. Ffolliott, Professor, School of Renewable Natural Resources, University of Arizona, Tucson, AZ 85721

Gerald R. Jacoby, Professor Emeritus, Eastern New Mexico University, Portales, NM 88130

Jennifer D. Knoepp, Research Soil Scientist, Southern Research Station, U.S. Department of Agriculture, Forest Service, Otto, NC 28763

Johanna D. Landsberg, Research Soil Scientist (Retired), Pacific Northwest Research Station, U.S. Department of Agriculture, Forest Service, Wenatchee, WA 98801

Daniel G. Neary, Project Leader, Rocky Mountain Research Station, U.S. Department of Agriculture, Forest Service, Flagstaff, AZ 86001

James R. Reardon, Physical Science Technician, Rocky Mountain Research Station, U.S. Department of Agriculture, Forest Service, Missoula, MT 59807

John N. Rinne, Research Fisheries Biologist, Rocky Mountain Research Station, U.S. Department of Agriculture, Forest Service, Flagstaff, AZ 86001

Peter R. Robichaud, Research Engineer, Rocky Mountain Research Station, U.S. Department of Agriculture, Forest Service, Moscow, ID 83843

Kevin C. Ryan, Project Leader, Rocky Mountain Research Station, U.S. Department of Agriculture, Forest Service, Missoula, MT 59807

Arthur R. Tiedemann, Research Soil Scientist (Retired), Pacific Northwest Research Station, U.S. Department of Agriculture, Forest Service, Wenatchee, WA 98801

Malcolm J. Zwolinski, Assistant Director and Professor, School of Renewable Natural Resources, University of Arizona, Tucson, AZ 85721

Abstract

Neary, Daniel G.; Ryan, Kevin C.; DeBano, Leonard F., eds. 2005. (revised 2008). **Wildland fire in ecosystems: effects of fire on soils and water**. Gen. Tech. Rep. RMRS-GTR-42-vol.4. Ogden, UT: U.S. Department of Agriculture, Forest Service, Rocky Mountain Research Station. 250 p.

This state-of-knowledge review about the effects of fire on soils and water can assist land and fire managers with information on the physical, chemical, and biological effects of fire needed to successfully conduct ecosystem management, and effectively inform others about the role and impacts of wildland fire. Chapter topics include the soil resource, soil physical properties and fire, soil chemistry effects, soil biology responses, the hydrologic cycle and water resources, water quality, aquatic biology, fire effects on wetland and riparian systems, fire effects models, and watershed rehabilitation.

Keywords: ecosystem, fire effects, fire regime, fire severity, soil, water, watersheds, rehabilitation, soil properties, hydrology, hydrologic cycle, soil chemistry, soil biology, fire effects models

Cover photo—Left photo: Wildfire encroaching on a riparian area, Montana, 2002. (Photo courtesy of the Bureau of Land Management, National Interagency Fire Center, Image Portal); Right photo: BAER team member, Norm Ambos, Tonto National Forest, testing for water repellancy, Coon Creek Fire 2002, Sierra Ancha Experimental Forest, Arizona.

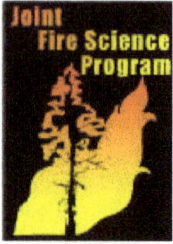

Preface

In 1978, a national workshop on fire effects in Denver, Colorado provided the impetus for the "Effects of Wildland Fire on Ecosystems" series. Recognizing that knowledge of fire was needed for land management planning, state-of-the-knowledge reviews were produced that became known as the "Rainbow Series." The series consisted of six publications, each with a different colored cover (frequently referred to as the Rainbow series), describing the effects of fire on soil (Wells and others 1979), water (Tiedemann and others 1979), air (Sandberg and others 1979), flora (Lotan and others 1981), fauna (Lyon and others 1978), and fuels (Martin and others 1979).

The Rainbow Series proved popular in providing fire effects information for professionals, students, and others. Printed supplies eventually ran out, but knowledge of fire effects continued to grow. To meet the continuing demand for summaries of fire effects knowledge, the interagency National Wildfire Coordinating Group asked Forest Service research leaders to update and revise the series. To fulfill this request, a meeting for organizing the revision was held January 46, 1993 in Scottsdale, AZ. The series name was then changed to "The Rainbow Series." The five-volume series covers air, soil and water, fauna, flora and fuels, and cultural resources.

The Rainbow Series emphasizes principles and processes rather than serving as a summary of all that is known. However, it does provide a lot of useful information and sources for more detailed study of fire effects. The five volumes, taken together, provide a wealth of information and examples to advance understanding of basic concepts regarding fire effects in the United States and Canada. While this volume focuses on the United States and Canada, there are references to information and examples from elsewhere in the world (e.g. Australia, South Africa, Spain, Zimbabwe, and others) to support the statements made. As conceptual background, they provide technical support to fire and resource managers for carrying out interdisciplinary planning, which is essential to managing wildlands in an ecosystem context. Planners and managers will find the series helpful in many aspects of ecosystem-based management, but they also have the responsibility to seek out and synthesize the detailed information needed to resolve specific management questions.

— The Authors

Acknowledgments

The Rainbow Series was completed under the sponsorship of the Joint Fire Sciences Program, a cooperative fire science effort of the U.S. Department of Agriculture (USDA) Forest Service and the U.S. Department of the Interior Bureau of Indian Affairs, Bureau of Land Management, Fish and Wildlife Service, and National Park Service. We thank Marcia Patton-Mallory and Louise Kingsbury for support and persistence to make this revision possible. We would like to dedicate this particular volume to a former USDA Forest Service Pacific Southwest Research Station research scientist, Steve S. Sackett, for his career-long commitment to research on the effects of fire on soils, to a former Rocky Mountain Research Station research scientist Roger D. Hungerford for his contributions to organic soil consumption and soil heating, and to the late Benee F. Swindel, a former USDA Forest Service, Southern Research Station scientist and Project Leader, for his counseling in the art of "perseverance".

The lead editor wishes to thank the following for insightful and helpful reviews, suggestions, information, and assistance that led to substantial technical and editorial improvements in the manuscript: Leonard DeBano, Peter Ffolliott, Randy Gimblett, George Ice, Marty Jurgensen, Karen Koestner, Dave Maurer, Kevin Ryan, Peter Robichaud, and Larry Schmidt.

Contents

Page

Summary vii

Chapter 1: Introduction 1
by Daniel G. Neary
Kevin C. Ryan
Leonard F. DeBano
Johanna D. Landsberg
James K. Brown
Background 1
   Importance of Fire to Soil and Water 2
   Scope 3
Fire Regimes 4
Fire Severity 5
   Fire Intensity versus Fire Severity 7
   Fire Intensity Measures 8
   Depth of Burn Measures 9
   Fire Severity Classification 13
   A Conceptual Model 15
Fire-Related Disturbances 17
   Fire 17
   Fire Suppression 17
   Type of Effects 17
This Book's Objective 17

Part A Effects of Fire on Soil 19

Part A—The Soil Resource: Its Importance,
   Characteristics, and General Responses
   to Fire 21
by Leonard F. DeBano
Daniel G. Neary
Introduction 21
The General Nature of Soil 22
Soil Properties—Characteristics, Reactions, and
   Processes 22
Soil Profile 22
Importance of Organic Matter 23
Fire Effects—General Concepts and Relationships 24
   Severity and Fire Intensity 24
   Combustion 24
   Heat Transfer 25
   Depth and Magnitude of Soil Heating 25
   Temperature Thresholds of Soil Properties 26
   Location of Soil Properties 26
Assessing Fire Effects on Soils 27
Management Implications 27
Summary 27

Chapter 2: Soil Physical Properties 29
by Leonard F. DeBano
Daniel G. Neary
Peter F. Ffolliott
Introduction 29
Soil Physical Characteristics 29
   Soil Texture and Mineralogy 29
   Soil Structure 30
   Bulk Density and Porosity 31
Physical Processes 31
   Heat Transfer in Soils 31
   Water Repellency 37

Page

Soil Erosion 41
   Processes and Mechanics 41
   Postfire Sediment Yields 46
Management Implications 51
Summary 51

Chapter 3: Soil Chemistry 53
by Jennifer D. Knoepp
Leonard F. DeBano
Daniel G. Neary
Introduction Soil Chemical Characteristics 53
   Organic Matter and Carbon 55
   Cation Exchange Capacity 59
   Cations 62
   Soil pH and Buffer Capacity 62
   Nitrogen 63
   Phosphorus 64
   Sulfur 64
Soil Chemical Processes 65
   Nutrient Cycling 65
   Nutrient Loss Mechanisms 65
   Nutrient Availability 66
Ash-Bed Effect 69
Management Implications 70
Summary 71

Chapter 4: Soil Biology 73
by Matt D. Busse
Leonard F. DeBano
Introduction 73
Biological Components of Soils 73
   Soil Microorganisms 74
   Soil Meso- and Macrofauna 78
   Roots and Reproductive Structures 78
   Amphibians, Reptiles, and Small Mammals 78
Biologically Mediated Processes in Soils 78
   Decomposition and Mineralization 79
   Nitrogen Cycling Processes 79
Fire Effects on Organisms and Biological Processes ... 80
   Soil Microorganisms 80
   Soil Meso- and Macrofauna 86
   Roots and Reproductive Structures 87
   Amphibians, Reptiles, and Small Mammals 87
   Biologically Mediated Processes 88
   Management Implications 89
Summary 91

Part B Effects of Fire on Water 93

Part B—The Water Resource: Its Importance,
   Characteristics, and General Responses
   to Fire 95
by Daniel G. Neary
Peter F. Ffolliott
Introduction 95
Hydrologic Cycle 95
   Interception 97
   Infiltration 98
   Evapotranspiration 99

Soil Water Storage 100
Snow Accumulation and Melt Patterns 101
Overland Flow 101
Baseflows and Springs 102
Pathways and Processes 102
Watershed Condition 103
Streamflow Discharge 104
Water Quality 105
Aquatic Biology 105

Chapter 5: Fire and Streamflow Regimes 107
by Daniel G. Neary
Peter F. Ffolliott
Johanna D. Landsberg
Introduction 107
Soil Water Storage 108
Baseflows and Springs 109
Streamflow Regimes 109
Effects of Wildfires 109
Effects of Prescribed Burning 110
Peakflows 111
Peakflow Mechanisms 114
Fire Effects 115
Management Implications 117
Summary 117

Chapter 6: Water Quality 119
by Daniel G. Neary
Johanna D. Landsberg
Arthur R. Tiedemann
Peter F. Ffolliott
Introduction 119
Water Quality Characteristics and Standards 119
Soil Erosion and Sedimentation Processes 120
Physical Characteristics of Water 121
Sediment 121
Water Temperature 123
pH of Water 124
Chemical Characteristics of Water 124
Dissolved Chemical Constituents 124
Nitrogen 125
Phosphorus 127
Sulfur 128
Chloride 128
Bicarbonate 128
Total Dissolved Solids 128
Nutrients and Heavy Metals 128
Biological Quality of Water 129
Fire Retardants 129
Rodeo-Chediski Fire, 2002: A Water Quality
Case History 131
Management Implications 132
Summary 134

Chapter 7: Aquatic Biota 135
by John N. Rinne
Gerald R. Jacoby
Fire Effects on Fish 135
Direct Fire Effects 136
Indirect Fire Effects 136
Other Anthropogenic Influencing Factors 138
Temporal-Spatial Scales 138

Species Considerations 139
Summary and General Management
Implications for Fish 139
Birds 140
Reptiles and Amphibians 140
Mammals 140
Invertebrates 140
Response to Fire 141
Invertebrate Summary 143

Part C Other Topics 145

Part C—Other Topics 147
by Daniel G. Neary

Chapter 8: Wetlands and Riparian Systems 149
by James R. Reardon
Kevin C. Ryan
Leonard F. DeBano
Daniel G. Neary
Introduction 149
Wetlands 149
Wetland and Hydric Soil Classification 151
Wetland Hydrology and Fire 153
Wetland Fire Effects and Soil Nutrient Responses . 162
Wetland Management Considerations 166
Riparian Ecosystems 166
Riparian Definition and Classification 167
Hydrology of Riparian Systems 167
Riparian Fire Effects 167
Role of Large Woody Debris 168
Riparian Management Considerations 168
Summary 169

Chapter 9: Fire Effects and Soil Erosion Models 171
by Kevin Ryan
William J. Elliot
Introduction 171
First Order Fire Effects Model (FOFEM) 171
Description, Overview, and Features 172
Applications, Potential Uses, Capabilities, and
Goals 172
Scope and Primary Geographic Applications 172
Input Variables and Data Requirements 172
Output, Products, and Performance 172
Advantages, Benefits, and Disadvantages 172
System and Computer Requirements 172
Models for Heat and Moisture Transport in Soils 173
WEPP, WATSED, and RUSLE Soil Erosion Models 173
Model Selection 176
DELTA-Q and FOREST Models 176
Models Summary 176

Chapter 10: Watershed Rehabilitation 179
by Peter R. Robichaud
Jan L. Beyers
Daniel G. Neary
Burned Area Emergency Rehabilitation (BAER) 179
BAER Program Analysis 181
Postfire Rehabilitation Treatment Decisions 183
The BAER Team and BAER Report 183
Erosion Estimates from BAER Reports 183
Hydrologic Response Estimates 183

Hillslope Treatments and Results 185
    Hillslope Treatments 185
    Hillslope Treatment Effectiveness 188
    Research and Monitoring Results 189
Channel Treatments and Results 191
    Channel Treatments 191
    Channel Treatment Effectiveness 192
Road and Trail Treatments and Results 194
    Road and Trail Treatments 194
    Road Treatment Effectiveness 195
Summary, Conclusions, and Recommendations 196
    Recommendations: Models and Predictions 196
    Recommendations: Postfire Rehabilitation
    Treatment 196
    Recommendations: Effectiveness monitoring 197

Chapter 11: Information Sources 199
   by Malcolm J. Zwolinski
    Daniel G. Neary
    Kevin C. Ryan
Introduction 199
Databases 200
    U.S. Fire Administration 200
    Current Wildland Fire Information 200
    Fire Effects Information System 200
    National Climatic Data Center 200
    PLANTS 200
    Fire Ecology Database 200
    Wildland Fire Assessment System 201
    National FIA Database Systems 201

Web Sites 201
    USDA Forest Service, Rocky Mountain
    Research Station Wildland Fire Research
    Program, Missoula, Montana 201
    Fire and Fire Surrogate Program 201
    National Fire Coordination Centers 201
    Other Web Sites 202
Textbooks 203
Journals and Magazines 204
Other Sources 205

Chapter 12: Summary and Research Needs 207
   by Daniel G. Neary
    Kevin C. Ryan
    Leonard F. DeBano
Volume Objective 207
Soil Physical Properties Summary 207
Soil Chemistry Summary 208
Soil Biology Summary 208
Fire and Streamflow Regimes Summary 208
Water Quality Summary 208
Aquatic Biota Summary 209
Wetlands Summary 210
Models Summary 210
Watershed Rehabilitation Summary 211
Information Sources Summary 211
Research Needs 211
References 213
Appendix A: Glossary 235

Summary

Fire is a natural disturbance that occurs in most terrestrial ecosystems. It is also a tool that has been used by humans to manage a wide range of natural ecosystems worldwide. As such, it can produce a spectrum of effects on soils, water, riparian biota, and wetland components of ecosystems. Fire scientists, land managers, and fire suppression personnel need to evaluate fire effects on these components, and balance the overall benefits and costs associated with the use of fire in ecosystem management. This publication has been written to provide up-to-date information on fire effects on ecosystem resources that can be used as a basis for planning and implementing fire management activities. It is a companion publication to the recently published book, *Fire's Effects on Ecosystems* by DeBano and others (1998).

In the late 1970s, the USDA Forest Service published a series of state-of-knowledge papers about fire effects on vegetation, soils, water, wildlife, and other ecosystem resources. These papers, collectively called "The Rainbow Series" because of their covers, were widely used by forest fire personnel. This publication updates both the Tiedemann and others (1979) paper on fire's effects on water and the Wells and others (1979) paper on soils.

This publication is divided into three major parts (A, B, C) and an introductory chapter that provides discussions of fire regimes, fire severity and intensity, and fire related disturbances. Part A describes the nature of the soil resource, its importance, characteristics and the responses of soils to fire and the relationship of these features to ecosystem functioning and sustainability. Part A is divided into three main chapters (2, 3, and 4) that describe specific fire effects on the physical, chemical, and biological properties of the soil, respectively. Likewise, Part B discusses the basic hydrologic processes that are affected by fire, including the hydrologic cycle, water quality, and aquatic biology. It also contains three chapters which specifically discuss the effect of fire on the hydrologic cycle, water quality, and aquatic biology in chapters 5, 6, and 7, respectively. Part C has five chapters that cover a wide range of related topics. Chapter 8 analyzes the effects of fire on the hydrology and nutrient cycling of wetland ecosystems along with management concerns. The use of models to describe heat transfer throughout the ecosystem and erosional response models to fire are discussed in chapter 9. Chapter 10 deals with important aspects of watershed rehabilitation and implementation of the Federal Burned Area Emergency Rehabilitation (BAER) program. Chapter 11 directs the fire specialists and managers to important information sources including data bases, Web sites, textbooks, journals, and other sources of fire effects information. A summary of the important highlights of the book are provided in chapter 12. Last, a glossary of fire terms is included in the appendix. The material provided in each chapter has been prepared by individuals having specific expertise in a particular subject.

This publication has been written as an information source text for personnel involved in fire suppression and management, planners, decisionmakers, land managers, public relations personnel, and technicians who routinely and occasionally are involved in fire suppression and using fire as a tool in ecosystem management. Because of widespread international interest in the previous and current "Rainbow Series" publications, the International System of Units (Systeme International d'Unites, SI), informally called the metric system (centimeters, cubic meters, grams), is used along with English units throughout the volume. In some instances one or the other units are used exclusively where conversions would be awkward or space does not allow presentation of both units.

Daniel G. Neary
Kevin C. Ryan
Leonard F. DeBano
Johanna D. Landsberg
James K. Brown

Chapter 1: Introduction

Background

At the request of public and private wildland fire managers, who recognize a need to assimilate current fire effects knowledge, the Rocky Mountain Research Station has produced a state-of-the-art integrated series of documents relative to management of ecosystems (Smith 2000, Brown and Smith 2000, Sandberg and others 2002, Neary and others this volume, and Jones and Ryan in preparation). The series covers our technical understanding of fire effects, an understanding that has grown considerably since the first version of this series, the "Rainbow Series," was published in 1979. Since that time our awareness has grown that fire is a fundamental process of ecosystems that must be understood and managed to meet resource and ecosystem management goals. The volumes in the current series are intended to be useful for land management planning, development of environmental assessments and environmental impact statements, training and education, informing others such as conservation groups and regulatory agencies, and accessing technical literature. Knowledge of fire effects has risen in importance to land managers because fire, as a disturbance process, is an integral part of the concept of ecosystem management and restoration ecology. Fire initiates changes in ecosystems that affect the composition, structure, and patterns of vegetation on the landscape. It also affects the soil and water resources of ecosystems that are critical to overall functions and processes.

Fire is a dynamic process, predictable but uncertain, that varies over time and landscape space. It has shaped plant communities for as long as vegetation and lightning have existed on earth (Pyne 1982). Recycling of carbon (C) and nutrients depends on biological decomposition and fire. In regions where decay is constrained either by dry or cold climates or saturated (in other words, anaerobic) conditions, fire plays a dominant role in recycling organic matter (DeBano and others 1998). In warmer, moist climates, decay plays the dominant role in organic matter recycling (Harvey 1994), except in soils that are predominantly saturated (in other words, hydric soils).

The purpose of this volume, *Effects of Fire on Soils and Water*, is to assist land managers with ecosystem restoration and fire management planning responsibilities in their efforts to inform others about the impacts of fire on these ecosystem resources. The geographic coverage in this volume is North America, but the principles and effects can be applied to any ecosystem in which fire is a major disturbance process.

This publication is divided into three major parts and an introductory chapter that provides discussions

USDA Forest Service Gen. Tech. Rep. RMRS-GTR-42-vol. 4. 2005

1

of fire regimes, fire severity and intensity, and fire related disturbances. Part A describes the nature of the soil resource, its importance, characteristics and the responses of soils to fire, and the relationship of these features to ecosystem functioning and sustainability. Part A begins with a general overview and then is divided into three chapters (2, 3, and 4). Likewise, part B begins with a general overview then is divided into three chapters (5, 6, and 7) that discuss the basic hydrologic processes that are affected by fire, including the hydrologic cycle, water quality, and aquatic biology. Part C has five chapters that cover a wide range of related topics. Chapter 8 analyzes the effects of fire on the hydrology and nutrient cycling of wetland ecosystems along with management concerns. The use of models to describe heat transfer throughout the ecosystem and erosional response models to fire are discussed in chapter 9. Chapter 10 deals with important aspects of watershed rehabilitation and implementation of the Federal Burned Area Emergency Rehabilitation (BAER) program. Chapter 11 directs the fire specialists and mangers to important information sources including databases, Web sites, textbooks, journals, and other sources of fire effects information. A summary of the important highlights of the book are provided in chapter 12. The book concludes with a list of references used in the volume, and a glossary of fire terms.

Importance of Fire to Soil and Water

Soil is the unconsolidated, variable-thickness layer of mineral and organic matter on the Earth's surface that forms the interface between the geosphere and the atmosphere. It has formed as a result of physical, chemical, and biological processes functioning simultaneously on geologic parent material over long periods (Jenny 1941, Singer and Munns 1996). Soil is formed where there is continual interaction between the soil system and the biotic (faunal and floral), climatic (atmospheric and hydrologic), and topographic components of the environment. Soil interrelates with other ecosystem resources in several ways. It supplies air, water, nutrients, and mechanical support for the sustenance of plants. Soil also receives and processes rainfall. By doing so, it partly determines how much becomes surface runoff, and how much is stored for delivery slowly from upstream slopes to channels where it becomes streamflow, and by how much is stored and used for soil processes (for example, transpiration, leaching, and so forth). When the infiltration capacity of the soil for rainfall is exceeded, organic and inorganic soil particles are eroded from the soil surface and become a major source of sediment, nutrients, and pollutants in streams that affect water quality. There is also an active and ongoing exchange of gases between the soil and the surrounding atmosphere. Soil also

provides a repository for many cultural artifacts, which can remain in the soil for thousands of years without undergoing appreciable change.

Fire can produce a wide range of changes in landscape appearance (fig. 1.1ABC; DeBano and others

Figure 1.1—Fire produced a wide range of changes in the forest landscape of a ponderosa pine forest in Arizona where their appearance ranged from (A) unburned ponderosa pine to those burned at (B) low-to-moderate severity and those burned at (C) high severity. (Photos by Peter Ffolliott).

2

USDA Forest Service Gen. Tech. Rep. RMRS-GTR-42-vol. 4. 2005

1998). The fire-related changes associated with different severities of burn produce diverse responses in the water, soil, floral, and faunal components of the burned ecosystems because of the interdependency between fire severity and ecosystem response. Both immediate and long-term responses to fire occur (fig. 1.2). Immediate effects also occur as a result of the release of chemicals in the ash created by combustion of biomass. The response of biological components (soil microorganisms and ecosystem vegetation) to these changes is both dramatic and rapid. Another immediate effect of fire is the release of gases and other air pollutants by the combustion of biomass and soil organic matter. Air quality in large-scale airsheds can be affected during and following fires (Hardy and others 1998, Sandberg and others 2002). The long-term fire effects on soils and water which are usually subtle, can persist for years following the fire, or be permanent as occurs when cultural resources are damaged (DeBano and others 1998, Jones and Ryan in preparation).

Other long-term fire effects arise from the relationships between fire, soils, hydrology, nutrient cycling, and site productivity (Neary and others 1999).

In the previous "Rainbow" series published after the 1978 National Fire Effects Workshop that reviewed the state-of-knowledge of the effects of fire, separate reports were published on soil (Wells and others 1979) and water (Tiedemann and others 1979). Because of the intricate linkage between soil and water effects, this volume combines both.

Scope

The scope of this publication covers fire and disturbances in forest, woodland, and shrubland, and grassland ecosystems of the United States and Canada. However, it is applicable to any area in the world having similar forest types and fire regimes. In some instances, research information from ecosystems outside of North America will be used to

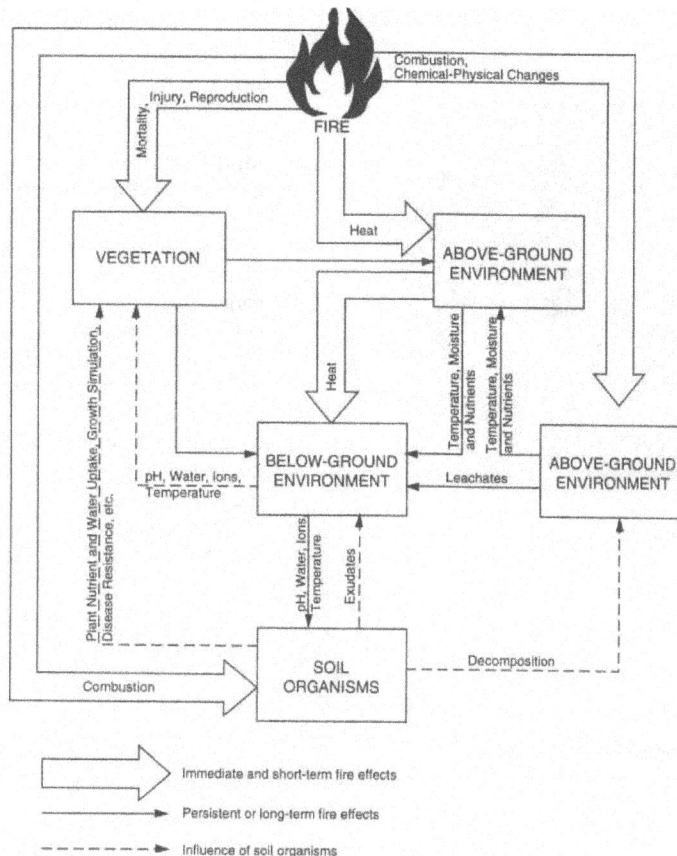

Figure 1.2—Immediate and long-term ecosystem responses to fire. (Adapted from Borchers and Perry 1990. In: *Natural and Prescribed Fire in Pacific Northwest Forests*, edited by J.D. Walstad, S.R. Radosevich, and D.V. Sandberg. Copyright © 1990 Oregon State University Press. Reproduced by permission).

USDA Forest Service Gen. Tech. Rep. RMRS-GTR-42-vol. 4. 2005

3

elucidate specific fire effects on soils and water. Fire effects on ecosystems can be described at several spatial and temporal scales (Reinhardt and others 2001). In this chapter we will describe fire relationships suitable to each spatial and temporal scale.

The fire-related disturbances included in this review include wildland fires and prescribed fires, both in natural and management activity fuels. It also includes disturbances from fire suppression such as fire lines and roads, and fire retardant applications.

Fire Regimes

The general character of fire that occurs within a particular vegetation type or ecosystem across long successional time frames, typically centuries, is commonly defined as the characteristic fire regime. The fire regime describes the typical or modal fire severity that occurs. But it is recognized that, on occasion, fires of greater or lesser severity also occur within a vegetation type. For example, a stand-replacing crownfire is common in long fire-return-interval forests (fig. 1.3).

Figure 1.3—High severity, stand replacing wildfire. (Photo by USDA Forest Service).

The fire regime concept is useful for comparing the relative role of fire between ecosystems and for describing the degree of departure from historical conditions (Hardy and others 2001, Schmidt and others 2002). The fire regime classification used in this volume is the same as that used in the volume of this series (Brown 2000) on the effects of fire on flora. Brown (2000) contains a discussion of the development of fire regime classifications based on fire characteristics and effects (Agee 1993), combinations of factors including fire frequency, periodicity, intensity, size, pattern, season, and depth of burn (Heinselman 1978), severity (Kilgore 1981), and fire periodicity, season, frequency, and effects (Frost 1998). Hardy and others (1998, 2001) used modal severity and frequency to map fire regimes in the Western United States (table 1.1).

The fire regimes described in table 1.1 are defined as follows:

- **Understory Fire Regime:** Fires are generally nonlethal to the dominant vegetation and do not substantially change the structure of the dominant vegetation. Approximately 80 percent or more of the aboveground dominant vegetation survives fires. This fire regime applies to certain fire-resistant forest and woodland vegetation types.
- **Stand Replacement:** Fires are lethal to most of the dominant aboveground vegetation. Approximately 80 percent or more of the aboveground dominant vegetation is either consumed or dies as a result of fire, substantially changing the aboveground vegetative structure. This regime applies to fire-susceptible forests and woodlands, shrublands, and grasslands.
- **Mixed:** The severity of fires varies between nonlethal understory and lethal stand replacement fires with the variation occurring in space or time. First, spatial variability occurs when fire severity varies, producing a spectrum from

Table 1.1—Comparison of fire regime classifications according to Hardy and others (1998, 2001) and Brown (2000).

	Hardy and others 1998, 2001		Brown 2000	
Fire regime group	Frequency (years)	Severity	Severity and effects	Fire regime
I	0-35	Low	Understory fire	1
II	0-35	Stand replacement	Stand replacement	2
III	35-100+	Mixed	Mixed	3
IV	35-100+	Stand replacement	Stand replacement	2
V	Greater than 200	Stand replacement	Stand replacement	2
			Non-fire regime	4

4

USDA Forest Service Gen. Tech. Rep. RMRS-GTR-42-vol. 4. 2005

understory burning to stand replacement within an individual fire. This results from small-scale changes in the fire environment (fuels, terrain, or weather) and random changes in plume dynamics. Within a single fire, stand replacement can occur with the peak intensity at the head of the fire while a nonlethal fire occurs on the flanks. These changes create gaps in the canopy and small to medium sized openings. The result is a fine pattern of young, older, and multiple-aged vegetation patches. While this type of fire regime has not been explicitly described in previous classifications, it commonly occurs in some ecosystems because of fluctuations in the fire environment (DeBano and others 1998, Ryan 2002). For example, complex terrain favors mixed severity fires because fuel moisture and wind vary on small spatial scales. Secondly, temporal variation in fire severity occurs when individual fires alternate over time between infrequent low-intensity surface fires and long-interval stand replacement fires, resulting in a variable fire regime (Brown and Smith 2000, Ryan 2002). Temporal variability also occurs when periodic cool-moist climate cycles are followed by warm dry periods leading to cyclic (in other words, multiple decade-level) changes in the role of fire in ecosystem dynamics. For example in an upland forest, reduced fire occurrence during the cool-moist cycle leads to increased stand density and fuel build-up. Fires that occur during the transition between cool-moist and warm-dry periods

can be expected to be more severe and have long-lasting effects on patch and stand dynamics (Kauffman and others 2003).
• **Nonfire Regime:** Fire is not likely to occur.

Subsequently, Schmidt and others (2002) used these criteria to map fire regimes and departure from historical fire regimes for the contiguous United States. This coarser-scale assessment was incorporated into the USDA Forest Service's Cohesive Strategy for protecting people and sustaining resources in fire adapted ecosystems (Laverty and Williams 2000). It is explicitly referred to in the United States as the 2003 Healthy Forest Restoration Act (HR 1904). The classification system used by Brown (2000) found in the *Effects of Fire on Flora* volume (Brown and Smith 2000) is also based on fire modal severity, emphasizes fire effects, but does not use frequency. Examples of vegetation types representative of each of the fire regime types are listed in tables 1.2a,b,c. These vegetation types are described in the Brown and Smith (2000) volume.

Fire Severity

At finer spatial and temporal scales the effects of a specific fire can be described at the stand and community level (Wells and others 1979, Rowe 1983, Turner and others 1994, DeBano and others 1998, Feller 1998, Ryan 2002). The commonly accepted term for describing the ecological effects of a specific fire is *fire severity*. Fire severity describes the magnitude of the disturbance and, therefore, reflects the degree of change in ecosystem components. Fire affects both the aboveground and belowground components of the

Table 1.2a—Examples of vegetation types associated with understory fire regimes in the United States and Canada. Some widely distributed vegetation types occur in more than one fire regime. Variations in fire regime result from regional differences in terrain and fire climate.

Communities	Source
Longleaf pine	Wade and others 2000
Slash pine	Wade and others 2000
Loblolly pine	Wade and others 2000
Shortleaf pine	Wade and others 2000
Pine flatwoods and pine rocklands	Myers 2000
Pondcypress wetlands	Myers 2000
Cabbage palmetto savannas and forests	Myers 2000
Oak-hickory forests	Wade and others 2000
Live oak forests	Myers 2000
Ponderosa pine	Arno 2000, Paysen and others 2000
Ponderosa pine-mixed conifer	Arno 2000
Jeffrey pine	Arno 2000
Redwood	Arno 2000
Oregon oak woodlands	Arno 2000

USDA Forest Service Gen. Tech. Rep. RMRS-GTR-42-vol. 4. 2005

5

Table 1.2b—Examples of vegetation types associated with mixed severity fire regimes in the United States and Canada. Some widely distributed vegetation types occur in more than one fire regime. Variations in fire regime result from regional differences in terrain and fire climate.

Communities	Source
Aspen	Duchesne and Hawkes 2000
Eastern white pine	Duchesne and Hawkes 2000
Red pine	Duchesne and Hawkes 2000
Jack pine	Duchesne and Hawkes 2000
Virginia pine	Wade and others 2000
Pond pine	Wade and others 2000
Mixed mesophytic hardwoods	Wade and others 2000
Northern hardwoods	Wade and others 2000
Bottomland hardwoods	Wade and others 2000
Coast Douglas-fir and Douglas-fir/hardwoods	Arno 2000
Giant sequoia	Arno 2000
California red fir	Arno 2000
Sierra/Cascade lodgepole pine	Arno 2000
Rocky Mountain lodgepole pine	Arno 2000
Interior Douglas-fir	Arno 2000
Western larch	Arno 2000
Whitebark pine	Arno 2000
Ponderosa pine	Arno 2000
Pinyon-juniper	Paysen and others 2000
Texas savanna	Paysen and others 2000
Western oaks	Paysen and others 2000

Table 1.2c—Examples of vegetation types associated with stand replacement fire regimes in the United States and Canada. Some widely distributed vegetation types occur in more than one fire regime. Variations in fire regime result from regional differences in terrain and fire climate.

Communities	Source
Boreal spruce-fir	Duchesne and Hawkes 2000
Conifer bogs	Duchesne and Hawkes 2000
Tundra	Duchesne and Hawkes 2000
Wet grasslands	Wade and others 2000
Prairie	Wade and others 2000
Bay forests	Wade and others 2000
Sand pine	Wade and others 2000
Table mountain pine	Wade and others 2000
Eastern spruce-fir	Wade and others 2000
Atlantic white-cedar	Wade and others 2000
Salt and brackish marshes	Wade and others 2000
Fresh and oligohaline marshes and wet prairie	Myers 2000, Wade and others 2000
Florida coastal prairies	Myers 2000
Florida tropical hardwood forests	Myers 2000
Hawaiian forests and grasslands	Myers 2000
Forests of Puerto Rico and the Virgin Islands	Myers 2000
Coast Douglas-fir	Arno 2000
Coastal true fir/mountain hemlock	Arno 2000
Interior true fir-Douglas-fir-western larch	Arno 2000
Rocky Mountain lodgepole pine	Arno 2000
Western white pine-cedar-hemlock	Arno 2000
Western spruce-fir-whitebark pine	Arno 2000
Aspen	Arno 2000
Grasslands (annual and perennial)	Paysen and others 2000
Sagebrush	Paysen and others 2000
Desert shrublands	Paysen and others 2000
Southwestern shrubsteppe	Paysen and others 2000
Chaparral-mountain shrub	Paysen and others 2000

6

USDA Forest Service Gen. Tech. Rep. RMRS-GTR-42-vol. 4. 2005

ecosystem. Thus severity integrates both the heat pulse above ground and the heat pulse transferred downward into the soil. It reflects the amount of energy (heat) that is released by a fire that ultimately affects resources and their functions. It can be used to describe the effects of fire on the soil and water system, ecosystem flora and fauna, the atmosphere, and society (Simard 1991). It reflects the amount of energy (heat) that is released by a fire that ultimately affects resource responses. Fire severity is largely dependent upon the nature of the fuels available for burning, and the combustion characteristics (in other words, flaming versus smoldering) that occur when these fuels are burned. This chapter emphasizes the relationship of fire severity to soil responses because the most is known about this relationship, and because soil responses (see chapters 2, 3, 4, and 10) are closely related to hydrologic responses (see chapters 5 and 6) and ecosystem productivity (see chapters 4 and 8).

Fire Intensity versus Fire Severity

Although the literature historically contains confusion between the terms *fire intensity* and *fire severity*, a fairly consistent distinction between the two terms has been emerging in recent years. Fire managers trained in the United States and Canada in fire behavior prediction systems use the term fire intensity in a strict thermodynamic sense to describe the rate of energy released (Deeming and others 1977, Stocks and others 1989). Fire intensity is concerned mainly with the rate of aboveground fuel consumption and, therefore the energy release rate (Albini 1976, Alexander 1982). The faster a given quantity of fuel burns, the greater the intensity and the shorter the duration (Byram 1959, McArthur and Cheney 1966, Albini 1976, Rothermel and Deeming 1980, Alexander 1982). Because the rate at which energy can be transmitted through the soil is limited by the soil's thermal properties, the duration of burning is critically important to the effects on soils (Frandsen and Ryan 1986, Campbell and others 1995). Fire intensity is not necessarily related to the total amount of energy produced during the burning process. Most energy released by flaming combustion of aboveground fuels is not transmitted downward (Packham and Pompe 1971, Frandsen and Ryan 1985). For example, Packham and Pompe (1971) found that only about 5 percent of the heat released by a surface fire was transmitted into the ground. Therefore, fire intensity is not necessarily a good measure of the amount of energy transmitted downward into the soil, or the associated changes that occur in physical, chemical, and biological properties of the soil. For example, it is possible that a high intensity and fast moving crown fire will consume little of the surface litter because only a small amount

of the energy released during the combustion of fuels is transferred downward to the litter surface (Rowe 1983, VanWagner 1983, Ryan 2002). In this case the surface litter is blackened (charred) but not consumed. In the extreme, one author of this chapter has seen examples in Alaska and North Carolina where fast spreading crown fires did not even scorch all of the surface fuels. However, if the fire also consumes substantial surface and ground fuels, the residence time on a site is greater, and more energy is transmitted into the soil. In such cases, a "white ash" layer is often the only postfire material left on the soil surface (Wells and others 1979, Ryan and Noste 1985) (fig. 1.4).

Because one can rarely measure the actual energy release of a fire, the term fire intensity can have limited practical application when evaluating ecosystem responses to fire. Increasingly, the term fire severity is used to indicate the effects of fire on the different ecosystem components (Agee 1993, DeBano and others 1998, Ryan 2002). Fire severity has been used describe the magnitude of negative fire impacts on natural ecosystems in the past (Simard 1991), but a wider usage of the term to include all fire effects is proposed. In this context severity is a description of the magnitude of change resulting from a fire and does not necessarily imply negative consequences. Thus, a low severity fire may restore and maintain a variety of ecological attributes that are generally viewed as positive, as for example in a fire-adapted longleaf pine (*Pinus palustris*) or ponderosa pine (*P. ponderosa*) ecosystem. In contrast a high severity fire may be a dominant, albeit infrequent, disturbance in a non-fire-adapted ecosystem, for example, spruce (*Picea* spp.) whereas it is abnormal in a fire-adapted ecosystem.

Figure 1.4—Gray to white ash remaining after a pinyon-juniper slash pile was burned at high temperatures for a long duration, Apache-Sitgreaves National Forest, Arizona. (Photo by Steve Overby).

USDA Forest Service Gen. Tech. Rep. RMRS-GTR-42-vol. 4. 2005

7

While all high severity fires may have significant negative social impacts, only in the latter case is the long-term functioning of the ecosystem significantly altered.

Fire Intensity Measures

Byram's (1959) definition of *fireline intensity* has become a standard quantifiable measure of intensity (Van Wagner 1983, Agee 1993, DeBano and others 1998). It is a measure of the rate of energy release in the flaming front of the spreading fire. It does not address the residual flaming behind the front nor subsequent smoldering combustion (Rothermel and Deeming 1980, Alexander 1982). Fireline intensity can be written as a simple equation:

$$I = Hwr$$

where

I = fireline intensity (BTU/ft/sec or kW/m/sec)
H = heat yield (BTU/lb or kW/kg of fuel)
w = mass of available fuel burned (lb/ft^2 or kg/m^2)
r = rate of spread (ft/sec or m/sec)

Fireline intensity is proportional to the flame length in a spreading fire and is a useful measure of the potential to cause damage to aboveground structures (Van Wagner 1973, Rothermel and Deeming 1980, Alexander 1982, Ryan and Noste 1985). The Canadian forest fire danger rating system calculates the intensity of surface fires and crown fires based on Byram's equation (Stocks and others 1989). Rothermel (1972) defined a somewhat different measure of fire intensity, heat per unit area, which is commonly used in fire behavior prediction in the United States (Albini 1976, Rothermel and Deeming 1980, Andrews 1986, Scott 1998, Scott and Reinhardt 2001). One problem with using current fire behavior prediction systems in ecological studies is that they focus on flaming combustion of fine fuels and do not predict all of the combustion and fuel consumed, or quantify all of the energy released, during a fire (Johnson and Miyanishi 2001). The intensity of residual combustion of large woody fuels is modeled in the BURNUP Model (Albini and Reinhardt 1995, Albini and others 1996) and the energy release rate from duff consumption is modeled in the First Order Fire Effects Model (FOFEM) v.5.0 (Reinhardt 2003).

Fires burn throughout a continuum of energy release rates (table 1.3) (Artsybashev 1983, Rowe 1983, Van Wagner 1983, Rothermel 1991). Ground fires burn in compact fermentation and humus layers and in organic muck and peat soils where they spread predominantly by smoldering (glowing) combustion and typically burn for hours to weeks (fig. 1.5). Forward rates of spread in ground fires range on the order of several inches (decimeters) to yards (meters) per day. Temperatures are commonly in excess of 572 °F (300 °C) for several hours (Frandsen and Ryan 1986, Ryan and Frandsen 1991, Hartford and Frandsen 1992, Agee 1993). The conditions necessary for ground fires are organic soil horizons greater than about 1.6 to 2.4 inches (4 to 6 cm) deep and extended drying (Brown and others 1985, Reinhardt and others 1997, Johnson and Miyanishi 2001). Surface fires spread by flaming combustion in loose litter, woody debris, herbaceous plants and shrubs and trees roughly less than 6 feet

Table 1.3—Representative ranges for fire behavior characteristics for ground, surface, and crown fires (From Ryan 2002).

Fire type	Dominant Combustion	General Description	Fire Behavior Characteristics		
			Rate of Spread (meters/minute)	Flame Length (meters)	Fireline Intensity (kW/meter)
Ground	Smoldering	Creeping	3.3E-4 to 1.6E-2	0.0	< 10
Surface	Flaming	Creeping	<3.0E-1	0.1 to 0.5	1.7E0 to 5.8E1
		Active/ Spreading Intense/ Running	3.0E-1 to 8.3E0 8.3E0 to 5.0E1	0.5 to 1.5 1.5 to 3.0	5.8E1 to 6.3E2 6.3E2 to 2.8E3
Transition	Flaming	Passive Crowning (Intermittent Torching)	variable[1]	3.0 to 10.0	variable[1]
Crowning	Flaming	Active Crowning Independent Crowning	1.5E1 to 1.0E2 Up to ca. 2.0E2	5.0 to 15[2] Up to ca. 70[2]	1.0E4 to 1.0E5 Up to ca. 2.7E6

[1]Rates of spread, flame length and fireline intensity vary widely in transitional fires. In subalpine and boreal fuels it is common for surface fires to creep slowly until they encounter conifer branches near the ground, then individual trees or clumps of trees torch sending embers ahead of the main fire. These embers start new fires, which creep until they encounter trees, which then torch. In contrast, as surface fires become more intense, torching commonly occurs prior to onset of active crowning.
[2]Flame lengths are highly variable in crown fires. They commonly range from 1.5 to 2 times canopy height. Fire managers commonly report much higher flames but these are difficult to verify or model. Such extreme fires are unlikely to result in additional fire effects within a stand but are commonly associated with large patches of continuous severe burning.

Figure 1.5—Smoldering ground fire. (Photo by Kevin Ryan).

(about 2 m) tall. Under marginal burning conditions surface fires creep along the ground at rates of 3 feet/hour (less than 1 m/hr) with flames less than 19 inches (less than 0.5 m) high (table 1.3). As fuel, weather, and terrain conditions become more favorable for burning, surface fires become progressively more active with spread rates ranging on the order of from tens of yards (meters) to miles (kilometers) per day. The duration of surface fires is on the order of 1 to a few minutes (Vasander and Lindholm 1985, Frandsen and Ryan 1986, Hartford and Frandsen 1992) except where extended residual burning occurs beneath logs or in concentrations of heavy woody debris. Here flaming combustion may last a few hours resulting in substantial soil heating (Hartford and Frandsen 1992). However, the surface area occupied by long-burning woody fuels is typically small, less than 10 percent and often much less (Albini 1976, Ryan and Noste 1985, Albini and Reinhardt 1995). If canopy fuels are plentiful and sufficiently dry, surface fires begin to transition into crown fires (Van Wagner 1977, Scott and Reinhardt 2001). Crown fires burn in the foliage, twigs, and epiphytes of the forest or shrub canopy located above the surface fuels. Such fires exhibit the maximum energy release rate but are typically of short duration, 30 to 80 seconds.

Fires burn in varying combinations of ground, surface, and crown fuels depending on the local conditions at the specific time a fire passes a given point. Ground fires burn independently from surface and crown fires and often occur some hours after passage of the flaming front (Artsybashev 1983, Rowe 1983, Van Wagner 1983, Hungerford and others 1995a, Hungerford and others 1995b). Changes in surface and ground fire behavior occur in response to subtle changes in the microenvironment, stand structure, and weather leading to a mosaic of fire treatments at multiple scales in the ground, surface, and canopy strata (Ryan 2002).

Depth of Burn Measures

The relationship of fire intensity to fire severity remains largely undefined because of difficulties encountered in relating resource responses to the burning process (Hungerford and others 1991, Hartford and Frandsen 1992, Ryan 2002). While quantitative relationships have been developed to describe changes in the thermal conductivity of soil, and changes in soil temperature and water content beneath surface and ground fires, these relationships have not been thoroughly extrapolated to field conditions (Campbell and others 1994, 1995). It is not always possible to estimate the effects of fire on soil, vegetation, and air when these effects are judged by only fire intensity measurements because other factors overwhelm fire behavior. The range of fire effects on soil resources can be expected to vary directly with the *depth of burn* as reflected in the amount of duff consumed and degree of large woody fuel consumption (Ryan 2002). Thus, for example, the depth of lethal heat (approximately 140 °F or 60 °C) penetration into the soil can be expected to increase with the increasing depth of surface duff that is burned (fig. 1.6).

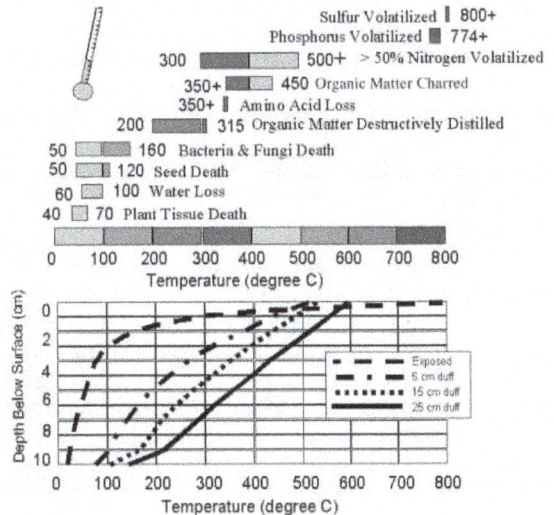

Figure 1.6—Temperature ranges associated with various fire effects (top) (from Hungerford and others 1991) compared to the depth of heat penetration into mineral soil (bottom) for a crown fire over exposed mineral soil (observed in jack pine *Pinus banksiana* in the Canadian Northwest Territories) or for ground fire burning in 5-, 15-, and 25-cm of duff (predicted via Campbell and others 1994, 1995). Conditions are for coarse dry soil, which provides the best conduction (i.e., a worst-case scenario). (From Ryan 2002).

Numerous authors have used measures of the depth of burn into the organic soil horizons or visual observation of the degree of charring and consumption of plant materials to define fire severity for interpreting the effects of fire on soils, plants, and early succession (Conrad and Poulton 1966, Miller 1977, Viereck and Dyrness 1979, Viereck and Schandelmeier 1980, Dyrness and Norum 1983, Rowe 1983, Zasada and others 1983, Ryan and Noste 1985, Morgan and Neuenschwander 1988, Schimmel and Granström 1996, DeBano and others 1998, Feller 1998). Depth of burn is directly related to the duration of burning in woody fuels (Anderson 1969, Albini and Reinhardt 1995) and duff (Frandsen 1991a, 1991b, Johnson and Miyanishi 2001). In heterogeneous fuels, depth of burn can vary substantially over short distances (for example, beneath a shrub or tree canopy versus the inter-canopy area, or beneath a log versus not (Tunstall and others 1976, Ryan and Frandsen 1991). At the spatial scale of a sample plot within a given fire, depth of burn can be classified on the basis of visual observation of the degree of fuel consumption and charring on residual plant and soil surfaces (Ryan and Noste 1985, Ryan 2002).

Ryan and Noste (1985) summarized literature on the relationships between depth of burn and the charring of plant materials. An adaptation of their table 2, updated to reflect subsequent literature (Moreno and Oechel 1989, Pérez and Moreno 1998, DeBano and others 1998, and Feller 1998) and experience, particularly in peat and muck soils, is presented in table 1.4. This table can be used as a field guide to classifying depth of burn on small plots (for example, quadrats). A brief description of depth of burn characteristics is provided for clarification of subsequent discussion of fire effects:

- *Unburned:* Plant parts are green and unaltered, there is no direct effect from heat. The extent of unburned patches (mosaics) varies considerably within and between burns as the fire environment (fuels, weather, and terrain) varies. Unburned patches are important rufugia for many species and are a source of plants and animals for recolinization of adjacent burned areas.

- *Scorched:* Fire did not burn the area, but radiated or convected heat from adjacent burned areas caused visible damage. Mosses and leaves are brown or yellow but species characteristics are still identifiable. Soil heating is negligible. Scorched areas occur to varying degrees along the edges of more severely burned areas. As it occurs on edges, the area within the scorched class is typically small (Dyrness and Norum 1983) and effects are typically similar to those in light burned

areas. The scorched class may, however, have utility in studies of microvariation of fire effects.

- *Light:* In forests the surface litter, mosses, and herbaceous plants are charred-to-consumed but the underlying forest duff or organic soil is unaltered. Fine dead twigs up to 0.25 inches (0.6 cm) are charred or consumed, but larger unburned branches remain. Logs may be blackened but are not deeply charred except where two logs cross. Leaves of understory shrubs and trees are charred or consumed, but fine twigs and branches remain. In nonforest vegetation, plants are similarly charred or consumed, herbaceous plant bases are not deeply burned and are still identifiable, and charring of the mineral soil is negligible. Light depth of burn is associated with short duration fires either because of light fuel loads (mass per unit area), high winds, moist fuels, or a combination of these three factors. Typical forest-floor moisture contents associated with light depth of burn are litter (O_i) 15 to 25 percent and duff (O_e+O_a) greater than 125 percent.

- *Moderate:* In forests the surface litter, mosses, and herbaceous plants are consumed. Shallow duff layers are completely consumed, and charring occurs in the top 0.5 inch (1.2 cm) of the mineral soil. Where deep duff layers or organic soils occur, they are deeply burned to completely consumed, resulting in deep char and ash deposits but the texture and structure of the underlying mineral soil are not visibly altered. Trees of late-successional, shallow-rooted species are often left on root pedestals or topple. Fine dead twigs are completely consumed, larger branches and rotten logs are mostly consumed, and logs are deeply charred. Burned-out stump holes and rodent middens are common. Leaves of understory shrubs and trees are completely consumed. Fine twigs and branches of shrubs are mostly consumed (this effect decreases with height above the ground), and only the larger stems remain. Stems of these plants frequently burn off at the base during the ground fire phase leaving residual aerial stems that were not consumed in the flaming phase lying on the ground. In nonforest vegetation, plants are similarly consumed, herbaceous plant bases are deeply burned and unidentifiable. In shrublands, average char-depth of the mineral soil is on the order of less than 0.4 inch (1 cm), but soil texture and structure are not noticeably altered. Charring may extend 0.8 to 1.2 inches (2.0 to 3.0 cm) beneath shrubs with deep leaf

10

USDA Forest Service Gen. Tech. Rep. RMRS-GTR-42-vol. 4. 2005

Table 1.4—Visual characteristics of depth of burn in forests, shrublands, and grasslands from observations of ground surface characteristics, charring, and fuel consumption for unburned and light (part A), moderate (part B), and deep (part C) classes (Modified from Ryan and Noste 1985).

Table 1.4 Part A

Depth of burn class	Vegetation type		
	Forests	Shrublands	Grasslands
Unburned			
Surface:	Fire did not burn on the surface.	See Forests	See Forests
Fuels:	Some vegetation injury may occur from radiated or convected heat resulting in an increase in dead fuel mass.	See Forests	See Forests
Occurrence:	A wide range exists in the percent unburned in natural fuels. Under marginal surface fire conditions the area may be >50 percent. Under severe burning conditions <5 percent is unburned. Commonly 10 to 20 percent of the area in slash burns is unburned. Unburned patches provide refugia for flora and fauna.	See Forests	See Forests
Light			
Surface:	Leaf litter charred or consumed. Upper duff charred but full depth not altered. Gray ash soon becomes inconspicuous leaving a surface that appears lightly charred to black.	Leaf litter charred or consumed, but some leaf structure is discernable. Leaf mold beneath shrubs is scorched to lightly charred but not altered over its entire depth. Where leaf mold is lacking charring is limited to <0.2 cm into mineral soil. Some gray ash may be present but soon becomes inconspicuous leaving a blackened surface beneath shrubs.	Leaf litter is charred or consumed but some plant parts are discernable. Herbaceous stubble extends above the soil surface. Some plant parts may still be standing, bases not deeply burned, and still recognizable. Surface is black after fire but this soon becomes inconspicuous. Charring is limited to <0.2 cm into the soil.
Fuels:	Herbaceous plants and foliage and fine twigs of woody shrubs and trees are charred to consumed but twigs and branches >0.5 cm remain. Coarser branches and woody debris are scorched to lightly charred but not consumed. Logs are scorched to blackened but not deeply charred. Rotten wood scorched to partially burned.	Typically, some leaves and twigs remain on plants and <60 percent of brush canopy is consumed. Foliage is largely consumed whereas fine twigs and branches >0.5 cm remain.	Typically, 50 to 90 percent of herbaceous fuels are consumed and much of the remaining fuel is charred.
Occurrence:	Light depth of burn commonly occurs on 10 to 100 percent of the burned area in natural fuels and 45 to 75 percent in slash fuels. Low values are associated with marginal availability of fine fuels whereas high values are associated with continuous fine fuels or wind-driven fires.	In shrublands where fine fuels are continuous, light depth of burn occurs on 10 to 100 percent depending on fine fuel moisture and wind. Where fine fuels are limited, burns are irregular and spotty at low wind speeds. Moderate to high winds are required for continuous burns.	Burns are spotty to uniform, depending on grass continuity. Light depth of burn occurs in grasslands when soil moisture is high, fuels are sparse, or fires burn under high wind. This is the dominant type of burning in most upland grasslands.

(con.)

Table 1.4 Part B

Depth of burn class	Vegetation type		
	Forests	**Shrublands**	**Grasslands**
Moderate			
Surface:	In upland forests litter is consumed and duff deeply charred or consumed, mineral soil not visibly altered but soil organic matter has been partially pyrolized (charred) to a depth >1.0 cm. Gray or white ash persists until leached by rain or redistributed by rain or wind. In forests growing on organic soils moderate depth of burn fires partially burn the root-mat but not the underlying peat or muck.	In upland shrublands litter is consumed. Where present, leaf mold deeply charred or consumed. Charring 1 cm into mineral soil, otherwise soil not altered. Gray or white ash quickly disappears. In shrub-scrub wetlands growing on organic soils moderate depth of burn fires partially burn the root-mat but not the underlying peat or muck.	In upland grasslands litter is consumed. Charring extends to <0.5 cm into mineral soil, otherwise soil not altered. Gray or white ash quickly disappears. In grasslands, sedge meadows and prairies growing on organic soils moderate depth of burn fires partially burn the root-mat but not the underlying peat or muck.
Fuels:	Herbaceous plants, low woody shrubs, foliage and woody debris <2.5 cm diameter consumed. Branch-wood 2.5 to 7.5 cm 90+ percent consumed. Skeletons of larger shrubs persist. Logs are deeply charred. Shallow-rooted, late successional trees and woody shrubs are typically left on pedestals or topple. Burned-out stump holes are common.	Herbaceous plants are consumed to the ground-line. Foliage and branches of shrubs are mostly consumed. Stems <1 cm diameter are mostly consumed. Stems >1 cm mostly remain.	Herbaceous plants are consumed to the ground-line.
Occurrence:	Moderate depth of burn occurs on 0 to 100 percent of natural burned areas and typically 10 to 75 percent on slash burns. High variability is due to variability in distributions of duff depth and woody debris.	Moderate depth of burn varies with shrub cover, age, and dryness. It typically occurs beneath larger shrubs and increases with shrub cover. Typically burns are more uniform than in light depth of burn fires.	Moderate depth of burn tends to occur when soil moisture is low and fuels are continuous. Then burns tend to be uniform. In discontinuous fuels high winds are required for high coverage in moderate depth of burn.
Table 1.4 Part C **Deep**			
Surface:	In forests growing on mineral soil the litter and duff are completely consumed. The top layer of mineral soil visibly altered. Surface mineral soil structure and texture are altered and soil is oxidized (reddish to yellow depending on parent material). Below oxidized zone, >1 cm of mineral soil appears black due to charred or deposited organic material. Fusion of soil may occur under heavy woody fuel concentrations. In forests growing on organic soils deep depth of burn fires burn the root-mat and the underlying peat or muck to depths that vary with the water table.	In shrublands growing on mineral soil the litter is completely consumed leaving a fluffy white ash surface that soon disappears. Organic matter is consumed to depths of 2 to 3 cm. Colloidal structure of surface mineral soil is altered. In shrub-scrub wetlands growing on organic soils deep depth of burn fires burn the root-mat and the underlying peat or muck to depths that vary with the water table.	In grasslands growing on mineral soil the litter is completely consumed leaving a fluffy white ash surface that soon disappears. Charring to depth of 1 cm in mineral soil. Soil structure is slightly altered. In grasslands growing on organic soils deep depth of burn fires burn the root-mat and the underlying peat or muck to depths that vary with the water table.
Fuels:	In uplands twigs and small branches are completely consumed. Few large, deeply charred branches remain. Sound logs are deeply charred and rotten logs are completely consumed. In wetlands twigs, branches, and stems not burned in the surface fire may remain even after subsequent passage of a ground fire.	In uplands twigs and small branches are completely consumed. Large branches and stems are mostly consumed. In wetlands twigs, branches, and stems not burned in the surface fire may remain even after subsequent passage of a ground fire.	All above ground fuel is consumed to charcoal and ash.
Occurrence:	In uplands deep depth of burn occurs under logs, beneath piles, and around burned-out stump holes, and typically occupies <10 percent of the surface except under extreme situations (e.g., extensive blow-down). In forested wetlands deep depth of burn can occur over large areas when the water table is drawn down during drought.	In uplands deep depth of burn typically is limited to small areas beneath shrubs where concentrations of deadwood burn-out. In shrub-scrub wetlands – see forests	In uplands deep depth of burn is limited to areas beneath the occasional log or anthropogenic features (e.g., fences, corrals). In wetlands – see forests.

litter. Typical forest-floor moisture contents associated with moderate depth of burn are litter (O_i) 10 to 20 percent and duff (O_e+O_a) less than 75 percent.

- *Deep:* In forests growing on mineral soil the surface litter, mosses, herbaceous plants, shrubs, and woody branches are completely consumed. Sound logs are consumed or deeply charred. Rotten logs and stumps are consumed. The top layer of the mineral soil is visibly oxidized, reddish to yellow. Surface soil texture is altered and in extreme cases fusion of particles occurs. A black band of charred organic matter 0.4 to 0.8 inch (1 to 2 cm) thick occurs at variable depths below the surface. The depth of this band increases with the duration of extreme heating. The temperatures associated with oxidized mineral soil are typical of those associated with flaming (greater than 932 °F or 500 °C) rather than smoldering (less than 932 °F or 500 °C). Thus, deep depth of burn typically only occurs where woody fuels burn for extended duration such as beneath individual logs or in concentrations of woody debris and litter-filled burned out stump holes. Representative forest-floor moisture contents associated with deep depth of burn are litter (O_i) less than 15 percent and duff (O_e+O_a) less than 30 percent. In areas with deep organic soils deep depth-of-burn occurs when ground fires consume the root-mat or burn beneath the root-mat. Trees often topple in the direction from which the smoldering fire front approached (Artsybashev 1983, Wein 1983, Hungerford and others 1995a,b).

Depth of burn varies continuously and, as is typical of classifications, there is some ambiguity at the class boundaries. The moderate depth of burn class is a broad class. Some investigators have chosen to divide the class into two classes (Morgan and Neuenschwander 1988, Feller 1998). The most common criteria for splitting the moderate class are between areas with shallow versus deep duff. Partial consumption of a deep layer may be more severe than complete consumption of a shallow layer for some effects but not others. For example, consumption of 8 inches (20 cm) of a 12-inch (30-cm) duff layer represents greater fuel consumption, smoke production, energy release, and nutrient release than complete consumption of a 4-inch (10-cm) layer, but because organic matter (duff) is a good insulator, heat effects are limited to less than an inch (1 to 2 cm) below the duff-burn boundary. In contrast, complete consumption of the 4-inch (10-cm) layer can be expected to have similar thermal effects at three to five times greater

depth (fig. 1.6). Duff consumption is a complex process (Johnson and Miyanishi 2001). Depth, bulk density, heat content, mineral content, moisture content, and wind speed all affect the energy release rate and soil heating. As these factors cannot be readily determined after a fire, it is difficult to describe postburn criteria that can be used to consistently split the class. While postfire examination of ground charring alone may not be adequate for classifying depth of burn, the actual depth can be inferred from the preponderance of the evidence, which includes reconstructing the prefire vegetative structure. Careful postfire observations of soil characteristics, fuel consumption, and the depth of charring of residual plant materials can be used to classify the depth of burn by using the descriptive characteristics provided in table 1.4.

Fire Severity Classification

Judging fire severity solely on ground-based processes ignores the aboveground dimension of severity implied in the ecological definition of the severity of a disturbance (White and Pickett 1985). This is especially important because soil heating is commonly shallow even when surface fires are intense (Wright and Bailey 1982, Vasander and Lindholm 1985, Frandsen and Ryan 1986, Hartford and Frandsen 1992, Ryan 2002). Ryan and Noste (1985) combined fire intensity classes with depth of burn (char) classes to develop a two-dimensional matrix approach to defining fire severity. Their system is based on two components of fire severity: (1) an aboveground heat pulse due to radiation and convection associated with flaming combustion, and (2) a belowground heat pulse due principally to conduction from smoldering combustion where duff is present or radiation from flaming combustion where duff is absent—in other words, bare mineral soil. Fire-intensity classes qualify the relative peak energy release rate for a fire, whereas depth of burn classes qualify the relative duration of burning. Their concept of severity focuses on the ecological work performed by fire both above ground and below ground. Ryan (2002) combined surface fire characteristic classes (table 1.3) and depth of burn classes (table 1.4) to revise the Ryan and Noste (1985) fire severity matrix (table 1.5). By this nomenclature two burned areas would be contrasted as having had, for example, an active spreading-light depth of burn fire versus an intense-moderate depth of burn fire. The matrix provides an approach to classifying the level of fire treatment or severity for ecological studies at the scale of the individual plant, sampling quadrat, and the community. The Ryan and Noste (1985) approach has been used to interpret differences in plant survival and regeneration (Willard and others 1995, Smith and

USDA Forest Service Gen. Tech. Rep. RMRS-GTR-42-vol. 4. 2005

13

Table 1.5—Two-dimensional Fire Severity Matrix (modified from Ryan and Noste 1985) and relative one-dimensional severity rating.

Characteristic Fire Behavior	Characteristic Depth of Burn			
	Unburned/Scorched	Light	Moderate	Deep
Crowning	**Low** (common edge effect) – when radiation and convection from nearby burning scorch foliage but surface is unburned. **Moderate** (occurs) – when fire burns over snow or water (wetlands).	**Moderate** – when crown-fire occurs over wet duff/soil, or thin (<4cm) duff.[1] **High** – when crownfire occurs over bare mineral soil.	**Moderate** – when residual duff (uplands) or root mat (wetlands) are present. **High** – when duff or root mat is completely consumed.	**High**
Intense/Running & Torching	See Above	See Above	See Above	**High**
Active/Spreading	**Low** – See Above	**Low**	**Low** – when residual duff (uplands) or root mat (wetlands) are present. **High** – when duff or root mat is completely consumed.	**Moderate** – when forest canopy remains. **High** – in forests shrublands and grasslands where above-ground vegetation is consumed.
Creeping	**Low** – boundary condition of no practical significance, except as noted below.	**Low**	**Low** – when residual duff (uplands) or root mat (wetlands) are present. **Moderate** – when duff or root mat is completely consumed.	See Above
Unburned	**Refugia** – flora and fauna not directly affected by fire but microenvironment may be altered.	NA	NA	NA

[1] Duff insulates the mineral soil from intense heat associated with flaming and thin duff does not burn independently by smoldering combustion. As a result maximum temperatures and soil heat flux are reduced.

Fischer 1997, Feller 1998) and to field-validate satellite-based maps of burned areas (White and others 1996). The depth of burn characteristics are appropriate for quadrat-level descriptions in species response studies and for describing fire severity on small plots within a burned area.

In the literature there is common usage of a one-dimension rating of fire severity (Wells and others 1979, Morrison and Swanson 1990, Agee 1993, DeBano and others 1998, and many others). The single-adjective rating describes the overall severity of the fire and usually focuses primarily on the effects on the soil resource. The fire severity rating in table 1.5 provides guidance for making comparisons to the two-dimensional severity rating of Ryan and Noste (1985) and Ryan (2002), and for standardizing the use of the term. At the spatial scale of the stand or community, fire severity needs to be based on a sample of the distribution of fire severity classes. In the original Rainbow volume on the effects of fire on soils, Wells and others 1979 (see also

14

USDA Forest Service Gen. Tech. Rep. RMRS-GTR-42-vol. 4. 2005

Ryan and Noste 1985, DeBano and others 1998) developed the following criteria to do this:

- Low severity burn—less than 2 percent of the area is severely burned, less than 15 percent moderately burned, and the remainder of the area burned at a low severity or unburned.
- Moderate severity burn—less than 10 percent of the area is severely burned, but more than 15 percent is burned moderately, and the remainder is burned at low severity or unburned.
- High severity burn—more than 10 percent of the area has spots that are burned at high severity, more than 80 percent moderately or severely burned, and the remainder is burned at a low severity.

The Wells and others (1979) criteria for defining the burn severity class boundaries are somewhat arbitrary but were selected on the basis of experience recognizing that even the most severe of fires has spatial variation due to random variation in the fire environment (fuels, weather, and terrain), and particularly localized fuel conditions. Recently, Key and Benson (2004) have developed a series of procedures for documenting fire severity in the context of field validation of satellite images of fire severity. Their procedures result in a continuous score, called the Composite Burn Index (CBI), which is based on visual observation of fuel consumption and depth of burn in several classes of fuels and vegetation.

In most situations, depth of burn is the primary factor of concern when assessing the impacts of fire on soil and water resources (fig. 1.7). Depth of burn relates directly to the amount of bare mineral soil exposed to rain-splash, the depth of lethal heat penetration, the depth at which a hydrophobic layer will form, the depth at which other chemical alterations occur, and the depth to which microbial populations will be affected. As such it affects many aspects of erodability and hydrologic recovery (Wright and Bailey 1982, DeBano and others 1998, Gresswell 1999, Pannkuk and others 2000). However, depth of burn is not the only controlling factor (Ryan 2002). For example, the surface microenvironment, shaded versus exposed, in surface fires versus crown fires can be expected to affect postfire species dynamics regardless of the depth of burn (Rowe 1983). In a surface fire, needles are killed by heat rising above the fire (Van Wagner 1973, Dickinson and Johnson 2001), thereby retaining their nutrients. Thus, litterfall of scorched needles versus no litterfall in crown fire areas can be expected to affect postfire nutrient cycling. Further, rainfall simulator experiments have shown that needle cast from underburned trees reduced erosion on sites where duff was completely consumed in contrast to crown fire areas with similar

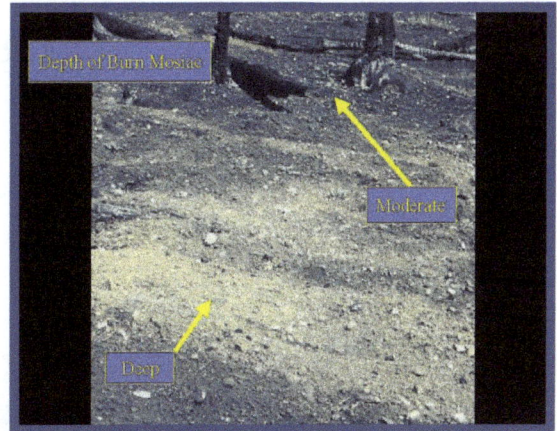

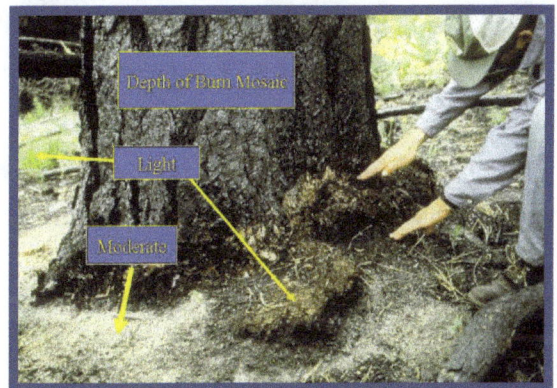

Figure 1.7—Depth of burn mosaics showing: (A) moderate to deep burn depths; and (B) light to moderate depths. (Photos by USDA Forest Service).

depth of burn (Pannkuk and others 2000). Thus, a lethal stand replacement crown fire (in other words, a fire that kills the dominant overstory; Brown 2000) represents a more severe fire treatment than a lethal stand replacement surface fire even when both have similar depth of burn (fig. 1.8). Thus for many ecological interpretations it is desirable to use the two-dimensional approach to rating fire severity to account for the effects of both the aboveground and belowground heat pulses.

A Conceptual Model

The previous discussion leads to development of a conceptual model to help planners, managers, and decisionmakers appreciate the spectrum of watershed responses to fire severity. The conceptual model describes fire severity as ranging from low water resource

USDA Forest Service Gen. Tech. Rep. RMRS-GTR-42-vol. 4. 2005

15

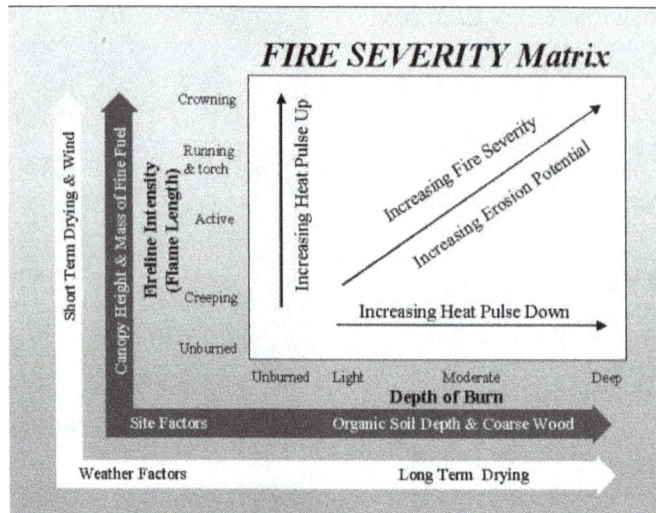

Figure 1.8—Site and weather factors associated with increasing fire severity and erosion potential.

responses likely to be experienced with a prescribed fire and no accompanying hydrologic events to high resource responses that could be expected from stand-replacing wildfire in forests and major storm events (fig. 1.9). Once again, fire severity is not directly related to fire intensity for the reasons discussed above.

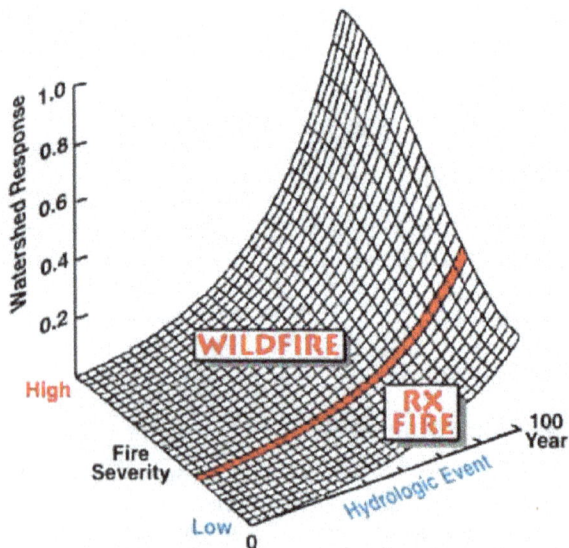

Figure 1.9—A conceptual model of watershed responses to fire severity.

Prescribed fire conditions (generally low fire severity) are depicted in figure 1.9 on the lower portion of the fire severity and resource response curve. These conditions are typically characterized by lower air temperature, higher relative humidity, and higher soil moisture burning conditions, where fuel loading is low and fuel moisture can be high. These conditions produce lower fire intensities and, as a consequence, lower fire severity leading to reduced potential for subsequent damage to soil and water resources. Prescribed fire, by its design, usually has minor impacts on these resources.

Fire at the other end of the severity spectrum (left side of fig. 1.9) more nearly represents conditions that are present during a wildfire, where temperatures, wind speeds, and fuel loadings are high, and humidity and fuel moisture are low. In contrast to prescribed burning, wildfire often has a major effect on soil and watershed processes, leading to increased sensitivity of the burned site to vegetative loss, increased runoff, erosion, reduced land stability, and adverse aquatic ecosystem impacts (Agee 1993, Pyne and others 1996, DeBano and others 1998).

Differences in watershed response along the spectrum between prescribed burning and wildfire or within an individual fire depend largely upon fire severity and the magnitude of hydrologic events following fire. Although significant soil physical, chemical, and biological impacts would occur with a stand-replacing wildfire, there would be no immediate watershed response in the absence of a hydrologic event. As indicated in figure 1.9, the magnitude of watershed response is keyed to the size (return period) of the hydrologic event or storm and the timing relative to fire. In addition to climate interactions with fire severity, topography also has a major influence on watershed response. Forested watersheds in mountainous regions of the West or East respond to storm events very differently than those in the Coastal Plain or Great Lakes where relief is at a minimum.

Fire-Related Disturbances _____

Fire

As discussed above, the primary disturbance to soil and water resources is a function of fire severity. Removal of vegetation alone is sufficient to produce significant soil and watershed responses (Anderson and others 1976, Swank and Crossley 1988, Neary and Hornbeck 1994). In areal extent and severity level, the disturbance produced by fire can be small or large, and uniform or a chaotic matrix. Beyond the initial vegetation and watershed condition impacts of the physical process of fire, suppression activities can add further levels of disturbance to both soils and water. These disturbances need to be evaluated along with those produced by fire.

Fire Suppression

Fire managers need to be aware of the soil and water impacts of suppression activities. The major soil and water disturbances associated with fire suppression are fire lines, roads, and fire retardants. Fire control lines created by hand or large equipment disturb the soil, alter infiltration, become sources of sediment, and can alter runoff patterns (see chapter 2). Suppression activities in boreal forests have been known to lead to decreasing permafrost depth resulting in downcutting and slope failures (Viereck 1982). Roads are already major sources of sediment in forest watersheds (Brown and Binkley 1994; Megahan 1984, Swift 1984). Temporary roads built during suppression operations increase the size of road networks and erosion

hazard areas within watersheds. In addition, traffic from heavy equipment and trucks can deteriorate the surfaces of existing roads, making them more prone to erosion during rainfall. Although fire retardants are basically fertilizers, they can produce serious short-term water quality problems if dropped into perennial streams.

Type of Effects

The effects of fire on soil and water resources can be direct and indirect. They occur at many scales (microsite to ecosystem level), in different patterns, and over variable periods. In most watersheds of fire dependent or dominated ecosystems, fire impacts to soils and water are significant components and variable backgrounds of cumulative watershed effects. An understanding of these effects is important for land managers who deal with wildfire and use prescribed fire to accomplish ecosystem management objectives.

This Book's Objective _____

The objective of this volume in the Rainbow series is to provide an overview of the state-of-the-art understanding of the effects of fire on soils and water in wildland ecosystems. It is meant to be an information guide to assist land managers with fire management planning and public education, and a reference on fire effects processes, pertinent publications, and other information sources. Although it contains far more information and detailed site-specific effects of fire on soils and water than the original 1979 Rainbow volumes, it is not designed to be a comprehensive research-level treatise or compendium. That challenge is left to several textbooks (Chandler and others 1991, Agee 1993, Pyne and others 1996, DeBano and others 1998). The challenge in developing this volume was in providing a meaningful summary for North American fire effects on soils and water resources despite enormous variations produced by climate, topography, fuel loadings, and fire regimes.

Notes

Part A

Effects of
Fire on Soil

Leonard F. DeBano
Daniel G. Neary

Part A—The Soil Resource: Its Importance, Characteristics, and General Responses to Fire

Introduction

Soil is a heterogeneous mixture of mineral particles and organic matter that is found in the uppermost layer of Earth's crust. The soil is formed as a product of the continual interactions among the biotic (faunal and floral), climatic (atmospheric and hydrologic), topographic, and geologic features of the environment over long periods (Jenny 1941, Singer and Munns 1996). Soils are important components of ecosystem sustainability because they supply air and water, nutrients, and mechanical support for the sustenance of plants. Soils also absorb water during infiltration. By doing so, they provide storage for water as well as acting as a conduit that delivers water slowly from upstream slopes to channels where it contributes to streamflow. There is also an active and ongoing exchange of gases between the soil and the surrounding atmosphere. When the infiltration capacity of the soil is exceeded, organic and inorganic soil materials are eroded and become major sources of sediment, nutrients, and pollutants in streams. These water and

erosional processes are described in part B and chapters 5, 6, and 7 of this publication.

To fully evaluate the effects of fire on a soil, it is first necessary to quantitatively describe the soil and then to discuss the movement of heat through the soil during a fire (wildfires or prescribed burns). During the process of soil heating, significant changes can occur in the physical, chemical, and biological properties that are relevant to the future productivity and sustainability of sites supporting wildland ecosystems. This introductory part A presents a general discussion on the properties of soils and the heating processes occurring in soils during fires, and provides some general information on the physical, chemical, and biological responses to fire. This part is also intended as an extended executive summary for those interested in the general concepts concerning fire effects on soils. Readers who are interested in more detailed information on fire effects on soils are directed to indepth discussions of the individual physical, chemical, and biological properties and processes in soil that are affected by fire (see chapters 2, 3, and 4, respectively).

USDA Forest Service Gen. Tech. Rep. RMRS-GTR-42-vol. 4. 2005

21

The General Nature of Soil _____

The features and the importance of soils are usually inconspicuous to the average person. Soil is simply the substrate that is walked on, that is used to grow trees or a garden, that creates a source of dust when the wind blows, that provides material that washes down the hillslope during runoff, or that is bared when the firefighter builds a fireline. However, closer examination shows that soil is a complex matrix made up of variable amounts of mineral particles, organic matter, air, and water. The inorganic constituents of soils contain a wide array of primary (for example, quartz) and secondary (for example, clays) minerals.

Organic matter is the organic portion of the soil and is made up of living and dead biomass that contains a wide range of plant nutrients. The living biomass in the soil consists of plant roots, microorganisms, invertebrates, and small and large vertebrate fauna that burrow in the soil. The nonliving organic matter is made up primarily of dead bark, large woody debris (dead trees, limbs, and so forth), litter, duff, and finely decomposed humus materials. Organic matter is broken down and decomposed by the actions of animals and microorganisms living in the soil. An important component of organic matter is humus, the colloidal soil organic matter (particles smaller than 3.9 to 20 x 10^{-6} inches [0.001 to 0.005 mm] in diameter) that decomposes slowly. Humus provides negative adsorption sites similar to clay minerals and also acts as an organic glue that helps to hold mineral soil particles together to form aggregates. This contributes to soil structure that creates pore space soil and provides passageways for the movement of air and water. The decomposition of organic matter also plays a central role in the cycling and availability of nutrients essential for plant growth. Organic matter also provides a source of energy necessary to support microbial populations in the soil.

Water and air occupy the empty spaces (pore space) created by the mineral-organic matter matrix in the soil. A delicate balance exists between the amount of pore space filled with water and that filled with air, which is essential for root respiration by living plants. Too much soil water can limit plant growth if the pore space is saturated with water (for example, waterlogged, anaerobic soils). In contrast, when too little water is available, plant growth can be limited by the lack of water necessary for transpiration and other physiological functions necessary for the growth of plants. Soil water also contains dissolved ions (cations and anions), and this is called the "soil solution." Many of the ions in the soil solution are absorbed by plant roots and used for plant growth.

A combination of the inorganic materials described above, along with variable amounts of finely divided and partially decomposed organic matter (humus), provides structure to the soil. Soil structure is the arrangement of the inorganic components of the soil into aggregates having distinctive patterns (for example, columnar, prismatic, blocky). These aggregates are stabilized by organic matter that provides an overall porous structure to the soil.

Soil Properties—Characteristics, Reactions, and Processes _____

The reader needs to be aware of some general definitions and terminology that are used in reference to soil properties when reading chapters 2, 3, and 4. The term "soil properties" is collectively used to include the characteristics, reactions, and processes that occur in soils. Traditionally, soils have been described in terms of physical, chemical, and biological properties. This classification is arbitrary, and in many cases the three classes of soil properties are not mutually exclusive but are so closely interconnected that it is impossible to clearly place a soil property in any one of the three categories. This interrelationship is particularly apparent in the discussions on organic matter and the different processes responsible for nutrient cycling. Because of this interdependency, we attempted to discuss the physical, chemical, and biological dimensions of nutrient cycling and organic matter separately, and then to cross-reference these discussions among the three general categories.

Soil Profile _____

Variable amounts and combinations of minerals, organic matter, air, and water produce a wide range of physical, chemical, and biological properties of a soil. However, these properties are not randomly distributed but occur in an orderly arrangement of horizontal layers called *soil horizons*. The arrangement of these layers extending from the surface litter downward to bedrock is referred to as the *soil profile*. A schematic profile is shown in figure A.1, and a real profile in figure A.2. Some profiles have distinct horizons as shown in figure A.2, but some soils have horizons that are not so distinct. The uppermost layers consist mainly of organic matter in various stages of decomposition. The surface litter layer (L-layer) is made up of undecomposed organic material that retains the features of the original plant material (leaves, stems, twigs, bark, and so forth). Immediately below the undecomposed layer is another organic layer that is in various stages of decomposition. It is called the fermentation layer (F-layer). In the F-layer, some of the original plant structure may still be discernable depending on the extent of decomposition. The lowermost surface organic matter layer is the humus layer (H-layer) that is completely

SOIL PROFILE SCHEMATIC

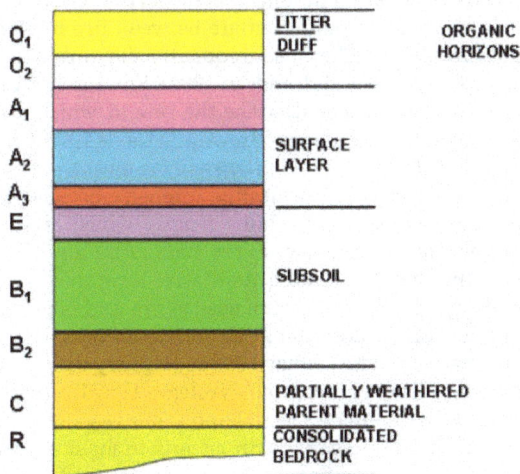

Figure A.1—A schematic of a well-developed (mature) soil profile showing a complete suite of the organic and inorganic soil horizons. (Figure courtesy of the USDA Forest Service, National Advanced Fire and Resource Institute, Tucson, AZ).

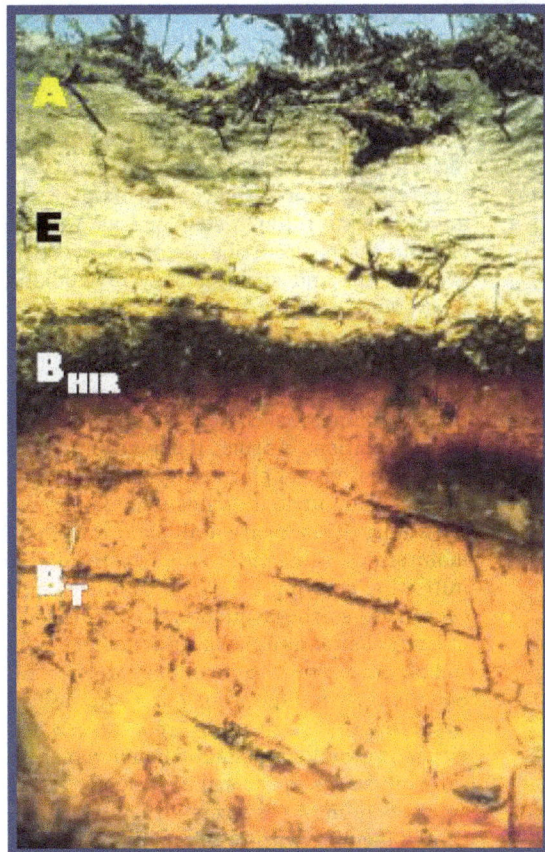

Figure A.2—Profile of a Pomona fine sand (Ultic Haplaquod, sandy, siliceous, hyperthermic family) from a slash pine (*Pinus elliottii*) stand in the flatwoods of northern Florida. (Photo by Daniel Neary).

decomposed organic matter. The H-layer is an important site for nutrient availability and storage. The finely decomposed organic matter in the H-layer is also the source of aggregating substances that combine with the mineral soil particles in the upper inorganic horizons to produce soil structure. The original plant structure is no longer identifiable in the H-layer. The combined F-and H-layer is commonly referred to as the duff.

More recent designations have been developed for the L-, F-, and H-layers described above. Current taxonomic terminology refers to the organic horizon as the O horizon. The L-layer is referred to as the O_i or O_1 horizon. The F-layer is designated as the O_e, or part of the O_2 horizon and the H-layer is denser than the L- and F-layers and is designated as the O_a or O_2 horizon.

The mineral soil horizons begin with the uppermost part of the A-horizon and extend downward to bedrock. Depending upon the age and development of the soil profile, there can be several intermediate mineral horizons (for example, E-, B-, and C-horizon). The A-horizon is the top mineral layer, and the upper part of this horizon often contains large quantities of finely decomposed organic matter (humus). The mineral E-horizon is located immediately below the A-horizon. It is the site where substantial amounts of silicate, clay, iron, aluminum, carbonate, gypsum, or silicon are lost by weathering and leaching that occurs during soil development. Materials leached downward from the

E-horizon accumulate mainly in the B-horizon. In well-developed (mature) soil profiles, the original rock structure can no longer be recognized in the B-horizon. The C-horizon is unconsolidated parent rock material remaining above the R-horizon that is made up of hard consolidated bedrock.

Importance of Organic Matter

The effects of fire on soils cannot be fully evaluated unless the role of organic matter in the functioning and sustainability of soil ecosystems is understood. Organic matter is the most important soil constituent that is found in soils. Although it is concentrated on the soil surface where it makes up most of the L-, F-, and H-layers, it also plays an important role in the

USDA Forest Service Gen. Tech. Rep. RMRS-GTR-42-vol. 4. 2005

23

properties of the underlying mineral soil horizons. Ecologically, organic matter plays three major roles:

- Organic matter enhances the structure of soils. The most important physical function of organic matter is its role in creating and stabilizing soil aggregates. A porous well-structured soil is essential for the movement of water, air, and nutrients through soils and as a result organic matter contributes directly to the productivity and sustainability of wildland ecosystems.
- Chemically, organic matter maintains and regulates the biogeochemical cycling of nutrients by providing an active medium for sustaining numerous chemical and biological transformations. As such, it plays a key role in the productivity of plant ecosystems. Its specific roles in nutrient cycling include: providing a storage reservoir for all plant nutrients, maintaining a balanced supply of available nutrients, creating a large cation exchange capacity for storing available nutrients in soils, and functioning as a chelating agent for essential plant micronutrients (for example, iron).
- Soil provides a habitat for plant and animal organisms that range in size from bacteria and viruses to small mammals. The microbiological populations in the soil are usually inconspicuous to most observers although the soil can be teeming with hundreds of millions of microorganisms in each handful of forest soil. Their activity and diversity far exceeds that of the other biological components of forest ecosystems (for example, vegetation, insects, wildlife, and so forth). Soil organic matter is of particular importance to soil microorganisms because it provides the main source of energy for sustaining soil microorganisms. Soil microorganisms are involved in nearly all of the processes responsible for the cycling and availability of nutrients such as decomposition, mineralization, and nitrogen (N) fixation.

Fire Effects—General Concepts and Relationships

The effects of fire on soil properties must be evaluated within the concept of a complex organic and inorganic matrix of the soil profile described above. The magnitude of change occurring during a fire depends largely upon the level of fire severity, combustion and heat transfer, magnitude and depth of soil heating, proximity of the soil property to the soil surface, and the threshold temperatures at which the different soil properties change.

Severity and Fire Intensity

When discussing the effects of fire on the soil resource it is important to differentiate between fire intensity and fire severity because frequently they are not the same (Hartford and Frandsen 1992). Fire intensity is a term that is used to describe the rate at which a fire produces thermal energy (Brown and Davis 1973, Chandler and others 1991). *Fire intensity* is most frequently quantified in terms of fireline intensity because this measure is related to flame length, which is easily measured (DeBano and others 1998). *Fire severity*, on the other hand, is a more qualitative term that is used to describe ecosystem responses to fire and is particularly useful for describing the effects of fire on the soil and water system (Simard 1991). Severity reflects the amount of energy (heat) that is released by a fire and the degree that it affects the soil and water resources. It is classified according to postfire criteria on the site burned and has been classified into low, moderate, and high fire severity. Detailed descriptions of the appearances of these three severity levels for timber, shrublands, and grasslands are presented in table 1.4, Parts A,B,C respectively. (Note: this table and the parts can be found on pages 11 and 12 in Chapter 1).

The level of fire severity depends upon:

- Length of time fuel accumulates between fires and the amount of these accumulated fuels that are combusted during a fire (Wells and others 1979).
- Properties of the fuels (size, flammability, moisture content, mineral content, and so forth) that are available for burning
- The effect of fuels on fire behavior during the ignition and combustion of these fuels.
- Heat transfer in the soil during the combustion of aboveground fuels and surface organic layers.

High intensity fires can produce high severity changes in the soil, but this is not always the case. For example, low intensity smoldering fires in roots or duff can cause extensive soil heating and produce large changes in the nearby mineral soil. In contrast, high severity crown fires may not cause substantial heating at the soil surface because they sweep so rapidly over a landscape that not much of the heat generated during combustion is transferred downward to the soil surface.

Combustion

Energy generated as heat during the combustion of aboveground and surface fuels provides the driving force that causes a wide range of changes in soil properties during a fire (DeBano and others 1998). *Combustion* is the rapid physical-chemical destruction of organic matter that releases the large amounts of energy stored in fuels as heat. These fuels consist of dead and live standing biomass, fallen logs, surface litter (including bark, leaves, stems, and twigs), humus, and sometimes roots. During the combustion process heat and a mixture of gaseous and particulate byproducts are released. Flames are the most visual characteristic of the combustion process.

Three components are necessary in order for a fire to ignite and initiate the combustion process (Countryman 1975). First, burnable fuel must be available. Second, sufficient heat must be applied to the fuel to raise its temperature to the ignition point. And last, sufficient oxygen (O_2) is needed to be present to keep the combustion process going and to maintain the heat supply necessary for ignition of unburned fuel. These three components are familiar to fire managers as the fire triangle.

A common sequence of physical processes occurs in all these fuels before the energy contained in them is released and transferred upward, laterally, or downward where it heats the underlying soil and other ecosystem components. There are five physical phases during the course of a fire, namely: preignition, ignition, flaming, smoldering, and glowing (DeBano and others 1998). These different phases have been described in more detail by several authors (Ryan and McMahon 1976, Sandberg and others 1979, Pyne and others 1996). Preignition is the first phase when the fuel is heated sufficiently to cause dehydration and start the initial thermal decomposition of the fuels (pyrolysis). After ignition, the three phases of combustion that occur are flaming, smoldering, and glowing combustion. When active flaming begins to diminish, smoldering increases, and combustion diminishes to the glowing phase, which finally leads to extinction of the fire.

Heat Transfer

Heat produced during the combustion of aboveground fuels (for example, dead and live vegetation, litter, duff) is transferred to the soil surface and downward through the soil by several heat transfer processes (radiation, convection, conduction, vaporization, and condensation). *Radiation* is the transfer of heat from one body to another, not in contact with it, by electromagnetic wave motion; it increases the molecular activity of the absorbing substance and causes the temperature to rise (Countryman 1976b). *Conduction* is the transfer of heat by molecular activity from one part of a substance to another, or between substances in contact, without appreciable movement or displacement of the substance as a whole (Countryman 1976a). *Convection* is a process whereby heat is transferred from one point to another by the mixing of one portion of a fluid with another fluid (Chandler and others 1991). Vaporization and condensation are important in fire behavior and

serve as a coupled reaction facilitating more rapid transfer of heat through soils. *Vaporization* of water occurs when it is heated to a temperature at which it changes from a liquid to a gas. *Condensation* occurs when water changes from a gas to a liquid with the simultaneous release of heat. The coupled reaction of vaporization and condensation provides a mechanism for the transfer of both water and organic materials through the soil during fires (DeBano and others 1998).

The mechanisms for the transfer of heat through different ecosystems components vary widely (table A.1). Although heat is transferred in all directions, large amounts of the heat generated during a fire are lost into the atmosphere (along with smoke, gases, and particulate matter generated by fire) by radiation, convection, and mass transfer (DeBano and others 1998). It has been estimated that only about 10 to 15 percent of the heat energy released during combustion of aboveground fuels is absorbed and transmitted directly downward to the litter and duff, or mineral soil if surface organic layers are absent. This occurs mainly by radiation (DeBano 1974, Raison and others 1986). Within the fuels themselves most of the heat transfer is by radiation, convection, and mass transfer. Meanwhile in the soil, convection and vaporization and condensation are the most important mechanisms for heat transfer in a dry soil. In a wet or moist soil, conduction can contribute significantly to heat transfer. The heat transfer processes occurring in the soil are described in more detail in chapter 2 of this publication. The transfer of heat through mineral soil is important because it causes soil heating and produces changes in the soil physical, chemical, and biological properties described in chapters 2, 3, and 4, respectively.

Depth and Magnitude of Soil Heating

As heat is transferred downward into and through the soil, it raises the temperature of the soil. The greatest increase in temperature occurs at, or near, the soil surface. Within short distances downward in the soil, however, the temperature increases quickly diminish so that within 2.0 to 3.9 inches (5 to 10 cm) of the soil surface the temperatures are scarcely above ambient temperature. A diagram of the heat increases with depth is called a temperature profile and is useful for determining the amount of change that occurs in a soil during a fire as the result of heating. The magnitude of these temperature increases depends on the severity of the fire as described above. Residence time of the fire (the duration of heating) is a particularly important feature of fires, affecting the depth and magnitude of soil heating. Detailed information on temperature profiles that can develop during grassland, shrubland, and forest fires is presented in chapter 2.

USDA Forest Service Gen. Tech. Rep. RMRS-GTR-42-vol. 4. 2005

25

Table A.1—Importance of different heat transfer mechanisms in the transfer of heat within different ecosystem components.

Heat transfer mechanism	Ecosystem component	Importance to heat transfer
Radiation	Air	Medium
	Fuel	High
	Soil	Low
Conduction	Air	Medium
	Fuel	Low
	Soil	Low (dry), high (wet)
Convection	Air	High
	Fuel	Medium
	Soil	Low
Mass transfer	Air	High
	Fuel	Low
	Soil	Low
Vaporization/condensation	Air	Low
	Fuel	Medium
	Soil	High

Temperature Thresholds of Soil Properties

An important feature when assessing the effect of fire on soil properties is the temperature at which nutrients are volatilized or that irreversible damage occurs to a particular soil property. This temperature is called the threshold temperature (DeBano and others 1998). Temperature thresholds have been identified for numerous physical, chemical, and biological properties. The ranges of temperatures over which some common soil properties change in response to soil heating are displayed in figure 1.6. These temperature thresholds have been classified into three general classes, namely:

- Relatively insensitive soil properties that do not change until temperatures have reached over about 842 °F (450 °C). This class includes clays, cations (calcium, magnesium, potassium) and other minerals such as manganese.
- Moderately sensitive soil properties that are changed at temperatures between 212 and 752 °F (100 and 400 °C). Materials belonging to this class include sulfur, organic matter, and soil properties dependent upon organic matter.
- Sensitive soil properties are those that are changed at temperatures less than 212 °F (100 °C). Examples of sensitive materials are living microorganisms (for example, bacteria, fungi, mychorrizae), plant roots, and seeds. This class also includes many of the biologically mediated nutrient cycling processes in soils.

Threshold values for specific physical, chemical, and biological soil properties are described in greater detail in chapters 2, 3, and 4, respectively.

Location of Soil Properties

The natural differentiation of the soil profile into horizons creates a stratification of the physical, chemical, and biological soil properties discussed above. Understanding this stratified arrangement is necessary in order to accurately assess the effects of fire and soil heating on the different soil properties. The soil properties near or on the soil surface are the most directly exposed to heat that is radiated downward during a fire. Soil heating generally decreases rapidly with soil depth in a dry soil because dry soil is a poor conductor of heat.

The organic horizons that make up the forest floor are particularly important when discussing fire effects, because they are directly subjected to heat produced by burning of surface fuel, and they contain a large proportion of the organic matter found in soil profiles (DeBano and others 1998). Although some of the individual nutrients contained in the organic matter may not be volatilized, others such as N are vaporized in direct proportion to the amount of organic material lost. Most of the fire effects produced during surface fires occur in the upper organic horizons, or in the top part of the A-horizon. Heating of the B-horizon and deeper in the soil profile occurs only when roots are ignited and create localized subsurface heating.

Assessing Fire Effects on Soils

The above general information on soils along with the detailed information given in chapters 2, 3, and 4 on physical, chemical, and biological soil properties, respectively, can be used to assess fire effects on soils. This assessment requires being able to quantify the effect of fire and associated soil heating on soil properties and includes three main steps:

- First, the amount of energy radiated downward during combustion of fuels must be estimated. This energy is the driving force responsible for producing changes in soil properties. In general, the magnitude of change in individual soil properties is largely dependent upon the amount of energy radiated onto the soil surface, and subsequently transferred downward into the underlying duff and mineral soil. This radiated heat increases the temperature and causes changes in organic matter and other soil properties. Therefore, the postfire appearance of vegetation, litter, duff, and upper soil horizons can be used to estimate the amounts of surface heating and used to classify fire severity as low, moderate, or high. The basic assumption used in this technique is that as the amount of heat radiated downward increases, the severity increases from low to moderate to high (in other words, the magnitude of change in the soil property increases). An earlier discussion describing fire severity provides the necessary framework for establishing the severity of the fire in different ecosystems.
- After the fire severity has been established it can be used to estimate soil temperatures that develop when different ecosystems are burned (for example, grassland, shrubland, forests). Representative soil temperatures for different severities of burning for different ecosystems are discussed in chapter 2.
- Finally, once the approximate soil temperatures have been established, the changes in specific soil properties can be estimated using temperature threshold information. The percentage loss of different nutrients can be used along with estimates of the quantities of nutrients affected by fire to estimate the total nutrient losses, or gains, which occurred on a specific site during a fire. Specific information on the temperatures at which different physical, chemical, and biological soil properties changes are give in chapters 2, 3, and 4, respectively.

Management Implications

The condition of the soil is a key factor in the productivity of forest ecosystems and the hydrologic functioning of watersheds. Cumulative impacts that occur in soils as a result of fire can manifest themselves in significant changes in soil physical, chemical, or biological properties. These include breakdown in soil structure, reduced moisture retention and capacity, development of water repellency, changes in nutrient pools cycling rates, atmospheric losses of elements, offsite erosion losses, combustion of the forest floor, reduction or loss of soil organic matter, alterations or loss of microbial species and population dynamics, reduction or loss of invertebrates, and partial elimination (through decomposition) of plant roots. Although the most serious and widespread impacts on soils occur with stand-replacing wildfires, prescribed fires sometimes produce localized problems. Managers need to be aware of the impacts that fire can have on soil systems, and that these impacts can lead to undesired changes in site productivity, sustainability, biological diversity, and watershed hydrologic response.

Land managers need to be aware that some changes in soil systems after fire are quite obvious (for example, erosion, loss of organic matter), but others are subtle and can have equal consequences to the productivity of a landscape. For example, carbon and N are the key nutrients affected by burning. The significance of these changes is directly tied to the productivity of a given ecosystem. With a given change in N capital, the productivity of a nutrient-rich soil system might not significantly change following burning. A similar loss in N capital in a nutrient stressed system could result in a much greater change in productivity. Recovery of soil nutrient levels after fires can be fairly slow in some ecosystems, particularly those with limited N; and in semiarid regions such as the Southwestern United States and Northern Mexico nutrient fixation and turnover rates are slow.

Summary

This introductory section to the chapters in part A has provided information on the general nature of soil systems, some of the important soil properties, the character of soil profiles, and important constituents such as organic matter. It also introduces key concepts of heat transfer to soils and thresholds for important soil properties. The three chapters in part A address in greater detail fire effects on individual physical, chemical, and biological properties and processes in soil systems (see chapters 2, 3, and 4, respectively).

USDA Forest Service Gen. Tech. Rep. RMRS-GTR-42-vol. 4. 2005

27

Notes

Leonard F. DeBano
Daniel G. Neary
Peter F. Ffolliott

Chapter 2:
Soil Physical Properties

Introduction

Soil physical properties are those characteristics, processes, or reactions of a soil that are caused by physical forces that can be described by, or expressed in, physical terms or equations (Soil Science Society of America 2001). These physical properties (including processes) influence the mineral component of the soil and how it interacts with the other two components (chemical and biological). Plants depend on the physical characteristics of soils to provide the medium for growth and reproduction. Fire can produce significant changes in the soil that profoundly affect the ecology of plants (Whelan 1995). The effect of fire on individual soil physical properties depends on the inherent stability of the soil property affected and the temperatures to which a soil is heated during a fire. The physical mechanisms responsible for heat transfer into soils are also discussed in this chapter along with the temperatures that develop during different severities of burning in several wildland ecosystems. The relationships between soil physical properties affected by fire and erosional processes are also reviewed.

Soil Physical Characteristics

Important physical characteristics in soil that are affected by soil heating include: texture, clay content, soil structure, bulk density, and porosity (amount and size). The threshold temperatures for these soil physical characteristics are given in table 2.1. Physical properties such as wettability and structure are affected at relatively low temperatures, while quartz sand content, which contributes to texture, is affected least and only at the most extreme soil temperatures.

Soil Texture and Mineralogy

Soil texture is based on the relative proportion of different-sized inorganic constituents that are found in the 0.08 inch (less than 2 mm) mineral fraction of the mineral soil (DeBano and others 1998). Several soil textural classes have been specified according to the relative proportions of sand (0.05 to 2 mm in diameter), silt (0.002 to 0.05 mm in diameter), and clay (less than 0.002 mm in diameter) particles in the soil. Various proportions of the sand, silt, and clay

USDA Forest Service Gen. Tech. Rep. RMRS-GTR-42-vol. 4. 2005

29

Table 2.1—Temperature thresholds for several physical characteristics of soil.

Soil characteristic	Threshold temperature		Source
	°F	°C	
Soil wettability	482	250	DeBano and Krammes 1966
Soil structure	572	300	DeBano 1990
Calcite formation	572-932	300-500	Iglesias and others 1997
Clay	860-1,796	460-980	DeBano 1990
Sand (quartz)	2,577	1,414	Lide 2001

fractions are used as the basis for identifying 12 textural classes (for example, sand, sandy loam, clay loam, silt loam). Clays are small-diameter silicate minerals having complex molecular structures that contribute to both the physical and chemical properties of a soil.

The components of soil texture (sand, silt, and clay) have high temperature thresholds and are not usually affected by fire unless they are subjected to high temperatures at the mineral soil surface (A-horizon). The most sensitive textural fraction is clay, which begins changing at soil temperatures of about 752 °F (400 °C) when clay hydration and clay lattice structure begin to collapse. At temperatures of 1,292 to 1,472 °F (700 to 800 °C), the complete destruction of internal clay structure can occur. However, sand and silt are primarily quartz particles that have a melting point of 2,577 °F (1,414 °C; Lide 2001). Only under extreme heating do quartz materials at the soil surface become fused. When fusion does occur, soil texture becomes more coarse and erodible. As a result, temperatures are rarely high enough to alter clays beyond a couple centimeters below the mineral soil surface. The effect of soil heating on the stability of clays is further mitigated by the concentration of clays during soil development in the B-horizons. These horizons are usually far removed from heating at the soil surface and rarely increase above ambient surface temperatures unless heated by smoldering roots.

The effect of soil heating on soil minerals other than clays has been studied to a limited extent. For example, a study on the effect of burning logs and slash piles on soil indicated that substantial changes can occur in the mineralogy of the underlying soil during severe heating while burning in juniper (*Juniperus* spp.) and oak (*Quercus* spp.) woodlands (Iglesias and others 1997). Although changes in minerals occurred in the juniper stands, they did not occur in the soils under oak. Followup laboratory burning experiments were done on calcite formation and the alteration of vermiculite in the soils collected from the juniper and oak woodland sites. Temperatures required for calcite formation in oak soils in the laboratory were found to be 932 °F (500 °C) compared to 572 °F (300 °C) in juniper soils.

Soil Structure

Soil structure has long been recognized as an important soil characteristic that can enhance productivity and water relations in both agricultural and wildland soils (DeBano and others 1998). Improving soil structure facilitates the infiltration into and the percolation of water through the soil profile, thereby reducing surface runoff and erosion (see chapter 5). The interaction of organic matter with mineral soil particles that create soil structure also increases the cation adsorption capacities of a soil and nutrient-supplying capabilities of the soil (see chapter 3).

Soil structure is the arrangement of primary soil particles into aggregates having distinctive patterns (columnar, prismatic, blocky). Humus is an important component of soil structure because it acts as a glue that helps hold mineral soil particles together to form aggregates and thus contributes to soil structure, particularly in the upper part of the mineral soil at the duff-upper A-horizon interface (see fig. A.1). However, further downward in the soil profile (in the B-horizon), soil structure is more dependent on clay minerals and the composition of the cations found in the soil solution.

Soil structure created as a result of organic matter in the soil can easily be affected by fire for two reasons. First, the organic matter in a soil profile is concentrated at, or near, the soil surface where it is directly exposed to heating by radiation produced during the combustion of aboveground fuels. Second, the threshold value for irreversible changes in organic matter is low. Living organisms can be killed by temperatures as low as 122 to 140 °F (50 to 60 °C). Nonliving organic matter begins changing at 224 °F (200 °C) and is completely lost at temperatures of 752 °F (400 °C) (DeBano 1990). The loss of soil structure reduces both the amount and size of soil pore space, as is described below.

Soil structure can also be changed by physical processes other than fire, such as deformation and compression by freezing and thawing, as well as by wetting and drying. The abundance of cations in saline and alkali soils can provide an aggregating effect, leading to a strong prismlike structure. Hydrophobic

30

USDA Forest Service Gen. Tech. Rep. RMRS-GTR-42-vol. 4. 2005

substances discussed later in this chapter also tend to improve the stability of soil aggregates by increasing their resistance to disintegration (slaking) when wetted (Giovannini and Lucchesi 1983, Giovannini and others 1983).

Bulk Density and Porosity

Bulk density is the mass of dry soil per unit bulk volume (expressed in g/cm^3) and is related to *porosity*, which is the volume of pores in a soil sample (nonsolid volume) divided by the bulk volume of the sample. Pore space in soils controls the rates of water (soil solution) and air movement through the soil. Well-aggregated soils contain a balance of macropores, which are greater than 0.02 inch (greater than 0.6 mm) in diameter, and micropores, which are less than 0.02 inch (less than 0.6 mm) in diameter (Singer and Munns 1996). This balance in pore sizes allows a soil to transmit both water and air rapidly through macropores and retain water by capillarity in micropores. Macropores in the surface soil horizons are especially important pathways for infiltration of water into the soil and its subsequent percolation downward through the soil profile.

Soil aggregation improves soil structure, creates macropore space, and improves aeration, and as a result decreases bulk density. Pore space not only influences the infiltration and percolation of water through the soil, but the presence of large pores also facilitates heat transfer by convection, and vaporization and condensation.

Fire and associated soil heating can destroy soil structure, affecting both total porosity and pore size distribution in the surface horizons of a soil (DeBano and others 1998). These changes in organic matter decrease both total porosity and pore size. Loss of macropores in the surface soil reduces infiltration rates and produces overland flow. Alteration of organic matter can also lead to a water repellent soil condition that further decreases infiltration rates. The scenario occurring during the destruction of soil structure by fire is:

- The soil structure collapses and increases the density of the soil because the organic matter that served as a binding agent has been destroyed.
- The collapse in soil structure reduces soil porosity (mainly macropores).
- The soil surface is further compacted by raindrops when surface soil particles and ash are displaced, and surface soil pores become partially or totally sealed.
- Finally, the impenetrable soil surface reduces infiltration rates into the soil and produces rapid runoff and hillslope erosion.

Physical Processes

The soil matrix provides the environment that controls several physical processes concerned with heat flow in soils during a fire. The results of heat transfer are manifested in the resulting soil temperatures that develop in the soil profile during a fire (Hartford and Frandsen 1992). Other soil physical processes affected by fire are infiltration rates and the heat transfer of organic substances responsible for water repellency.

Heat Transfer in Soils

The energy generated during the ignition and combustion of fuels provides the driving force that is responsible for the changes that occur in the physical, chemical, and biological properties of soils during a fire (Countryman 1975). Mechanisms responsible for heat transfer in soils include radiation, conduction, convection, mass transport, and vaporization and condensation (table A.1).

Radiation is defined as the transfer of heat from one body to another, not in contact with it, by electromagnetic wave motion (Countryman 1976b). Radiated energy flows outward in all directions from the emitting substance until it encounters a material capable of absorbing it. The absorbed radiation energy increases the molecular activity of the absorbing substance, thereby increasing its temperature.

Conduction is the transfer of heat by molecular activity from one part of a substance to another part, or between substances in contact, without appreciable movement or displacement of the substance as a whole (Countryman 1976a). Metals are generally good conductors in contrast to dry mineral soil, wood, and air that conduct heat slowly. Water as a liquid is a good conductor of heat up to the boiling point, and has an especially high capacity for storing heat until it evaporates.

Convection is a process whereby heat is transferred from one point to another by the mixing of one portion of a fluid with another fluid (Chandler and others 1991). Heat transfer by convection plays an important role in the rate of fire spread through aboveground fuels. In soils, however, the complicated air spaces and interconnections between them provide little opportunity for the movement of heat through the soil by convection.

Vaporization and condensation are important coupled heat transfer mechanisms that facilitate the rapid transfer of heat through dry soils. Vaporization is the process of adding heat to water until it changes phase from a liquid to a gas. Condensation occurs when a gas is changed into a liquid with heat being released during this process. Both water and organic materials can be moved through the soil by vaporization and condensation.

USDA Forest Service Gen. Tech. Rep. RMRS-GTR-42-vol. 4. 2005

31

Heat Transfer Pathways and Models—The heat that is generated by the combustion of surface and aboveground fuels is transferred to the mineral soil surface where it is transferred downward into the underlying soil by a series of complex pathways (fig. 2.1). Quantifying these different pathways for heat flow requires the mathematical modeling of fire behavior, duff ignition and combustion, and the transfer of heat downward to and through moist and dry mineral soil (Dimitrakopoulos and others 1994).

The heat radiated downward during the combustion of aboveground fuels is transferred either to the surface of the forest floor (path A), or directly to the surface of mineral soil if organic surface layers are absent (path B). In most forest ecosystems, heat is usually transferred to an organic layer of litter and duff (path A). When duff is ignited it can produce additional heat that is subsequently transferred to the underlying mineral soil (path D). More details concerning the influence of smoldering and burning duff

on soil heating is presented below. If duff does not ignite, it does not heat the underlaying mineral soil (path C). The heat reaching the mineral soil is either transferred through a dry soil (path E) or a moist soil (path F). Dry soils are common during wildfires, whereas prescribed fires can be planned so as to burn over wet or dry soils. The temperature profiles that develop during heat transfer into moist and dry mineral soil vary widely and as a result affect the physical, chemical, and biological soil properties differently (DeBano and others 1998).

The temperature in moist soils does not rise much above 203 °F (95 °C) until all the water in a given soil layer has been vaporized. As a result, most of the chemical and physical properties of soil are not greatly affected by heating until the soil becomes dry. However, irreversible damage to living organisms near the soil surface is likely to occur because soil temperatures can easily be elevated above lethal temperatures 140 °F (60 °C) for seeds and microorganisms (see part A). Also, the lethal temperatures for microorganisms are lower for moist soils than for those that are dry.

The depth that heat penetrates a moist soil depends on the water content of the soil, and on the magnitude and duration of the surface heating during the combustion of aboveground fuels, litter, and duff (Frandsen 1987). During long-duration heating, such as that occurring under a smoldering duff fire or when burning slash piles, substantial heating can occur 40 to 50 cm downward in the soil. This prolonged heating produces temperatures that are lethal to soil organisms and plant roots. Increased thermal conductivity of moist soil may also create lethal temperatures at much deeper soil depths than if the soil was dry, due to increased thermal conductivity.

Organic-Rich Soils—Organic matter-rich soils are created when the primary productivity exceeds decomposition. Organic matter accumulations can vary from thick surface duff layers located mainly on the soil surface to deep deposits of peat that have been accumulating for thousands of years. The ignition and combustion of these organic-rich soils is of global concern because of the magnitude and duration of these fires and because of the severity of soil heating that occurs during these types of fires (fig. 2.2). The primary combustion process during these fires is by smoldering. The role of fires in wetlands is discussed further in chapter 8 of this publication.

General fire relationships: Although peatland soils are usually saturated, they can dry out during drought periods and become highly combustible. The fires that occur in peatland soils can be extremely long lasting and cover extensive areas where contiguous deposits of peat are present. Such was the case for one of the largest and longest burning fires in the world that occurred in Kalimantan, Indonesia (Kilmaskossu 1988,

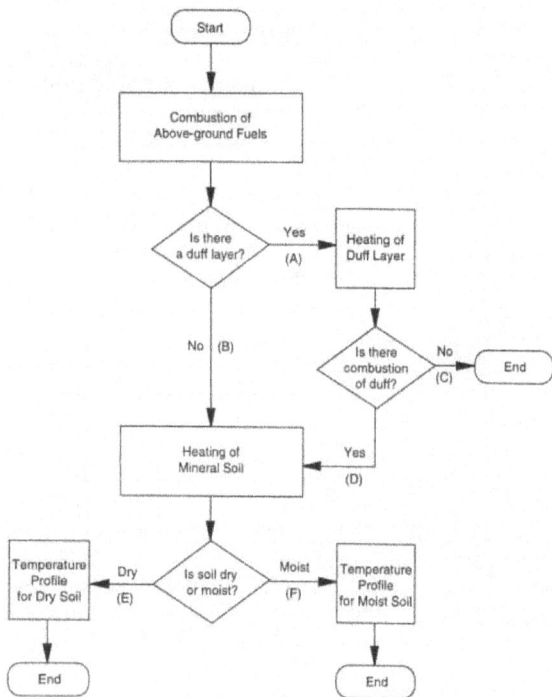

Figure 2.1—A conceptual model of heat flow pathways from combusting fuels downward through the litter into the underlying mineral soil. (Adapted from Dimitrakopoulos and others 1994. A simulation model of soil heating during wildfires. In Sala, M., and J. L. Rubio, editors. Soil Erosion as a Consequence of Forest Fires. Geoforma Edicones, Logrono, Spain, pp. 199-206. Copyright © 1994 Geoforma Edicones, Longro, Spain, ISBN 84-87779-14-X, DL 7353-1994).

32

USDA Forest Service Gen. Tech. Rep. RMRS-GTR-42-vol. 4. 2005

Figure 2.2—Burn out of surface organic matter in the Seney National Wildlife Refuge of Michigan. (Photo by Roger Hungerford)

DeBano and others 1998). This fire was started during drought season when the farmers were clearing an area of logging slash by burning before planting agricultural crops. The fire started in 1983 and burned unchecked more-or-less continuously until the later part of the 1990s. The entire burn covered an area of over 8.4 million acres (3.4 million ha).

Although a thick accumulation of organic matter is commonly associated with tropical and semitropical ecosystems, the largest areas of organic soils are actually found in the boreal regions of the world. Although conditions are usually cool and moist in boreal forests, fires can occur periodically in underlying wetland soils during low rainfall years, at which time these fires mostly burn only the drier surface layers (Wein 1983). The combustion of peatland soils in the boreal forests during wildfires that are started by lightning have been identified as a major source of CO^2 that is released into the atmosphere. It has been estimated that greater than 20 percent of the atmospheric emissions linked to global warming are caused by these fires in the boreal forests worldwide (Conard and Ivanova 1997). During the 1980s alone, more than 138.3 million acres (56 million ha) of boreal forest were estimated to have burned globally (Stocks 1991). Research is currently under way to develop better methods for quantifying the amounts of organic matter lost as the result of wildland fires in peatlands (Turetsky and Wieder 2001).

Soil heating pathways: When dealing with the combustion of organic soils it is important to understand the processes that sustain combustion. The heat produced by the combustion of aboveground fuels can be transferred to the duff (path A, fig. 2.1). Duff (or thick organic layers) can act as an insulating layer when it does not ignite (path C), or a heat source when it ignites, combusts, and continues smoldering (path D).

Therefore, the amount of heat transferred into the underlying mineral soil depends on whether the duff burns and whether the smoldering duff acts as a long-term source of heat. The ignition and combustion of the duff is complex, and attempts to correlate it with heat produced during slash burning have been largely unsuccessful (Albini 1975). Although the duff complicates the heat transfer from the burning aboveground fuels into the underlying mineral soil, some features controlling duff ignition and combustion are known. Important variables needed to describe heat production in duff include depth, total amount, density of packing, the amount of inorganic constituents present, and the moisture content.

If duff does not burn, it provides a barrier to heat flow because the thermal conductivity of organic matter is low (path C, fig. 2.1). The probability of ignition in organic soils (including duff) depends on both inorganic constituents and moisture content (Hungerford and others 1995b). Organic soils are not necessarily completely organic matter, but instead they may contain variable amounts of mineral soil as a result of mixing by surface disturbance. The chances of ignition in organic soils decrease as mineral content increases at any given moisture content. Likewise, the chances of ignition of organic soil decrease as moisture content increases at any given mineral content. In general, duff burns more efficiently when the moisture content is below 30 percent. Varying amounts will burn at moisture contents from 30 to 150 percent, and it is too wet to burn when moisture contents exceed 150 percent (Brown and others 1985). Moisture affects both the thermal conductivity and the heat capacity of the duff, which in turn affect its ignitability (Hungerford 1990).

Combustion of the duff and thick organic deposits such as peat soils involves a smoldering reaction that is initiated by the ignition of a spot or several spots by fire brands, hot ash material, or radiated and conducted heat from the fire front (Pyne and others 1996). After duff ignites, it can transfer large amounts of heat into the underlying soil by convection, conduction, and radiation, and can raise the mineral soil above 350 °C for several hours. Therefore, it becomes difficult to quantify this combined heat flow into the underlying mineral soil. When thick layers of organic materials ignite, glowing combustion can also create an ash layer on the surface of the glowing duff. This ash layer retards heat dissipation upward, thereby causing more heat to penetrate into the soil (Sackett and Haase 1992). As a result, organic layers can transfer 40 to 73 percent of the heat generated during the smoldering process into the underlying mineral soil (Hungerford and Ryan 1996). The ignition, smoldering, and combustion of thick duff layers can continue for hours, thereby allowing substantial time for heat to be transferred deeply into the soil.

USDA Forest Service Gen. Tech. Rep. RMRS-GTR-42-vol. 4. 2005

33

A duff burnout model: Combustion in duff (duff burnout) and organic soils was summarized by Hungerford and others (1995a,b). Ignition is initiated at a single point or several locations on the surface duff (fig. 2.3). Ignition can also occur in cracks or depressions in duff, or be caused by woody material that burns downward through the duff (fig. 2.3A). Fire burns both laterally and vertically after ignition (fig. 2.3B). Fire will burn laterally until it encounters incombustible conditions (moist organic matter, rocks, or the absence of duff). It burns vertically until it reaches mineral soil or moisture conditions that will

not support combustion. During the smoldering of the fire, a hole develops in the burned-out organic layer. Horizontal spread of the fire can leave a thin unburned top crust (Pyne and others 1996). As the smoldering zone moves laterally and vertically, it creates a drying zone caused by the heat from the glowing zone, which allows the glowing front to advance until it reaches incombustible conditions (fig. 2.3C).

Soil Temperature Profiles—Heat absorption and transfer in soils produce elevated temperatures throughout the soil. Temperature increases near the surface are greatest, and they are the least downward in the soil. These temperature regimes are called temperature profiles and can be highly variable depending mainly on the amount of soil water present. Dry soils are poor conductors of heat and thereby do not heat substantially below about 2 inches (5 cm) unless heavy long-burning fuels are combusted. In contrast, wet soils conduct heat rapidly via the soil water although temperatures remain at the boiling point of water until most of the water has been lost. The final soil temperatures reached vary considerably between fires (different fires may produce similar soil temperatures, and conversely, similar fires can produce widely different soil temperatures) and within fires because of heterogeneous surface temperatures.

Numerous reports describing soil temperatures during fire under a wide range of vegetation types and fuel arrangements are present in the literature. As a point of reference, some typical soil temperature profiles are presented for different severities of fire in grass, chaparral, and forests.

Soil temperature increases generated during a cool-burning prescribed fire in mixed conifer forests are low and of short duration (fig. 2.4). This type of fire would

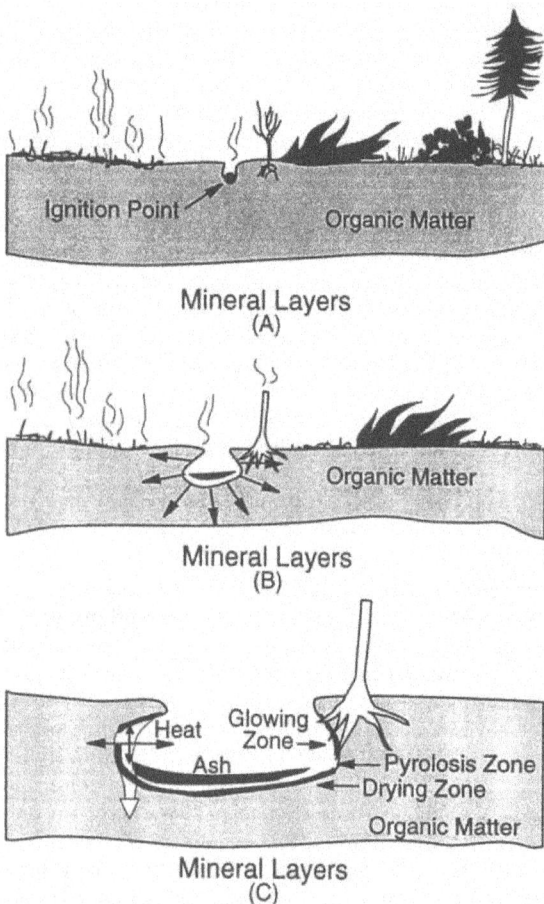

Figure 2.3—A schematic diagram of the smoldering process. The initial ignition point is created by the passing fire front. The fire spreads concentrically from the ignition point (A), develops concentric burned areas (B), and finally develops into a large burned out area (C). (Adapted from Hungerford and others 1995b).

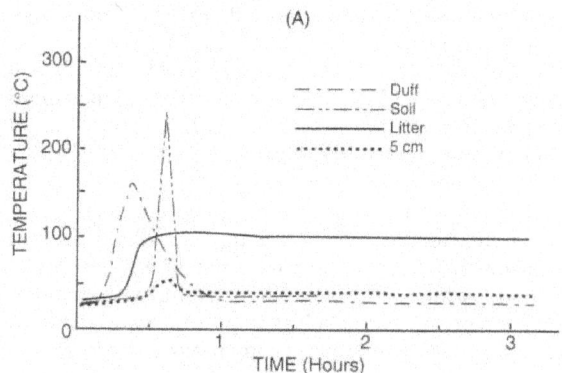

Figure 2.4—Surface and soil temperatures recorded under a cool-burning prescribed fire in mixed conifer forest. (Adapted from Agee 1973, University of California, Water Resources Center, Contribution 143).

34

USDA Forest Service Gen. Tech. Rep. RMRS-GTR-42-vol. 4. 2005

be carried by the surface litter and would probably not consume much standing vegetation, although it might affect some smaller seedlings.

Fire behavior during brush fires, however, differs widely from that occurring during prescribed burning in forests. Both wildfires and prescribed fires in brush fields need to be carried through the plant canopy. The difference between wild and prescribed fires is mainly the amount and rate at which the plant canopy is consumed. During wildfires, the entire plant canopy can be consumed within a matter of seconds, and large amounts of heat that are generated by the combustion of the aboveground fuels are transmitted to the soil surface and into the underlying soil. In contrast, brush can be prescribe-burned under cooler burning conditions (for example, higher fuel moisture contents, lower wind speeds, higher humidity, lower ambient temperatures, using northerly aspects) such that fire behavior is less explosive. Under these cooler burning conditions the shrub canopy may be not be entirely consumed, and in some cases a mosaic burn pattern may be created (particularly on north-facing slopes). The soil temperature profiles that were measured during low, medium, and high severity fires in chaparral vegetation in southern California are presented in figure 2.5A, B, and C.

The highest soil temperatures are reached when concentrated fuels such as slash piles and thick layers of duff burn for long periods (fig. 2.6A and B). The soil temperatures under a pile of burning eucalyptus logs (fig. 2.6A) reached lethal temperatures for most living biota at a depth of almost 22 cm in the mineral soil (after Roberts 1965). It must be kept in mind, however, that this extreme soil heating occurred on only a small fraction of the area, although the visual effects on plant growth were observed for several years. An

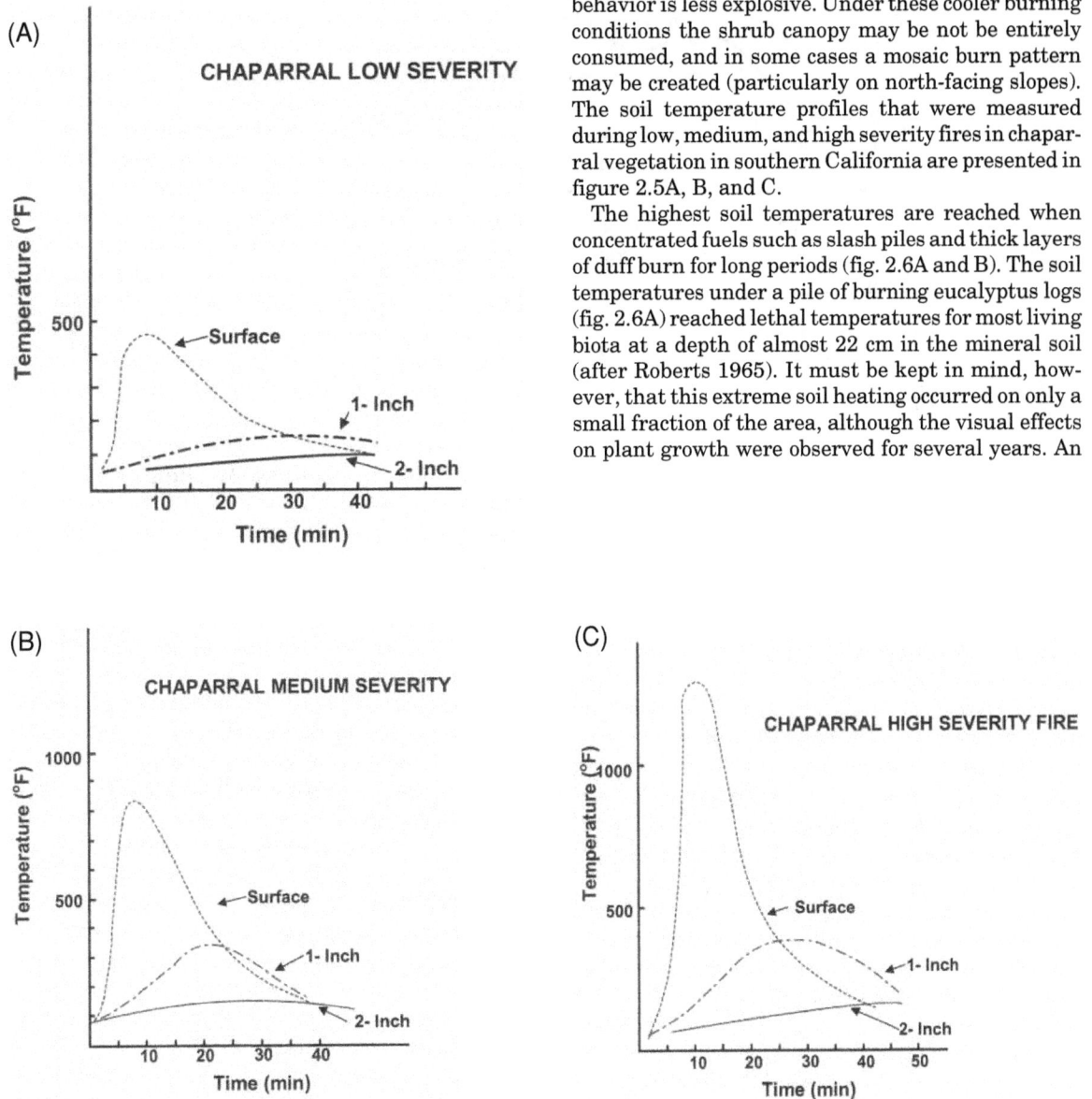

Figure 2.5—Soil temperature profiles during a low severity chaparral fire (A), a medium severity chaparral fire (B), and high severity chaparral fire (C). (After DeBano and others 1979)

USDA Forest Service Gen. Tech. Rep. RMRS-GTR-42-vol. 4. 2005

35

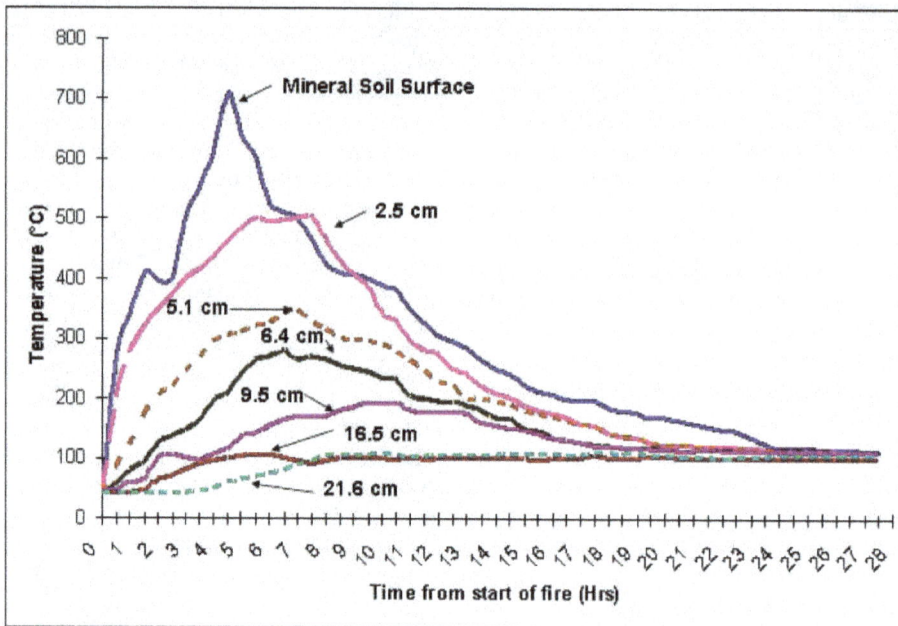

(A)

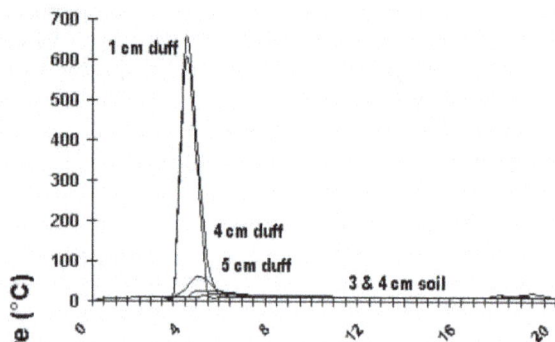

(B)

Low Soil Heating

Prescribed Burn (47T/A Slash)
Pre-burn duff = 7 cm Postburn = 2-3 cm

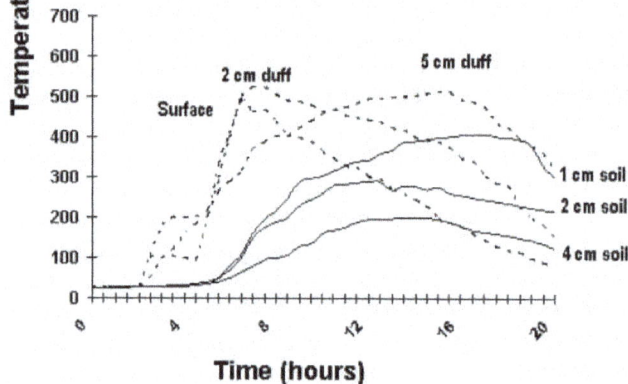

High Soil Heating

Larch Duff Burn
Pre-burn duff = 7 cm Postburn = 0

Figure 2.6—Soil temperatures profiles under (A) windrowed logs (After W.B. Roberts. 1965. Soil temperatures under a pile of burning eucalyptus logs. Australian Forest Research 1(3):21-25), and (B) under a 7-cm (18 inch) duff layer in a larch forest. (After Hungerford 1990).

example of extensive soil heating that can occur during the burning of areas having large accumulations of duff and humus (fig. 2.6B) was reported during the complete combustion of 2.8 inches (7 cm) of a duff layer found under larch (Hungerford 1990).

Based on the information available on the relationship between soil heating and type of fire, the following generalities can be made:

- Crown fires are fast-moving, wind-driven, large, impressive, and usually uncontrollable, and they have a deep flame front (fig. 1.3). Usually little soil heating results when a fire front passes mainly through the tree crowns.
- Surface fires, compared to crown fires, are slower moving, smaller, patchy, and are more controllable, and they may also have a deep flame front (fig. 2.7A). These fires usually ignite and combust a large portion of the surface fuels in forests and brushlands that can produce substantial soil heating.
- Grass fires are fast-moving and wind-driven, may be large, and have a narrow flame front (fig. 2.7B). The amount of fuel available for burning in grasslands is usually much less than that contained in brushlands and forests, and as a result, soil heating is substantially less than occurs during surface or smoldering fires.
- Smoldering fires do not have flames, are slow-moving and unimpressive, but frequently have long burnout times. They generally are controllable although they may have a deep burning front. Soil heating during this long duration smoldering process may be substantial. Temperatures within smoldering duff often are between 932 and 1,112 °F (500 and 600 °C). The duration of burning may last from 18 to 36 hours, producing high temperatures in the underlying mineral soil.

Water Repellency

The creation of *water repellency* in soils involves both physical and chemical processes. It is discussed within the context of physical properties because of its importance in modifying physical processes such as infiltration and water movement in soils. Although hydrophobic soils had been observed since the early 1900s (DeBano 2000a,b), fire-induced water repellency was first identified on burned chaparral watersheds in southern California in the early 1960s. Watershed scientists were aware of it earlier, but it had been referred to simply as the "tin roof" effect because of its effect on infiltration (fig. 2.8A, B, and C). In southern California both the production of a fire-induced water repellency and the loss of protective vegetative cover play a major role in the postfire runoff and erosion, and the area is particularly important because of the large centers of

Figure 2.7—Surface fire in (A) an uneven aged ponderosa pine forest, Mogollon Rim, Arizona, and (B) Alaskan grasslands. (Photos by USDA Forest Service).

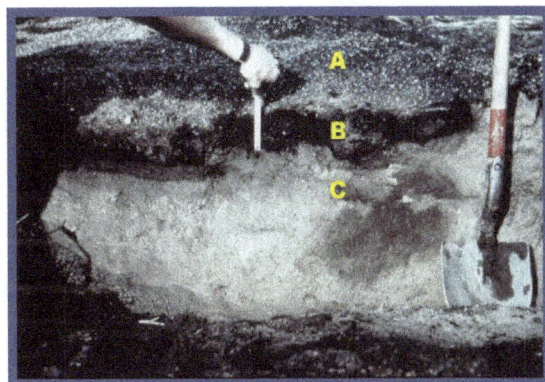

Figure 2.8—The "tin roof" effect on burned chaparral watersheds as described by earlier watershed researchers include (A) the wettable ash and carbon surface layer, (B) the discontinuous water repellent layer, and (C) the wettable subsoil. (After DeBano 1969).

populations located immediately below steep, unstable chaparral watersheds.

Nature of Water Repellency in Soils—Normally, dry soils have an affinity for adsorbing liquid and vapor water because there is strong attraction between the mineral soil particles and water. In water-repellent soils, however, the water droplet "beads up" on the soil surface where it can remain for long periods and in some cases will evaporate before being absorbed by the soil. Water, however, will not penetrate some soils because the mineral particles are coated with hydrophobic substances that repel water. Water repellency has been characterized by measuring the contact angle between the water droplet and the water-repellent soil surface. Wettable dry soils have a liquid-solid contact angle of nearly zero degrees. In contrast, water-repellent soils have liquid-solid contact angles around 90 degrees (fig. 2.9).

Causes of Water Repellency—Water repellency is produced by soil organic matter and can be found in both fire and nonfire environments (DeBano 2000a,b). Water repellency can result from the following processes involving organic matter:

- An irreversible drying of the organic matter (for example, rewetting dried peat).
- The coating of mineral soil particles with leachates from organic materials (for example, coarse-grained materials treated with plant extracts).
- The coating of soil particles with hydrophobic microbial byproducts (for example, fungal mycelium).
- The intermixing of dry mineral soil particles and dry organic matter.

Figure 2.9—Appearance of water droplets that are "balled up" on a water-repellent soil. (After DeBano 1981).

- The vaporization of organic matter and condensation of hydrophobic substances on mineral soil particles during fire (for example, heat-induced water repellency).

Formation of Fire-Induced Water-Repellent Soils— A hypothesis by DeBano (1981) describes how a water-repellent layer is formed beneath the soil surface during a fire, noting that organic matter accumulates on the soil surface under vegetation canopies during the intervals between fires. During fire-free intervals, water repellency occurs mainly in the organic-rich surface layers, particularly when they are proliferated with fungal mycelium (fig. 2.10A). Heat produced during the combustion of litter and aboveground fuels vaporizes organic substances, which are then moved downward into the underlying mineral soil where they condense in the cooler underlying soil layers (fig. 2.10B) The layer where these vaporized hydrophobic substances condense forms a distinct water-repellent layer below and parallel to the soil surface (fig. 2.10C).

The magnitude of fire-induced water repellency depends upon several parameters, including:

- The severity of the fire. The more severe the fire, the deeper the layer, unless the fire is so hot it destroys the surface organic matter.
- Type and amount of organic matter present. Most vegetation and fungal mycelium contain hydrophobic compounds that induce water repellency.
- Temperature gradients in the upper mineral soil. Steep temperature gradients in dry soil enhance the downward movement of volatilized hydrophobic substances.
- Texture of the soil. Early studies in California chaparral showed that sandy and coarse-textured soils were the most susceptible to fire-induced water repellency (DeBano 1981). However, more recent studies indicate that water repellency frequently occurs in soils other than coarse-textured ones (Doerr and others 2000).
- Water content of the soil. Soil water affects the translocation of hydrophobic substances during a fire because it affects heat transfer and the development of steep temperature gradients.

Effect of Water Repellency on Postfire Erosion—Fire affects water entering the soil in two ways. First, the burned soil surface is unprotected from raindrop impact that loosens and disperses fine soil and ash particles that can seal the soil surface. Second, soil heating during a fire produces a water-repellent layer at or near the soil surface that further impedes infiltration into the soil. The severity of the water repellency in the surface soil layer, however, decreases

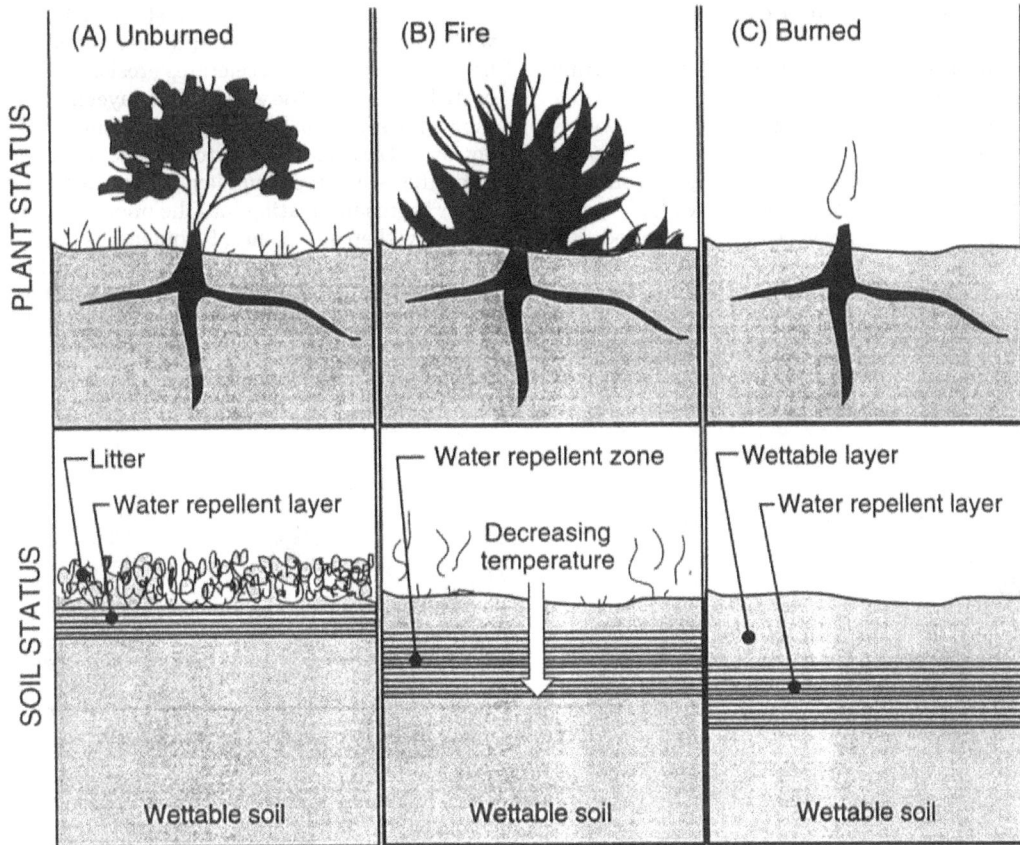

Figure 2.10—Formation of fire-induced water repellency. Water repellency before (A), during (B), and following (C) fire. (After DeBano 1981).

over time as it is exposed to moisture; in many cases, it does not substantially affect infiltration beyond the first year. More detailed effects of water repellency on the infiltration process are discussed in chapter 6. Water repellency has a particularly important effect on two postfire erosion processes, that of raindrop splash and rill formation.

Raindrop splash: When a water-repellent layer is formed at the soil surface, the hydrophobic particles are more sensitive to raindrop splash than those present on a wettable soil surface (Terry and Shakesby 1993). Consequently, raindrops falling on a hydrophobic surface produce fewer, slower moving ejection droplets that carry more sediment a shorter distance than in the case of a wettable soil. Further, the wettable surfaces have an affinity for water and thereby become sealed and compacted during rainfall, which makes them increasingly resistant to splash detachment. Conversely, the hydrophobic soil remains dry

and noncohesive; particles are easily displaced by splash when the raindrop breaks the surrounding water film.

Rill formation: A reduction in infiltration caused by a water-repellent layer quickly causes highly visible rainfall-runoff-erosion patterns to develop on the steep slopes of burned watersheds. The increased surface runoff resulting from a water-repellent layer quickly entrains loose particles of soil and organic debris, and produces surface runoff that rapidly becomes concentrated into well-defined rills. As a result, extensive rill networks develop when rainfall exceeds the slow infiltration rates that are characteristic of water-repellent soils.

The sequence of rill formation as a result of fire-induced water repellency has been documented to follow several well-defined stages (Wells 1987). First, the wettable soil surface layer, if present, is saturated during initial infiltration (fig. 2.11A). Water infiltrates

USDA Forest Service Gen. Tech. Rep. RMRS-GTR-42-vol. 4. 2005

39

rapidly into the wettable surface ash layer until it is impeded by a water-repellent layer. This process occurs uniformly over the landscape so that when the wetting front reaches the water-repellent layer, it can neither drain downward or laterally (fig. 2.11A). If the water-repellent soil layer is on the soil surface, runoff begins immediately after rain droplets reach the soil surface. As rainfall continues, water fills all available pores until the wettable soil layer becomes saturated. Because of the underlying water-repellent layer, the saturated pores cannot drain, which creates a positive pore pressure above the water-repellent layer. This increased pore pressure decreases the shear strength of the soil mass and produces a failure zone located at the boundary between the wettable and water-repellent layers

where pore pressures are greatest (fig. 2.11B). As the water flows down this initial failure zone, turbulent flow develops, which accelerates erosion and entrains particles from both the wettable ash layer if present and the water-repellent layer (fig. 2.11C and 2.11D). The downward erosion of the water-repellent rill continues until the water-repellent layer is eroded away and water begins infiltrating into the underlying wettable soil. Flow then diminishes, turbulence is reduced, and down-cutting ceases. The final result is a rill that has stabilized immediately below the water-repellent layer (fig.2.11E). On a watershed basis these individual rills develop into a well-defined network that can extend throughout a small watershed (fig. 2.11F).

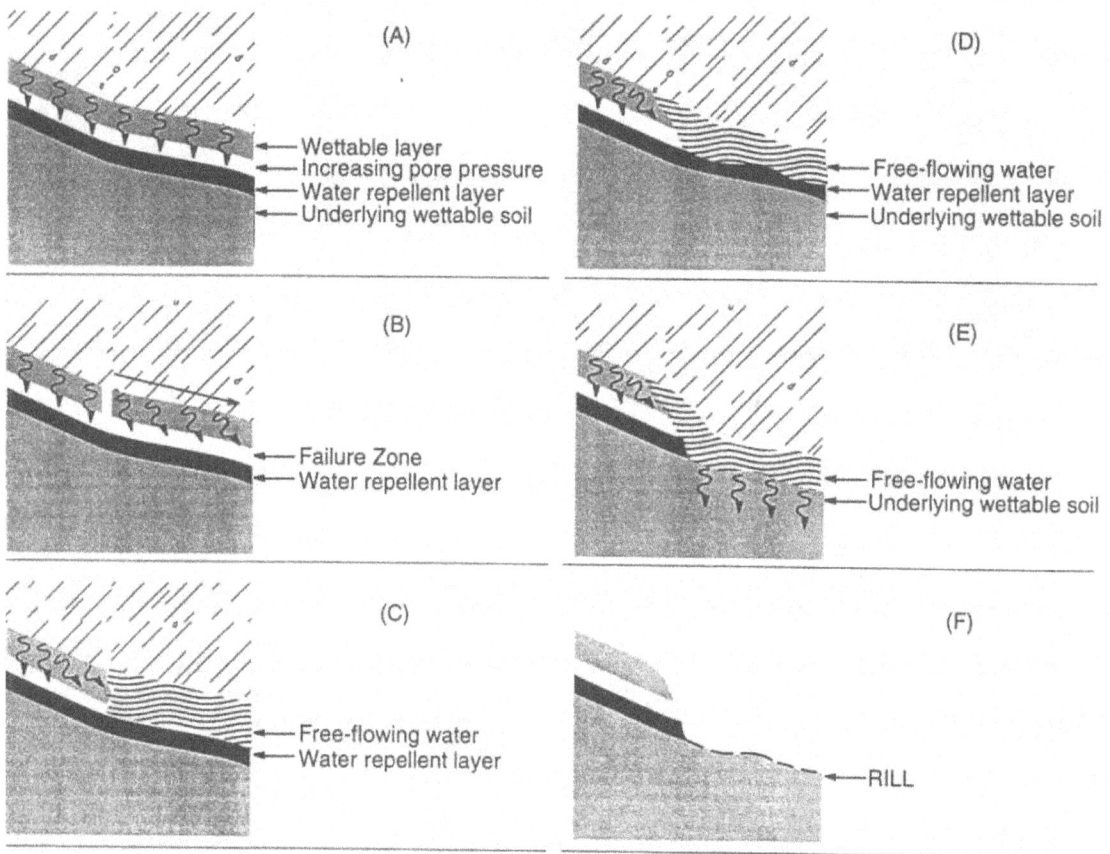

Figure 2.11—Sequence of rill formation on a burned slope with a water-repellent layer includes (A) saturation of wettable surface area, (B) development of a failure zone in wettable surface layer, (C) free flowing water over the water-repellent layer, (D) erosion of the water-repellent layer, (E) removal of the water repellent layer and infiltration into underlying wettable soil, and (F) resultant rill. (From Wells 1987. The effects of fire on the generation of debris flows in southern California. Reviews in Engineering Geology. 7:105-114. Modified with permission of the publisher, the Geological Society of America, Boulder, Colorado U.S.A. Copyright © 1987 Geological Society of America.)

Soil Erosion

Processes and Mechanics

The erosion process involves three separate components that are a function of sediment size and transport medium (water or air) velocity. These are: (1) detachment, (2) transport, and (3) deposition. Erosion occurs when sediments are exposed to water or air and velocities are sufficient to detach and transport the sediments. Table 2.2 gives a generalized breakdown of sediment classes and detachment/transport/deposition velocities in water.

Erosion is a natural process occurring on landscapes at different rates and scales depending on geology, topography, vegetation, and climate. Natural rates of erosion are shown in table 2.3. Geologic erosion rates were calculated on large basins so they are higher than those listed for forests. The data from forests come from much smaller watershed experiments. Natural erosion rates increase as annual precipitation increases, peaking in semiarid ecoregions when moving from desert to wet forest (Hudson 1981). This occurs because there is sufficient rainfall to cause natural erosion from the sparser desert and semiarid grassland covers. As precipitation continues to increase, the landscapes start supporting dry and eventually wet forests, which produce increasingly dense plant and litter covers that decrease natural erosion. However, if the landscapes are denuded by disturbance (for example, fire, grazing, timber harvesting, and so forth), then the rate of erosion continues to increase with increasing precipitation (fig. 2.12). Surface conditions after fire are important for determining where water moves and how much erosion is produced (table 2.4).

Table 2.2—Sediment size classes and some typical detachment/deposition velocities (varies with sediment characteristics and flow depth).

Sediment type	Size class		Detachment velocity		Deposition velocity	
	in	*cm*	*in/sec*	*cm/sec*	*in/sec*	*cm/sec*
Boulders	39.37	100.00				
Cobble	10.08	25.60	7.480	19.00	4.724	12.00
Gravel	2.520	6.40	5.906	15.00	3.150	8.00
Sand	0.078	0.19	1.378	3.50	0.591	1.48
Silt	0.007	0.02	0.669	1.70	0.059	0.15
Clay	<0.001	<0.01	9.843	25.00	0.004	0.01

Table 2.3—Natural sediment losses in the United States.

Location	Watershed condition	Sediment loss		Reference
		tons/ac	*Mg/ha*	
United States	Geologic erosion:			
	Natural, lower limit	0.26	0.58	Schumm and Harvey 1982
	Natural, upper limit	6.69	15.00	Schumm and Harvey 1982
Eastern U.S. Forests	Lower baseline	0.05	0.10	Patric 1976
	Upper baseline	0.11	0.22	
Western U.S.	Lower baseline	<0.01	<0.01	Biswell and Schultz 1965
	Upper baseline	2.47	5.53	DeByle and Packer 1972

USDA Forest Service Gen. Tech. Rep. RMRS-GTR-42-vol. 4. 2005

41

Figure 2.12—Erosion from a clearcut and burned *Pinus rigida* stand planted on degraded farmland, Southern Appalachian Mountains, Georgia. (Photo by Daniel Neary).

Table 2.4—Soil surface conditions affect infiltration, runoff, and erosion.

Soil surface condition	Infiltration	Runoff	Erosion
Litter charred	High	Low	Low
Litter consumed	Medium	Medium	Medium
Bare soil	Low	High	High
Water repellent layers	Very low	Very high	Severe

Erosion is certainly the most visible and dramatic impact of fire apart from the consumption of vegetation. Fire management activities (wildfire suppression, prescribed fire, and postfire watershed rehabilitation) can affect erosion processes in wildland ecosystems. Wildfire, fireline construction, temporary roads, and permanent, unpaved roads receiving heavy vehicle traffic will increase erosion. Increased stormflows after wildfires will also increase erosion rates. Burned Area Emergency Rehabilitation (BAER) work on watersheds will decrease potential postfire erosion to varying degrees depending on the timing and intensity of rainfall (see chapter 10; Robichaud and others 2000).

Sheet, Rill, and Gully Erosion: Progressive Erosion—In sheet erosion, slope surfaces erode uniformly. This type proceeds to rill erosion in which small, linear, rectangular channels cut into the surface of a slope. Further redevelopment of rills leads to the formation of deep, large, rectangular to v-shaped channels (gullys) cut into a slope (fig. 2.13).

Some special erosion conditions can be encountered. For instance, in ecoregions with permafrost, the progression of erosion from sheet to rill to gully interacts with the depth of permafrost thaw. Until thaw occurs, erosion is essentially frozen. Fire and fire control activities such as fireline construction will affect thaw depth after wildfires, and subsequent erosion of firelines can be substantial (fig. 2.14).

Dry Ravel—*Dry ravel* is the gravity-induced downslope surface movement of soil grains, aggregates, and rock material, and is a ubiquitous process in semiarid steepland ecosystems (Anderson and others 1959). Triggered by animal activity, earthquakes, wind, freeze-thaw cycles, and thermal grain expansion during soil heating and cooling, dry ravel may best be

42

USDA Forest Service Gen. Tech. Rep. RMRS-GTR-42-vol. 4. 2005

Figure 2.13—Incised gully after post-wildfire runoff on the Hondo Fire, 1996, Carson National Forest. (Photo by Russell Lafayette).

described as a type of dry grain flow (Wells 1981). Fires greatly alter the physical characteristics of hillside slopes, stripping them of their protective cover of vegetation and organic litter and removing barriers that were trapping sediment. Consequently, during and immediately following fires, large quantities of surface material are liberated and move downslope as

dry ravel (Krammes 1960, Rice 1974). Dry ravel can equal or exceed rainfall-induced hillslope erosion after fire in chaparral ecosystems (Krammes 1960, Wohlgemuth and others 1998).

In the Oregon Coast Range, Bennett (1982) found that prescribed fires in heavy slash after clearcutting produced noncohesive soils that were less resistant to the force of gravity. Dry ravel on steep slopes (greater than 60 percent) that were prescribe-burned produced 118 yard^3/acre (224 m^3/ha) of surface erosion compared to 15 yard^3/acre (29 m^3/ha) on moderate slopes with burning, and 9 yard^3/acre (17 m^3/ha) where burning was not done after clearcutting. Sixty-four percent of the erosion, as dry ravel, occurred within the first 24 hours after burning.

Mass Failures—This term includes slope creep, falls, topples, rotational and translational slides, lateral spreads, debris flows, and complex movements (Varnes 1978). *Slope creep* is a slow process that does not deliver large amounts of sediment to stream channels in the periods normally considered in natural resources management. The most important for forest management considerations are rotational and translational slides, flows, and complex movement (Ice 1985). *Slump-earthflows* and *debris avalanches* are more likely with increased water in the soil because of the decreased tension between soil particles, increased

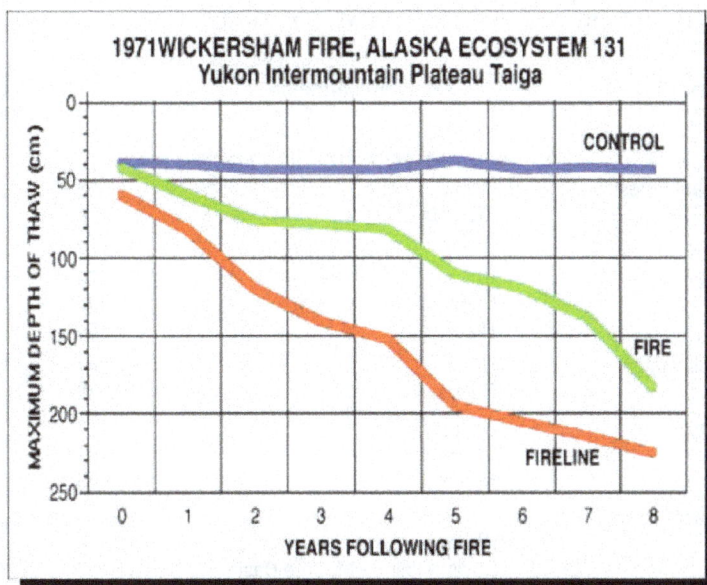

Figure 2.14—Postwildfire erosion following permafrost thaw in Alaska. (After Viereck 1982, Figure courtesy of the USDA Forest Service, National Advanced Fire and Resource Institute, Tucson, AZ).

loading on slopes produced by excess soil water, and a buoyant effect created by soil water along a failure plane. These types of failures are most often associated with clearcutting, tree death due to disease or insects, and road network construction. Slope failures associated with fires are the result of the loss of the forest floor, surface sealing, and the development of water repellency. These processes produce a diversion of rainfall from infiltration to surface runoff. The result is a dramatic increase in debris torrents in channels that greatly increase sediment delivery to channels and flooding (see chapter 5).

Debris avalanches are the largest, most dramatic, and main form of mass wasting that delivers sediment to streams (Benda and Cundy 1990). They can range from slow moving earth flows to rapid avalanches of soil, rock, and woody debris. Debris avalanches occur when the mass of soil material and soil water exceed the sheer strength needed to maintain the mass in place. The loss of root strength (for example, soil strength, anchoring, soil mass cohesion, and so forth) due to removal of trees or tree death caused by insects or disease aggravates the situation. Steep slopes, logging, road construction, and heavy rainfall aggravate debris avalanching potential (table 2.5).

Most fire-associated mass failures are debris flows associated with development of water repellency in soils (DeBano and others 1998). Chaparral occupying steep slopes in southern California has a high potential for mass failures, particularly when deep-rooted chaparral species are replaced with shallower-rooted grass species (Rice 1974). These mass failures are a large source of sediment delivered to stream channels (can be 50 percent of the total postfire sediment yield in some ecoregions; fig. 2.15). Wells (1981) reported that wildfire in chaparral vegetation in coastal southern California can increase average debris avalanche sediment delivery in large watersheds from 18 to 4,845 yard^3/mile^2/year (7 to 1,910 m^3/km^2/year). However, individual storm events in smaller basins can trigger much greater sediment yields (Gartner and others 2004; table 2.6). Rates as high as 221,026 yard^3/mile^2 (65,238 m^3/km^2) have been measured after single storms in California chaparral. Other ecoregions in the Western United States have postfire debris flows that have been larger (for example, 304,761 yard^3/mile^2 or 89,953 m^3/km^2 in ecoregion M331, Colorado) but not with the same frequency (table 2.6). The situation could change with the increasing severity, frequency, and distribution of forest wildfires that have characterized the past decade.

Table 2.5—Annual sediment yields from debris avalanches in undisturbed forests and those affected by clearcutting, roads, and wildfire. (From Ice 1985, Neary and Hornbeck 1994).

Ecoregion-location	Treatment/condition	Sediment yield	
		yd^3/mi^2/yr	m^3/km^2/yr
M242 CASCADE MIXED-CONIFER-MEADOW FOREST PROVINCE[1]			
Siuslaw National Forest, Oregon	Uncut	95	28
	Clearcut	376	111
	Roads	11,858	3,500
H.J. Andrews, Oregon	Uncut	122	36
	Clearcut	447	132
	Roads	3,964	1,170
Northwest Washington	Uncut	244	72
	Roads	39,978	11,800
British Columbia	Uncut	37	11
	Clearcut	81	24
	Roads	955	282
Entiat Experimental Forest, Washington	Wildfire – Fox Basin	10,164	3,600
	Wildfire – Burns	420	124
	BasinWildfire – McCree Basin	12,197	3,000
M262 CALIFORNIA COASTAL RANGE WOODLAND-SHRUB-CONIFER PROVINCE			
Southern California	Uncut	24	7
	Wildfire	6,461	1,907

[1]Bailey's (1995) descriptions of ecoregions of the United States.

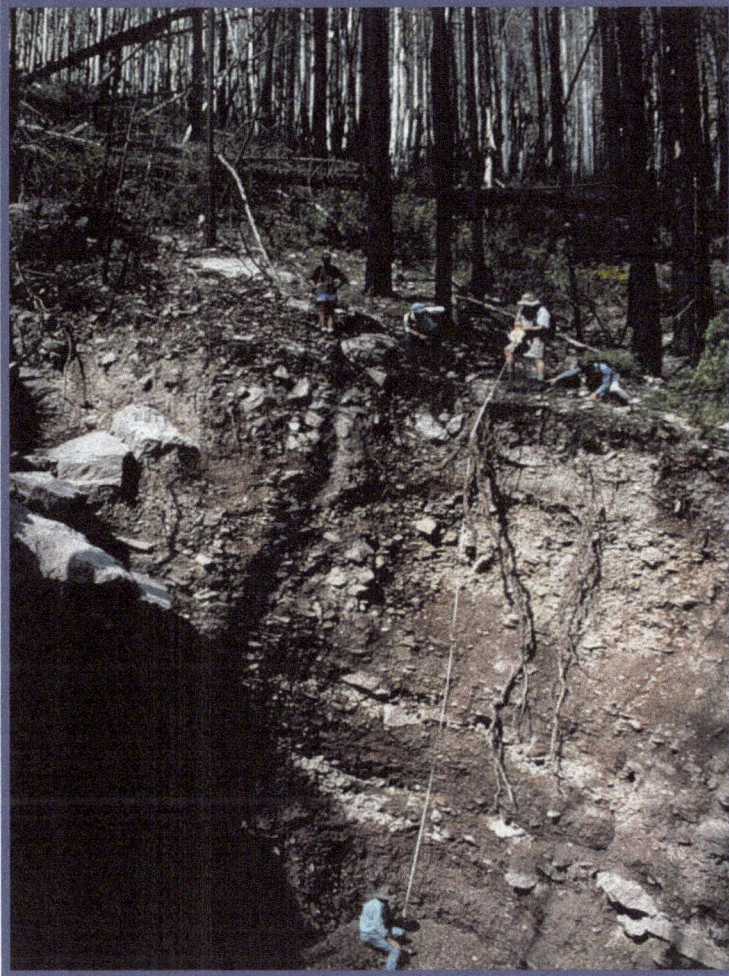

Figure 2.15—Deeply incised gully in the Chiricahua Mountains after the Rattlesnake Fire, 1996, Coronado National Forest, Arizona. (Photo courtesy of the Tree Ring Laboratory, University of Arizona).

Cannon (2001) describes several types of debris flow initiation mechanisms after wildfires in the Southwestern United States. Of these, surface runoff, which increases sediment entrainment, was the dominant triggering mechanism. Campbell and others (1977) reported a 416-fold increase in sediment yield after wildfire in southwestern ponderosa pine.

Channel Stability—Fire-related sediment yields vary, depending on fire frequency, climate, vegetation, and geomorphic factors such as topography, geology, and soils (Swanson 1981). In some regions, more than 60 percent of the total landscape sediment production over the long term is fire-related. Much of that sediment loss can occur the first year after a wildfire (Rice 1974, Agee 1993, DeBano and others 1996, DeBano

and others 1998, Wohlgemuth and others 1998, Robichaud and Brown 1999).

A stable stream channel reflects a dynamic equilibrium between incoming and outgoing sediment and streamflow (Rosgen 1996). Increased sideslope erosion after fires can alter this equilibrium by transporting additional sediment into channels (aggradation). Here it is stored until increased peakflows (see chapter 5) produced after fires erode the channel (degradation) and move the stored material downstream (Heede and others 1988). Sediment transported from burned areas as a result of increased peakflows can adversely affect aquatic habitat, recreation areas, roads, buildings, bridges, and culverts. Deposition of sediments alters habitat and can fill in lakes and reservoirs (Reid 1993, Rinne 1996).

USDA Forest Service Gen. Tech. Rep. RMRS-GTR-42-vol. 4. 2005

45

Table 2.6—Event-based sediment yields from debris flows due to wildfires. (From Gartner and others 2004).

Ecoregion-location	Treatment/condition		Sediment yield per event	
	Burn area	Rainfall		
	%	mm	yd³/mi²	m³/km²
M242 CASCADE MIXED-CONIFER-MEADOW FOREST PROVINCE[1]				
Entiat Valley, WA, 1972	100	335	1,355	400
M262 CALIFORNIA COASTAL RANGE WOODLAND-SHRUB-CONIFER PROVINCE				
Los Angeles County, CA, 1914	80	Unknown	60,069	17,730
Los Angeles County, CA, 1928	100	36	45,680	13,483
Los Angeles County, CA, 1933	100	356	67,943	20,054
San Dimas, W. Fork, CA, 1961	100	40	54,906	16,206
Glendora, Glencoe, CA, 1969	80	1,143	203,280	60,000
Glendora, Rainbow, CA, 1969	80	1,143	221,026	65,238
Big Sur, Pfiefer, CA, 1972	100	21	22,588	6,667
Sierra Madre, CA, 1978	100	38	7,650	2,258
San Bernardino, CA, 1980	NA	Unknown	160,432	47,353
Laguna Canyon, CA, 1993	85	51	73,303	21,636
Hidden Springs, CA, 1978	100	250	84,700	25,000
Sierra Madre, CA, 1978	100	38	7,650	2,258
Topanga, CA, 1994	100	66	783	231
Ventura, Slide Creek, CA, 1986	100	122	871	257
M331 ROCKY MOUNTAIN STEPPE-OPEN WOODLAND-CONIFEROUS FOREST				
Glenwood Springs, CO, 1994	97	17	41,537	12,260
Glenwood Springs, CO, 1994	58	17	7,247	2,139
M341 NV-UT SEMI-DESERT-CONIFEROUS FOREST-ALPINE MEADOW				
Santaquin, UT, 2001	29	12	304,761	89,953
Santaquin, UT, 2001	28	12	31,657	9,344
M313 AZ-NM MOUNTAINS SEMIDESERT-WOODLAND-CONIFER PROVINCE				
Huachuca Mountains, AZ, 1988	80	8	56,468	16,667

[1]Bailey's (1995) descriptions of ecoregions of the United States.

Postfire Sediment Yields

Baseline Yields—Some reference sediment yield baselines are presented in tables 2.3, 2.5, and 2.6. Natural erosion rates for undisturbed forests in the Western United States of less than 0.01 to 2.47 tons/acre/year (less than 0.01 to 5.53 Mg/ha/year) are higher than Eastern United States yields of 0.05 to 0.10 tons/acre/year (0.1 to 0.2 Mg/ha/year) but don't approach the upper limit of geologic erosion (Maxwell and Neary 1991). These differences are due to natural site factors such as soil and geologic erosivity, rates of geologic uplift, tectonic activity, slope, rainfall amount and intensity, vegetation density and percent cover, and fire frequency. Landscape-disturbing activities such as mechanical site preparation (6.7 tons/acre; 15 Mg/ha/year; Neary and Hornbeck 1994), agriculture (249.8 tons/acre; 560 Mg/ha/year; Larson and others 1983), and road construction (62.4 tons/acre; 140 Mg/ha/year; Swift 1984) produce the most sediment loss and can match or exceed the upper limit of natural geologic erosion.

Yields from Fires—Fire-related sediment yields vary considerably, depending on fire frequency, climate, vegetation, and geomorphic factors such as topography, geology, and soils (Anderson and others 1976, Swanson 1981). In some regions, over 60 percent of the total landscape sediment production over the long term is fire-related. Much of that sediment loss can occur the first year after a wildfire (Rice 1974, Agee 1993, DeBano and others 1996, DeBano and others 1998, Wohlgemuth and others 1998, Robichaud and Brown 1999). An example is the large amount of sediment that filled in a 10 acre (4 ha) lake on the Coronado National Forest after the Rattlesnake Fire of 1996 (fig. 2.16). Tables 2.8, 2.9, and 2.10 show the range of sediment yield increases from the first

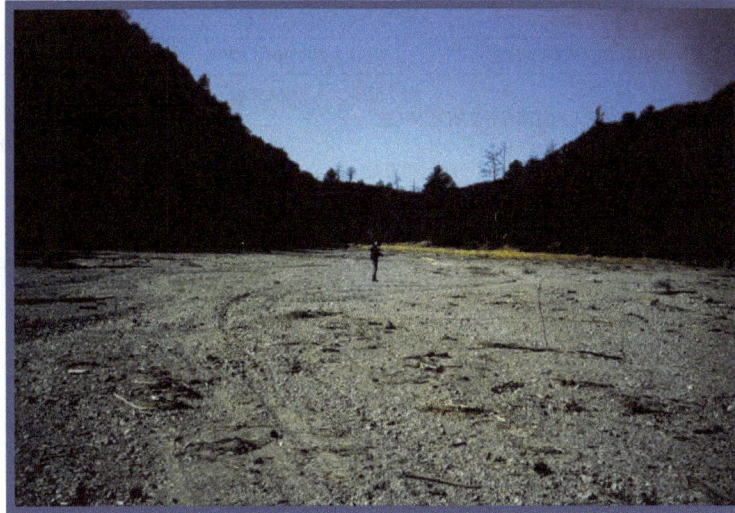

Figure 2.16—Rucker Lake, Coronado National Forest, filled in by erosion off of the Rattlesnake Fire, Chiricahua Mountains, Arizona. (Photo by Daniel Neary).

Table 2.7—Sediment losses produced by different land management activities.

Location	Management activity	1st year sediment loss		Reference
		tons/ac	*Mg/ha[1]*	
CONTINENTAL UNITED STATES				
USA	Cropland			Larson and others 1983
	Maximum tolerance	5.00	11.20	
	Maximum loss	249.98	560.50	
Eastern U.S.A.	Forest roadbuilding	62.44	140.00	Swift 1984
M212 ADIRONDACK-NEW ENGLAND MIXED FOREST PROVINCE[2]				
New Hampshire	Clearcut	0.16	0.37	
231 SOUTHEASTERN MIXED FOREST PROVINCE				
North Carolina	Cut, shear, disk	4.18	9.37	Douglass and Godwin 1980
Virginia	Cut, shear, disk	7.14	15.00	Fox and others 1983
Mississippi	Cut, disk, bed	6.36	14.25	Beasley 1979
232 COASTAL PLAIN MIXED FOREST PROVINCE				
Florida	Clearcut, windrow	0.02	0.04	Riekerk 1983
M231 OUCACHITA MIXED FOREST-MEADOW PROVINCE				
Arkansas	Clearcut, shear	0.11	0.54	Beasley and others 1986
M313 AZ-NM MOUNTAINS SEMIDESERT-WOODLAND-CONIFER PROVINCE				
Arizona	Clearcut, roading	0.06	0.14	Neary and Hornbeck 1994
M332 MIDDLE ROCKY MOUNTAIN STEPPE-CONIFER-MEADOW PROVINCE				
Montana	Clearcut	0.08	0.18	DeByle and Packer 1972
M242 CASCADE MIXED-CONIFER-MEADOW FOREST PROVINCE				
Oregon	Clearcut, roading	0.65	0 .46	Neary and Hornbeck 1994

[1]Mg/ha is metric tons per hectare.
[2]Bailey's (1995) descriptions of ecoregions of the United States.

Table 2.8—Sediment losses the first year after prescribed (Rx) fire and wildfires, part 1.

Location	Management activity	1st year sediment loss		Reference
		tons/ac	Mg/ha[1]	
231 SOUTHEASTERN MIXED FOREST PROVINCE[2]				
South Carolina	Loblolly pine			Van Lear and others 1985
	Control	0.012	0.027	
	Understory burn (R=2)	0.019	0.042	
	Burn, cut (R=2)	0.067	0.151	
North Carolina	Southern hardwoods			Copley and others 1944
	Control	0.002	0.004	
	Semi-annual Rx burn	3.077	6.899	
Mississippi	Scrub oak			Ursic 1970
	Control	0.210	0.470	
	Rx burn (R=3 yr)	0.509	1.142	
Mississippi	Scrub oak			Meginnis 1935
	Control	0.025	0.056	
	Rx burn	0.330	0.739	
Texas	Loblolly pine			Pope and others 1946
	Control	0.050	0.112	
	Annual Rx burning	0.359	0.806	
Texas	Loblolly pine			Ferguson 1957
	Control	0.100	0.224	
	Single Rx burn	0.210	0.470	
M231 OUACHITA MIXED FOREST-MEADOW PROVINCE				
Arkansas	Shortleaf pine			Miller and others 1988
	Control	0.016	0.036	
	Cut, slash Rx burn	0.106	0.237	
M222 OZARK BROADLEAF FOREST PROVINCE				
Oklahoma	Mixed hardwoods			Daniel and others 1943
	Control	0.010	0.022	
	Annual Rx burning	0.110	0.246	

[1]Mg/ha is metric tons per hectare.
[2]Bailey's (1995) descriptions of ecoregions of the United States.

year after prescribed burns and wildfires. Sediment yields 1 year after prescribed burns and wildfires range from very low, in flat terrain and in the absence of major rainfall events, to extreme, in steep terrain affected by high intensity thunderstorms. Erosion on burned areas typically declines in subsequent years as the site stabilizes, but the rate of recovery varies depending on burn or fire severity and vegetation recovery.

Soil erosion following fires can vary from under 0.1 tons/acre/year (0.1 Mg/ha/year) to 6.7 tons/acre/year (15 Mg/ha/year) in prescribed burns, and from less than 0.1 tons/acre (less than 0.1 Mg/ha/year) in low-severity wildfire, to more than 164.6 tons/acre/year (369 Mg/ha/year) in high-severity wildfires on steep slopes (Hendricks and Johnson 1944, Megahan and Molitor 1975, Neary and Hornbeck 1994, Robichaud and Brown 1999). For example, Radek (1996) observed erosion of 0.13 tons/acre/year (0.3 Mg/ha/year) to 0.76 tons/acre/year (1.7 Mg/ha/year) from several large wildfires that covered areas ranging from 494 to 4,370 acres (200 to 1,770 ha) in the northern Cascades Mountains. Three years after these fires, large erosional events occurred from spring rainstorms, not from snowmelt. Most of the sediment produced did not leave the burned area. Sartz (1953) reported an average soil loss of 1.5 inch (37 mm) after a wildfire on a north-facing slope in the Oregon Cascades. Raindrop splash and sheet erosion accounted for the measured soil loss. Annual precipitation was 42.1 inches (1,070 mm), with a maximum intensity of 3.54 inch/hour (90 mm/hour). Vegetation covered the site within 1 year after the burn. Robichaud and Brown (1999) reported first-year erosion rates after a wildfire from 0.5 to 1.1 tons/acre (1.1 to 2.5 Mg/ha/year), decreasing by an order of magnitude by the second year, and to no

Table 2.9—Sediment losses the first year after prescribed (Rx) fire and wildfires, part 2.

Location	Management activity	1st year sediment loss		Reference
		tons/ac	*Mg/ha[1]*	
M242 CASCADE MIXED-CONIFER-MEADOW FOREST PROVINCE[2]				
Washington	Mixed conifer			Helvey 1980
	Control	0.012	0.028	
	Wildfire	1.049	2.353	
M261 SIERRAN STEPPE-MIXED FOREST-CONIFER FOREST PROVINCE				
California	Ponderosa pine			Biswell and Schultz 1965
	Control	<0.001	<0.001	
	Understory Rx burn	<0.001	<0.001	
261 CALIFORNIA COASTAL CHAPARRAL				
California	Chaparral			Wells 1981
	Control	0.019	0.043	
	Wildfire (R=3 yr)	12.758	28.605	
California	Chaparral			Krammes 1960
	Control	2.466	5.530	
	Wildfire	24.664	55.300	
California	Chaparral			Wohlgemuth 2001
	Control			
	Rx burn	0.708	1.587	
	Reburn	0.389	0.872	
	Wildfire	9.058	20.309	
313 COLORADO PLATEAU SEMI-DESERT PROVINCE				
Arizona	Chaparral			Pase and Lindenmuth 1971
	Control	<0.001	<0.001	
	Prescribed fire	1.685	3.778	
Arizona	Chaparral			Pase and Ingebo 1965
	Control	0.043	0.096	
	Wildfire	12.798	28.694	
Arizona	Chaparral			Glendening and others 1961
	Control	0.078	0.175	
	Wildfire	90.984	204.000	

[1]Mg/ha is metric tons per hectare.
[2]Bailey's (1995) descriptions of ecoregions of the United States.

sediment by the fourth, in an unmanaged forest stand in eastern Oregon. DeBano and others (1996) found that following a wildfire in ponderosa pine, sediment yields from a low severity fire recovered to normal levels after 3 years, but moderate and severely burned watersheds took 7 and 14 years, respectively. Nearly all fires increase sediment yield, but wildfires in steep terrain produce the greatest amounts, 12.5 to164.8 tons/acre/year (28 to more than 369 Mg/ha/year). Noble and Lundeen (1971) reported an average annual sediment production rate of 2.5 tons/acre (5.7 Mg/ha) from a 902 acre (365 ha) burn on steep river breaklands in the South Fork of the Salmon River, Idaho. This rate was approximately seven times greater than hillslope sediment yields from similar, unburned lands in the vicinity.

Sediment yields usually are the highest the first year after a fire and then decline in subsequent years. However, if precipitation is below normal, the peak sediment delivery year might be delayed until years 2 or 3. In semiarid areas like the Southwest, postfire sediment transport is episodic in nature, and the delay may be longer. All fires increase sediment yield, but it is wildfire that produces the largest amounts. Slope is a major factor in determining the amount of sediment yielded during periods of rainfall following fire (see table 2.10). There is growing evidence that short-duration, high-intensity rainfall (greater than 50 mm/hour in 10- to-15 minute bursts) over areas of about 1 km^2 (247 acres) often produce the flood flows that result in large amounts of sediment transport (Neary and

USDA Forest Service Gen. Tech. Rep. RMRS-GTR-42-vol. 4. 2005

49

Table 2.10—Sediment losses the first year after prescribed (Rx) fire and wildfires, part 3

Location	Management activity	1st year sediment loss		Reference
		tons/ac	Mg/ha[1]	
M313 AZ-NM MOUNTAINS SEMIDESERT-WOODLAND-CONIFER PROVINCE[2]				
Arizona	Ponderosa Pine			Campbell and others 1977
	Control	0.001	0.003	DeBano and others 1996
	Wildfire low severity	0.036	0.080	
	Wildfire moderate severity	0.134	0.300	
	Wildfire high severity	0.559	1.254	
Arizona	Mixed conifer			Hendricks and Johnson 1944
	Control	<0.001	<0.001	
	Wildfire, 43% slope	31.969	71.680	
	Wildfire, 66% slope	89.914	201.600	
	Wildfire, 78% slope	164.842	369.600	
315 SOUTHWEST PLATEAU AND PLAINS STEPPE AND SHRUB PROVINCE				
Texas	Juniper and grass			Wright and others 1982
	Control	0.027	0.060	
	Rx burn (3 yr)	6.690	15.000	
	Rx burn, seed (1 yr)	1.338	3.000	
Texas	Juniper and grass			Wright and others 1976
	Control: level	0.011	0.025	
	Rx burn: level	0.013	0.029	
	Control: 15-20%	0.034	0.076	
	Rx: 15-20% slope	0.836	1.874	
	Control: 43-54%	0.006	0.013	
	Rx: 43-54% slope	3.766	8.443	
M332 MIDDLE ROCKY MTN STEPPE-CONIFER FOREST-MEADOW PROVINCE				
Montana	Larch, Douglas-fir			DeByle 1981
	Control	<0.001	<0.001	
	Slash Rx burn	0.067	0.150	
342 IINTERMOUNTAIN SEMIDESERT				
Idaho	Sagebrush, grass, forb			Pierson and others 2001b
	Control, interspace	<0.013	<0.030	
	Moderate severity fire	0.056	0.125	
	High severity fire	0.686	1.538	

[1]Mg/ha is metric tons per hectare.
[2]Bailey's (1995) descriptions of ecoregions of the United States.

Gottfried 2002, Gottfried and others 2003, Gartner and others 2004; see also chapter 5).

Best Management Practices certainly have value in reducing sediment losses from prescribed fires. O'Loughlin and others (1980) reported that a 66 foot (20 m) buffer strip in a steep watershed reduced sediment loss after prescribed fire from 800 percent of the control watershed to 142 percent. Mitigative techniques for reducing sediment losses after wildfires often are used as part of burned area emergency watershed rehabilitation, but they have their limitations (see chapter 10).

After fires, turbidity can increase due to the suspension of ash and silt-to-clay-sized soil particles in streamflow. Turbidity is an important water quality parameter because high turbidity reduces municipal water quality and can adversely affect fish and other aquatic organisms (see chapter 7). It is often the most easily visible water quality effect of fires (DeBano and others 1998). Less is known about turbidity than sedimentation in general because it is difficult to measure, highly transient, and extremely variable.

Extra coarse sediments (sand, gravel, boulders) transported off of burned areas or as a result of increased storm peakflows can adversely affect aquatic habitat, recreation areas, and reservoirs. Deposition of coarse sediments destroys aquatic and riparian habitat and fills in lakes or reservoirs (Reid 1993, Rinne 1996).

Management Implications _____

Resource managers need to be aware of the changes in the physical properties that occur in the soil during a fire. The most important physical process functioning during a fire is the transfer of heat into the soil. Soil heating not only affects soil physical properties but also changes many of the chemical and biological properties in soils described in chapters 3 and 4, respectively.

Wildfires present their own unique concerns. Although little can be done to modify soil heating during a wildfire, managers need to be aware of the susceptibility of severely burned areas to postfire runoff and erosion. Excessive heating of the underlying soils during these uncontrollable fires can change soil structure to such an extent that water infiltration is impeded, creating excessive runoff and erosion following fire. Formation of water repellent soils during these wildfires may present special concerns with erosion following wildfires and needs to be addressed when initiating postfire treatments. Postfire rehabilitation is an important activity on areas burned by wildfire where it is necessary to reestablish plant cover as soon as possible to protect the bare soil surface from raindrop impact. However, discretion is needed in order to apply the most effective and practical postfire rehabilitation treatments. Numerous revegetation techniques that are available for use on burned watersheds are discussed in chapter 10 of this publication and elsewhere (Robichaud and others 2000).

The use of prescribed fire, however, presents the manager with alternatives for minimizing the damage done to the soil. The least amount of damage occurs during cool-burning, low-severity fires. These fires do not heat the soil substantially, and the changes in most soil properties are only minor and are of short duration. However, the burning of concentrated fuels (for example, slash, large woody debris) can cause substantial damage to the soil resource, although these long-term effects are limited to only a small proportion of the landscape where the fuels are piled. These types of fire use should be avoided whenever possible. The burning of organic soils is also a special case where extensive damage can occur unless burning prescriptions are carefully planned.

Summary _____

The physical processes occurring during fires are complex and include both heat transfer and the associated change in soil physical characteristics. The most important soil physical characteristic affected by fire is soil structure because the organic matter component can be lost at relatively low temperatures. The loss of soil structure increases the bulk density of the soil and reduces its porosity, thereby reducing soil productivity and making the soil more vulnerable to postfire runoff and erosion. Although heat is transferred in the soil by several mechanisms, its movement by vaporization and condensation is the most important. The result of heat transfer in the soil is an increase in soil temperature that affects the physical, chemical, and biological properties of the soil. When organic substances are moved downward in the soil by vaporization and condensation they can cause a water-repellent soil condition that further accentuates postfire runoff and erosion. Water repellency accelerates postfire runoff, which in turn creates extensive networks of surface rill erosion. Water repellency also increases erosion by raindrop splash. The magnitude of change in soil physical properties depends on the temperature threshold of the soil properties and the severity of the fire. The greatest change in soil physical properties occurs when smoldering fires burn for long periods.

USDA Forest Service Gen. Tech. Rep. RMRS-GTR-42-vol. 4. 2005

51

Notes

Jennifer D. Knoepp
Leonard F. DeBano
Daniel G. Neary

Chapter 3:
Soil Chemistry

Introduction

The chemical properties of the soil that are affected by fire include individual chemical characteristics, chemical reactions, and chemical processes (DeBano and others 1998). The soil chemical characteristics most commonly affected by fire are organic matter, carbon (C), nitrogen (N), phosphorus (P), sulfur (S), cations, cation exchange capacity, pH, and buffer power. Some purely chemical reactions occur in soils. These include the exchange of cations adsorbed on the surface of mineral soil particles and humus with their surrounding solutions. Another predominately chemical reaction is the chemical weathering of rocks and their eventual transformation into secondary clay minerals during soil formation. During the chemical decomposition of rock material, the soil and its surrounding solution become enriched with several cations. Associated with the chemical interactions during weathering and soil formation are physical forces (freezing and thawing, wetting and drying) and biological activities (production of organic acids during the decomposition of humus) that also accelerate soil development. The most common chemical processes occurring in soils that are affected by fire, however, are those mechanisms that are involved in nutrient availability and the losses and additions of nutrients to the soil.

Soil Chemical Characteristics

The chemical characteristics of soils range from the inorganic *cations*—for example, calcium (Ca), sodium (Na), magnesium (Mg), potassium (K), and so forth—that are adsorbed on the surface of clay materials to those contained mainly within the organic matrix of the soil—for example, organic matter, C, N, P, S. All chemical characteristics are affected by fire, although the temperatures at which changes occur can vary widely. The best estimates available in the literature for the threshold temperatures of individual soil chemical characteristics are given in table 3.1.

The published information describing the effects of fire on changes in individual chemical constituents of soils and organic matter are contradictory and have often led to differing conclusions about the magnitude and importance of the chemical changes that actually occur during a fire. Different studies have concluded that soil chemical constituents increase, decrease, or remain the same (DeBano and others 1998). This has been particularly true for studies reporting changes in N and other nutrients that can volatilize readily during a fire (for example, organic matter, sulfur, and phosphorus). Differing conclusions arise primarily because of the method used for

USDA Forest Service Gen. Tech. Rep. RMRS-GTR-42-vol. 4. 2005

53

Table 3.1—Temperature thresholds for several soil chemical characteristics of soil.

Soil characteristic	Threshold temperature		Source
	°F	°C	
Organic matter	212	100	Hosking 1938
Nitrogen	414	200	White and others 1973
Sulfur	707	375	Tiedemann 1987
Phosphorus and potassium	1,425	774	Raison and others 1985a,b
Magnesium	2,025	1,107	DeBano 1991
Calcium	2,703	1,484	Raison and others 1985a,b
Manganese	3,564	1,962	Raison and others 1985a,b

calculating the chemical constituents. Chemical changes can be expressed in either percentages (or some other expression of concentrations, for example, ppm or mg/kg) or be based on the actual changes in total amounts of the constituent (for example, pounds/acre or kg/ha). Before fire, the percent of a given chemical constituent is usually based on the amount contained in a prefire sample that can contain variable amounts of organic matter. In contrast, following the fire the percent of the same chemical constituent is based on the weight of a burned sample that contains varying amounts of ash along with charred and unburned organic matter. Thus, the confusion in nutrient changes arises because different bases are used for calculating the change in a particular chemical constituent (that is, based on mainly organic matter before combustion as compared to ashy and unburned materials following a fire). This confusion between percentages and total amounts was first reported in a study on the effect of fire on N loss during heating (Knight 1966). This study indicated that the differences between percent N and total amount of N started at 212 °F (100 °C) and became greater until about 932 °F (500 °C). Because of these difficulties in interpreting concentration and percentage data, the following discussion on the fire-related

changes in chemical constituents will first focus on the more fundamental changes in chemical constituents in wildland ecosystems.

As a general rule, the total amounts of chemical elements are never increased by fire. The total amounts of different chemical elements on a particular burned site most likely decrease, although in some cases may remain the same (for example, elements with high temperature thresholds such as Mg, Ca, and others listed in table 3.1). The fire, however, does change the form of different elements and in many cases makes them more available for plants and other biological organisms. A classic example of this is total N contained in the ecosystem organic matter (table 3.2). When organic matter is combusted, total N on the site is always decreased, although increases in the available forms of N are likely to occur as is discussed in a later section, "Nitrogen." Therefore, managers must be alert when interpreting the significance of the sometimes contradictory changes in different nutrients during a fire that are reported in the literature. The following sections focus on describing these changes in terms of the underlying chemical processes and to indicate the management implications of these changes in terms of soil and ecosystem productivity and postfire management.

Table 3.2—Effect of burning at low, medium, and high severities on organic matter and mineralizable N in forest soils in northern Idaho. (Adapted from Niehoff 1985, and Harvey and others 1981).

| Treatment | Organic matter Mineral soil 2.5-7.5 cm | | Mineralizable N | | Total N |
			Mineral 2.5-7.5 cm	Organic 0-2.5 cm	
	Percent	Change (%)	ppm (mg/kg)	ppm (mg/kg)	Change (%)
Undisturbed	3.6	0	9.4	68	0
Clearcut					
No burn	3.9	+8	9.7	97	+22
Low	4.1	+12	9.5	75	+8
Medium	2.8	−22	9.3	5	−82
High	0.6	−83	0.7	0	−99

Organic Matter and Carbon

Many chemical properties and processes occurring in soils depend upon the presence of organic matter. Not only does it play a key role in the chemistry of the soil, but it also affects the physical properties (see chapter 2) and the biological properties (see chapter 4) of soils as well. Soil organic matter is particularly important for nutrient supply, cation exchange capacity, and water retention. However, burning consumes aboveground organic material (future soil organic matter, including large logs), and soil heating can consume soil organic matter (fig. 3.1). The purpose of the following discussion is to focus as much as possible on the purely chemical properties of organic matter and on changes that occur as the result of soil heating. Because organic C is one of the major constituents of organic matter, the changes in organic matter and organic C during soil heating are considered to be similar for all practical purposes.

Location of Organic Matter in Different Ecosystems—Organic compounds are found in both aboveground and belowground biomass where they make up the standing dead and live plants and dead organic debris (that is, leaves, stems, twigs, and logs) that accumulate on the soil surface and throughout the soil profile (DeBano and others 1998). Organic matter found in the soil consists of at least seven components, namely:

- The L-layer, (O_i) which is made up of readily identifiable plant materials.
- The F-layer, (O_e) which contains partially decomposed organic matter but can still be identified as different plant parts (needles, leaves, stems, twigs, bark, and so forth).
- The H-layer, (O_a) which is made up of completely decayed and disintegrated organic

materials, some of which is usually mixed with the upper mineral soil layers.
- Coarse woody debris that is eventually decayed but can remain on the soil surface or buried in the mineral soil for long periods.
- Charcoal or other charred materials that become mixed with the forest floor and uppermost layers of the mineral soil.
- The uppermost part of the A-horizon, which is composed mainly of a mixture of humus and mineral soil particles.
- A mixture of mineral soil, plant roots, and biomass (live and dead) that is concentrated primarily in the A-horizon but may extend downward into the B-horizon or deeper depending upon the type of vegetation growing on the site.

The amount of aboveground and belowground organic matter varies widely between different vegetation types depending upon on the temperature and moisture conditions prevailing in a particular area (DeBano and others 1998). In almost all ecosystems throughout the world, greater quantities of C (a measure of organic matter) are found belowground than aboveground (fig. 3.2). In grasslands, savannas, and tundra-covered areas, much greater quantities of organic C are found in the underground plant parts than aboveground (less than 10 percent of the total C in these herbaceous vegetation ecosystems is found

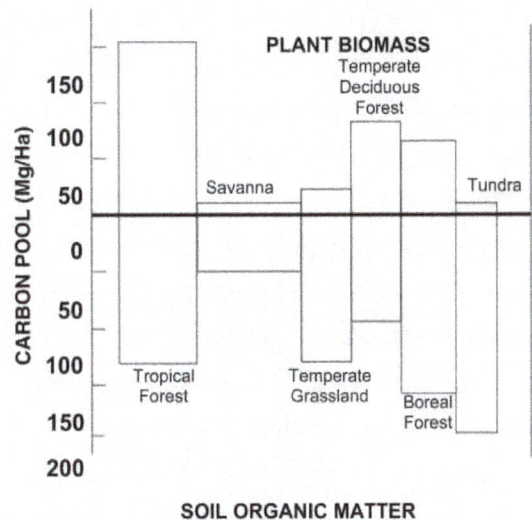

Figure 3.2—Distribution of C and soil organic matter (including litter) in major ecosystem types of the world. (Adapted from J. M. Anderson, 1991. The effects of climate change on decomposition processes in grassland and coniferous forest. Ecological Applications. 1: 326-347. Copyright © 1991 Ecological Society of America.)

Figure 3.1—Large logs combusting in a prescribed fire in a ponderosa pine stand. (Photo by Daniel Neary).

USDA Forest Service Gen. Tech. Rep. RMRS-GTR-42-vol. 4. 2005

55

aboveground). Tundra ecosystems are unique in that large amounts of organic matter accumulate on the soil surface because the low year-long temperatures severely limit decomposition. In forest ecosystems, C is more evenly distributed aboveground and belowground (for example, temperate deciduous and boreal forests). In general, soils with larger proportions of organic matter in the aboveground biomass and on their forest floors are more prone to disturbances (including fire) in their nutrient and C regimes than those in which most of the C in the ecosystem is located belowground.

Dynamics of Organic Matter Accumulation— In forests, the dynamics of the forest floor are responsible for the accumulation of organic matter, and the forest floor provides a major storage reservoir for nutrients that are cycled within natural ecosystems (fig. 3.3). An aggrading forest ecosystem sequesters nutrients and C aboveground in both the biomass and the forest floor (Knoepp and Swank 1994). Over many years this material forms the forest floor or the organic soil horizons (designated as Oi, Oe, and Oa horizons in part A, fig A.1). Depending on the soil type, organic matter may be concentrated in the forest floor or spread in decreasing amounts downward through the soil profile (fig. 3.4 and A.2).

The forest floor increases during forest development and aggradation when the rate of addition is greater than the rate of decomposition. For example, Knoepp and Swank (1994) found that forest floor mass

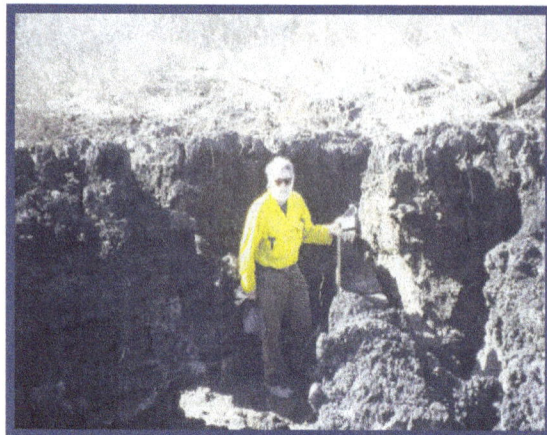

Figure 3.4—Burn out of high organic matter soil in the Barca Slough, Vandenburg Air Force Base, California. (Photo by U.S. Air Force).

increased by 28 and 45 percent over a 10-year period in an aggrading mixed oak (*Quercus* spp.) forest and white pine (*Pinus strobus*) plantation, respectively. In Arizona, Covington and Sackett (1992) examined forest floor accumulation in a ponderosa pine (*P. ponderosa*) stand having several different age "substands" within it and found several significant differences in forest floor mass among the substands. In general, the younger sapling areas had the smallest accumulations of forest floor materials compared to old growth pine substands that had the greatest. Over long periods the inputs and outputs of forest floor materials, coarse woody debris, fine woody debris, and leaf litter eventually reach a dynamic equilibrium depending upon the stand type (for example, coniferous versus hardwood) and forest management practices used (for example, uncut versus cut, log only versus whole-tree harvest).

Coarse Woody Debris—*Coarse woody debris* (including slash piles) is an important component of the organic matter pools found in forested ecosystems (fig. 3.5). In many cases it is partially or totally covered by soil and humus layers, and it has been found to comprise more than 50 percent of the total surface organic matter (this can amount to 16.5 to 22.3 tons/acre or 37 to 50 Mg/ha) in old growth forests in the Inland Northwest (Page-Dumroese and others 1991, Jurgensen and others 1997). Coarse woody debris, along with smaller organic matter, enhances the physical, chemical, and biological properties of the soil and thereby contributes directly to site productivity (Brooks and others 2003). It also provides a favorable microenvironment for the establishment

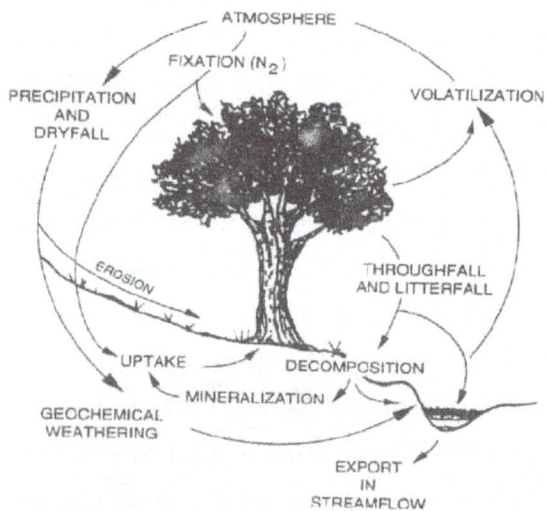

Figure 3.3—Nutrient cycling in natural environments. (Adapted from Brown 1980.)

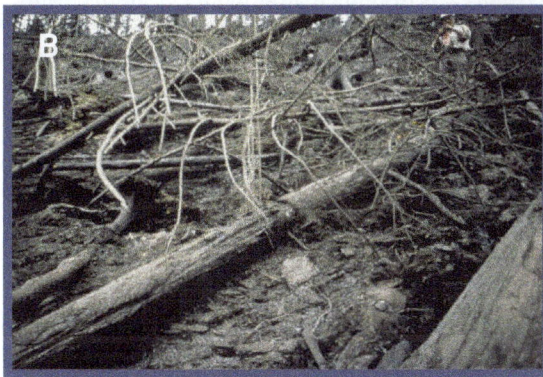

Figure 3.5—Coarse woody debris: (A) slash pile from pinyon-juniper thinning operation, Apache-Sitgreaves National Forest, Arizona; (B) logging slash in a Douglas-fir and mixed conifer, Mt. Baker-Snoqualmie National Forest, Cascade Mountains, Washington. (Photos by Malchus Baker and Kevin Ryan, respectively).

of seedlings and plant growth. The microbiological role of coarse woody debris is discussed further in chapter 4.

Fire Effects—Fire not only affects the organic matter by directly affecting its chemical composition but it also indirectly affects the subsequent decomposition rates. The magnitude of changes is related to the severity of the burn (for example, low, moderate, or high).

Chemical Changes: The low temperature threshold of organic matter makes it especially sensitive to soil heating during fire (table 3.1). Also, a major portion of the organic matter is located in the uppermost part of the soil profile (surface litter and humus layers) where it is exposed directly to the heat radiated downward

during a fire (see part A). The changes in organic matter during the course of heating have been of interest to scientists for more than 60 years (Hosking 1938), and the changes reported by these earlier studies showed that:

- Losses of organic matter can occur at temperatures below 212 °F (100 °C).
- Volatile constituents in organic matter are lost at temperatures up to 392 °F (200 °C).
- Destructive distillation destroys about 85 percent of the soil organic matter at temperatures between 392 and 572 °F (200 and 300 °C).
- Above 572 °F (300 °C), the greater part of the residual organic matter consists of carbonaceous material, which is finally lost upon ignition.
- Heating the soil to 842 °F (450 °C) for 2 hours, or to 932 °F (500 °C) for 1/2 hour, destroys about 99 percent of the organic matter.

Recent studies on the detailed heat-induced changes in organic matter have improved our understanding of the specific chemical changes that occur in organic matter during the course of heating (DeBano and others 1998). At low temperatures, the changes in organic matter affect the more sensitive functional groups; at higher temperatures the thermal decomposition of nuclei occurs (Schnitzer and Hoffman 1964). At lower temperatures between 482 and 752 °F (250 and 400 °C) both phenolic OH and carboxyl (COOH) groups are lost, although phenolic groups are the more stable of the two. Another study of the thermal changes occurring in the H-layer under an evergreen oak forest in Spain showed that oxygen-containing functional groups found in humic and fulvic acids were altered (Almendros and others 1990). Humic acids were converted into alkali-insoluble substances that contributed to soil humus, while fulvic acids were transformed into acid-insoluble polymers. Biomass that was not completely burned contained both alkali-soluble lignin materials and brown products formed by dehydration of carbohydrates. The lignin compounds formed were more resistant to further chemical and biological change.

In a separate study, Ross and others (1997) found a decrease in total soil C and potassium sulfate extractable C at 1.5 and 2.5 years after burning. Sands (1983) found that 24 years after an intense site preparation burn on sandy soils, soil C was still lower than on adjacent unburned sites. This was generally the case for all soil C components he examined, including total organic C, extractable C, water-soluble C, humic acids, and carbohydrates.

Decomposition rates: In unburned ecosystems, natural decomposition processes (most biologically mediated) slowly release nutrients to tree and plant roots

USDA Forest Service Gen. Tech. Rep. RMRS-GTR-42-vol. 4. 2005

57

growing within the forest floor and to the mineral soil below. These biological processes add organic matter and nutrients to soils in forest environments under moderate temperatures through the activity of insects and microbes (DeBano 1991). Burning, however, acts as an instantaneous physical decomposition process that not only volatilizes nutrients, such as N, from the site but also alters the remaining organic materials (St. John and Rundel 1976).

Burning not only rapidly accelerates the rates of organic matter decomposition during the fire itself but can also indirectly affect postfire decomposition rates. For example, Schoch and Binkley (1986) and Raison and others (1990) studied loblolly pine (*P. taeda*) and radiata pine (*P. radiata*) plantations following fire. They found that decomposition rates of the remaining forest floor increased after burning, releasing ammonium (NH_4) and other nutrients. Observed changes in nutrient release and availability may be due to the alteration of organic matter solubility as a result of soil heating during a fire. The change in nutrient release and availability along with increased soil temperature and moisture content may also increase biological activity. This response may be short lived, however, because this readily available organic matter often diminishes rapidly and decomposition rates decrease (Raison and others 1990). Conversely, some studies have noted decreases in the rates of organic matter decomposition following burning. Monleon and Cromack (1996) measured decreased rates of decomposition in ponderosa pine forests immediately following burning and for up to 12 years. They concluded that the lower decomposition rates may be due to the combination of increased temperature and decreased moisture in the postfire forest floor. Springett (1976) also measured slower decomposition in Australian plantations and native forests caused by the changes in soil temperature and moisture as well as a decrease in the diversity and density of soil fauna following burning. This suggested that burning on a frequent rotation could simplify the litter fauna and flora, but these changes may permanently alter patterns of organic matter decomposition and nutrient release.

Both fire severity and frequency of burning affect the amount of organic matter that is lost as a result of burning. These are two characteristics of fire effects that fire management specialists can alter within the context of a prescribed burning program.

Fire severity: The effect of severity of burning on the amount of organic matter burned was reported for a 350 acre (142 ha) wildfire that burned a table mountain pine stand in the Shenandoah Valley (Groeschl and others 1990, 1992, 1993). This wildfire left a mosaic pattern of areas that were burned at different severities. They reported that on areas burned by a low-severity fire, the forest floor Oi and Oe layers were completely combusted, but the Oa layer remained. High-severity burning also consumed the Oa layer. Of the 10.1 tons/acre (22.6 Mg/ha) of C present in the forest floor in the unburned areas, no C remained in the high-severity burned areas compared to 9.3 tons/acre (20.8 Mg/ha) C that was left on the burned areas at low severities.

The effect of prescribed burning on the C and N content of the forest floor on an area supporting a mixed pine-oak overstory with a ericaceous shrub layer was studied in the Southern Appalachian Mountains (Vose and Swank 1993, Clinton and others 1996). The study areas were treated with felling-and-burning of existing trees to stimulate pine regeneration. Carbon and N content of the forest floor in these systems was examined after felling, prior to burning, and it was found that the forest floor contained between 15 and 22 percent of the total aboveground C on the site and 44 to 55 percent of the total aboveground N. Prescribed burning of these sites consumed the entire Oi layer, but 75 to 116 percent of the total forest floor C and 65 to 97 percent of the N remained in the combined Oe and Oa layers. On these sites, about half of the total aboveground N pool was contained in the forest floor. Total aboveground N losses ranged from 0.086 tons/acre (0.193 Mg/ha) for the lowest severity fire to 0.214 tons/acre (0.480 Mg/ha) on the most severe burn. Under prescribed burning conditions it is frequently possible to select appropriate weather conditions prior to burning (for example, time since last rain, humidity, temperature) to minimize the effects of fire on the consumption of organic matter. Therefore, it has been recommended that weather conditions prior to burning (for example, time since rainfall) could be used as a predictor of forest floor consumption (Fyles and others 1991).

Responses of total C and N are variable and depend on site conditions and fire characteristics. For example, Grove and others (1986) found no change in organic C in the surface 0 to 1.2 inch (0-3 cm) of soil immediately following burning; percent total N, however, increased. Knoepp and Swank (1993a,b) found no consistent response in total N in the upper soil layer, but did find increases in NH_4 concentrations and N mineralization on areas where a burning treatment followed felling.

Fire frequency: As would be expected, frequency of burning can affect C accumulations. A study was carried out on tropical savanna sites in Africa having both clay and sandy soils that were burned repeatedly every 1, 3, or 5 years (Bird and others 2000). While sites with clay soils had greater total C than did the sandy soils, they responded similarly to burning. All unburned sites had 40 to 50 percent greater C than burned sites. Low frequency burning (every 5 years)

58

USDA Forest Service Gen. Tech. Rep. RMRS-GTR-42-vol. 4. 2005

resulted in an increase in soil C of about 10 percent compared to the mean of all burned areas. High frequency burning (every year) decreased C about 10 percent. In another study, Wells and others (1979) reported the results of a 20-year burning study in a pine plantation in South Carolina. They found that periodic burning over 20-year period removed 27 percent of the forest floor. Annual burning conducted in the summer removed 29 percent of the forest floor as compared to a 54 percent loss resulting from winter burning. The total organic matter content of the surface soil (0 to 2 inches or 0 to 5 cm) increased in all cases, but there was no effect on the 2 to 3.9 inches (5-10 cm) soil layer. Interestingly, when they summed the organic matter in the forest floor and in the surface 0 to 3.9 inches (0-10 cm) of soil they found that these low-severity periodic burns sites had not reduced but only redistributed the organic matter.

The incidence of fire has been found to also affect the organic matter composition of savannahlike vegetation (referred to as "cerrado") in central Brazil (Roscoe and others 2000). On plots exposed to more frequent fires (burned 10 times in 21 years) C and N were decreased in the litter by 1.652 and 0.046 tons/acre (3.703 and 0.104 Mg/ha), respectively, although no significant differences were noted in the upper 3 feet (1 meter) of the underlying soil. Interestingly, the increase in fire incidence replaced the C_3-C with C_4-C by about 35 percent throughout the soil profile. This suggested that a more rapid rate of soil organic matter turnover occurred in areas burned by frequent fires, and as a result the soil would not be able to replace sufficient C to maintain long-term productivity of the site.

Prescribed fire was returned into overstocked ponderosa pine stands on the Mogollon Rim of Arizona for the purpose of restoring fire into the ecosystem and removing fuel buildups (Neary and others 2003). Prescribed fires were ignited at intervals of 1, 2, 4, 6, 8, and 10 years to determine the best fire return interval for Southwest ponderosa pine ecosystems (Sackett 1980, Sackett and others 1996). Two sites were treated—one on volcanic-derived soils and the other on sedimentary-derived soils near Flagstaff, AZ. Soil total C and total N levels were highly variable and exhibited an increasing, but inconsistent, concentration trend related to burn interval. They ranged from 2.9 to more than 6.0 percent total C and 0.19 to 0.40 percent total N (fig. 3.6). High spatial variability was measured within treatments, probably due to microsite differences (location of samples in the open, under large old-growth trees, in small-diameter thickets, in pole-sized stands, next to downed logs, and so forth). Stratification of samples by microsite differences could possibly reduce the within-plot variability but add complexity to any sampling design. Although there were statistically significant differences between the total C levels in soils of the

unburned plots and the 8-year burning interval, there were no differences between burning intervals. There also was a statistically significant difference between unburned and 2-year burning interval and the 8-year burning interval in total soil N. This study determined that burning increased mineral soil C and N, which conflicted with Wright and Hart's (1997) contention that the 2-year burning interval could deplete soil N and C pools. This study did not examine the mineral fractions of the soil N pool, NH_4-N, and NO_3-N. Although the mineral forms of N are small (less than 2 percent of the total soil N pool), they are important for plant nutrition and microorganism population functions

Soil organic matter: Summary reports have described the effect of different management activities (including the effect of fire) on the organic matter and N found in the mineral portion of forest soils (D.W. Johnson 1992, Johnson and Curtis 2001). A meta-analysis on the results of 13 studies completed between 1975 and 1997 (table 3.3) show that the C and N contents of both the A-horizon and the underlying mineral soil layers change only a small amount (less than 10 percent) in the long term (fig. 3.7 and 3.8). These results agreed with the conclusions of another review (E. A. Johnson 1992) that indicated the overall effect of fire was not significant, although there was a significant effect of time since fire. It must be remembered, however, that although small changes in soil organic matter and C occurred in the soils during these studies, that substantial amounts of both organic litter and duff were most likely consumed during these fires. Organic matter and N losses from the forest floor could have a lasting effect on the long-term productivity and sustainability of forest sites, particularly when they occur on nutrient-deficient sites (see the later discussions in this chapter on N loss and ecosystem productivity).

Cation Exchange Capacity

Cation exchange is the interchange between cations in solution and different cations adsorbed on the surface of any negatively charged materials such as a clay or organic colloids (humus). *Cation exchange capacity* is the sum of the exchangeable cations found on organic and inorganic soil colloids (fig. 3.9). It arises from the negatively charged particles found on clay particles and colloidal organic matter in the soil. Cation exchange capacity sites are important storage places for soluble cations found in the soil. The adsorption of cations prevents the loss of these cations from the soils by leaching following fire. Although most of the exchange sites in soils are negative and attract cations, there are some positively charged sites that can attract anions (anion exchange has been reported to occur on clay particles).

USDA Forest Service Gen. Tech. Rep. RMRS-GTR-42-vol. 4. 2005

59

The relative contribution of clay particles and organic matter to the cation exchange capacity of the soil depends largely upon the proportion of the two components and the total quantities of each present (Tate 1987). Cation exchange capacity also depends upon the type of clay and organic matter present. Clay materials such as montmorillonites have large exchange capacities, and other clays such as kaolinite are much lower. Other mineral particles such as silt and sand contain few adsorption sites for cations. In organic matter, the degree of humification affects the cation exchange capacity, and the more extensive the decomposition of organic material, the greater the exchange capacity.

Soil heating during a fire can affect cation exchange capacity in at least two ways. The most common change is the destruction of humus compounds. The location of the humus layer at, or near, the soil surface

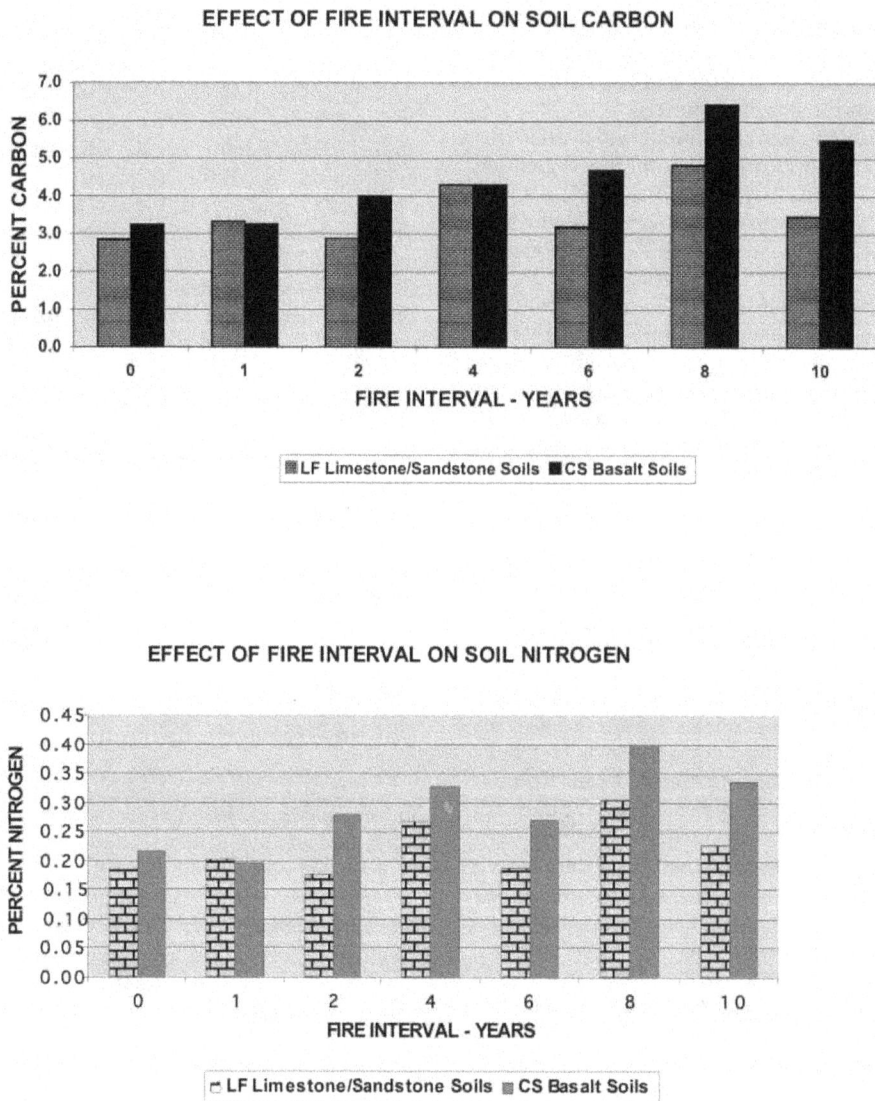

EFFECT OF FIRE INTERVAL ON SOIL CARBON

EFFECT OF FIRE INTERVAL ON SOIL NITROGEN

Figure 3.6—Effect of fire interval on (A) 0 to 2 inch (0-5 cm) soil total carbon, and (B) 0 to 2 inch (0-5 cm) soil nitrogen, Limestone Flats and Chimney Springs burning interval study, Arizona. (From Neary and others 2003).

makes it especially vulnerable to partial or total destruction during a fire because organic and humic materials start decomposing at about 212 °F (100 °C) and are almost completely destroyed at 932 °F (500 °C). These temperatures are easily reached during brushland and forest fires (see chapter 2 discussion on soil temperatures). In contrast, the cation exchange capacity of the clay materials is more resistant to change because heating and temperatures of 752 °F (400 °C) must be reached before dehydration occurs. The complete destruction of clay materials does not occur until temperatures of 1,292 to 1,472 °F (700 to 800 °C) are reached. In addition, clay material is seldom located on the soil surface but instead is located at least several centimeters below the soil surface in the B-horizon where it is well protected from

Table 3.3—References for the Johnson and Curtis (2001) meta-analysis of the effects of forest fires on soil C and N contents.

Location	Species	Fire type	Reference
Southern U.S.A.			
SC	Longleaf pine	PF[1]	Binkley and others 1992
SC, AL, FL, LA	Loblolly and other pines	PF	McKee 1982
Southwest U.S.A.			
AZ	Ponderosa pine	PF	Covington and Sackett 1986
AZ	Pinyon-juniper	PF	Klopatek and others 1991
Northwest U.S.A.			
WA	Mixed conifer	WF	Grier 1975
WA, OR	Douglas-fir, conifer mix	BB	Kraemer and Hermann 1979
OR	Ponderosa pine	PF	Monleon and others 1997
MT	Mixed conifer	PF	Jurgenson and others 1981
Alaska and Canada			
AL	Mixed spruce and birch	WF	Dyrness and others 1989
BC	Sub-boreal spruce	BB	Macadam 1987
World			
Australia	Eucalyptus	BB	Rab 1996
Algeria	Oak	WF	Rashid 1987
Sardinia	Chaparral	PF	Giovannini and others 1987

[1]PF: prescribed fire; WF: wildfire; BB: cut and broadcast burn.

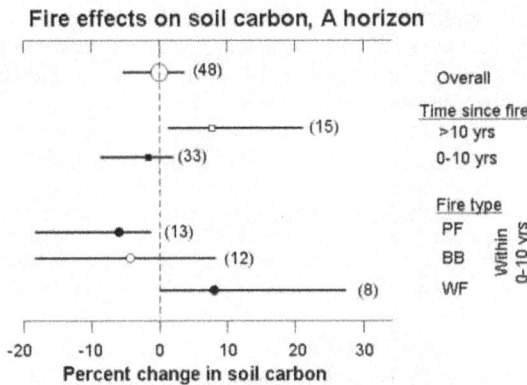

Figure 3.7—Fire effects on soil C, A-horizon nonparametric metaanalysis results; 99 percent confidence intervals (bars) and number of studies (in parentheses); PF = prescribed fire, WF = wildfire, and BB = broadcast burning of slash after harvest) (After Johnson and Curtis 2001, Forest Ecology and Management, Copyright © 2001, Elsevier B.V. All rights reserved).

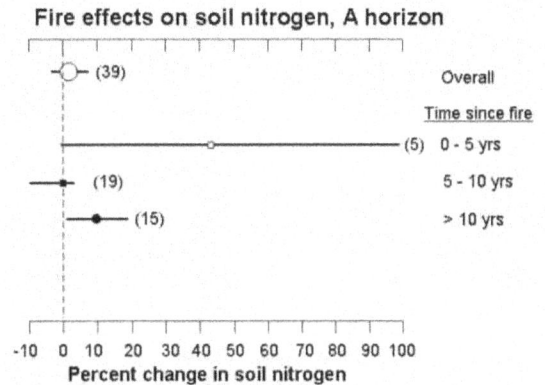

Figure 3.8—Fire effects on soil N, A-horizon nonparametric metaanalysis results; 99 percent confidence intervals (bars) and number of studies (in parentheses). (After Johnson and Curtis 2001).

USDA Forest Service Gen. Tech. Rep. RMRS-GTR-42-vol. 4. 2005

61

Figure 3.9—Cation exchange capacity in soil is provided by both organic matter (humus) found in the forest floor and upper soil horizons as well as inorganic mineral particles such as silts and clays found lower in the profile. (Soil profile from Appalachian Mountains, Nantahala National Forest; photo by Daniel Neary).

surface heating. In general, the reduction in exchange capacity as the result of a fire is proportional to the amount of the total cation capacity that is provided by the organic component (DeBano and others 1998). The amount of cation exchange capacity remaining after a fire affects the leaching losses of soluble nutrients released during the fire. For example, the prefire cation exchange capacity of sandy soils may consist mainly of exchange sites found on the humus portion of the soil. If large amounts of humus are destroyed in these sandy soils during burning, then no mechanisms are available to prevent large losses of soluble nutrients by leaching.

The loss of cation exchange capacity, as the result of organic matter destroyed by fire, has been reported on by several authors. Soto and Diaz-Fierros (1993) intensively monitored changes in soil cation exchange capacity for one of the six soils they exposed to increasing temperatures. They found that cation exchange capacity decreased from 28.4 meq/100 g (28.4 cmol/kg) at 77 °F (25 °C) down to 1 meq/100 g (1 cmol/kg) when exposed to 1,292 °F (700 °C). The largest decrease occurred between 338 and 716 °F (170 and 380 °C), dropping from 28.1 to 6.9 meq/100 g (28.1 to 6.9 cmol/kg). Sands (1983) examined two adjacent radiata pine sites on sandy soils in Southeastern Australia 24 years after cutting. He found that the sites that received an intense site preparation burning before planting had decreased cation exchange capacity downward in soils to 20 cm compared to no changes on unburned naturally regenerated sites.

A stand replacement fire in the Southern Appalachians that resulted in a mosaic burn pattern similar to a wildfire produced a slight but significant decrease in cation exchange capacity 3 months after burning (Vose and others 1999). Also associated with the change in cation exchange capacity on midslope areas (medium-severity burn) was a decrease in exchangeable K and Mg, along with an increase in soil pH.

Cations

Cations found in the soil that are affected by fire include Ca, Mg, Na, K, and ammonia (NH_4), although these cations are not usually deficient in most wildland soils (DeBano 1991). In many studies, a significant increase in soil cation concentration following either prescribed burning or a wildfire has been reported (Grove and others 1986, Raison and others 1990, Soto and Diaz-Fierros 1993). The NH_4 cation, which is an important component of N cycling and soil productivity, responds differently from the other cations. With the exception of NH_4, cations have high temperature thresholds and, as a result, are not easily volatilized and lost from burned areas. The ash deposited on the soil surface during a fire contains high concentrations of cations, and their availability is increased, including NH_4 (Marion and others 1991, DeBano and others 1998; fig. 1.4). The amount of NH_4 released by burning depends upon fuel loading and the quantity of fuel combusted (Tomkins and others 1991). Some of the cations can be lost through particulate transfer in the smoke (Clayton 1976).

Monovalent cations, such as Na and K, are present largely as chlorides and carbonates that are readily mobilized (Soto and Diaz-Fierros 1993). Divalent ions, such as Ca and Mg, are less mobile and are commonly present as oxides and carbonates. The formation of insoluble calcium carbonate can occur, which limits the availability of P following fire. Although these readily available monovalent and divalent cations probably do not materially affect plant growth directly, their amount and composition determines base saturation, which plays an important role in controlling the pH regimes in soils (DeBano and others 1998).

Soil pH and Buffer Capacity

Soil *pH* is a measure of the hydrogen ion activity in the soil and is determined at specified moisture contents. Neutral soils have a pH of 7, acidic soils have a pH less than 7, and basic soils are those with a pH greater than 7. *Buffer capacity* is the ability of ions associated with the solid phase to buffer changes in ion concentration of the soil solution.

The combustion of organic matter during a fire and the subsequent release of soluble cations tend to increase pH slightly because basic cations are released during combustion and deposited on the soil surface. The increase in soil pH, however, is usually temporary depending upon the original soil pH, amount of ash released, chemical composition of the ash, and wetness of the climate (Wells and others 1979). The ash-bed effect discussed later in this chapter is an example of these factors in which large amounts of nutrients are deposited, with pH values being measurably changed by fire.

The pH of the soil is an important factor affecting the availability of plant nutrients (fig. 3.10). The nutrients released during a fire that are most likely to be affected are P, iron, and copper. P is particularly important because it is a macronutrient that is frequently limiting in wildland ecosystems, and it can also become insoluble at both high or low pHs (see part A). At low pH, P forms insoluble compounds with iron and at high pH, Ca compounds tend to immobilize it.

Nitrogen

Nitrogen is considered the most limiting nutrient in wildland ecosystems and as such it requires special consideration when managing fire, particularly in N-deficient ecosystems (Maars and others 1983). Nitrogen is unique because it is the only soil nutrient that is not supplied to the soil by chemical weathering of parent rock material. Almost all N found in the vegetation, water, and soil of wildland systems has to be added to the system from the atmosphere. A rare exception is the addition of some synthetic N-fertilizers that have been produced industrially and used for fertilizing forested areas. The cycling of N involves a series of interrelated complex chemical and biological

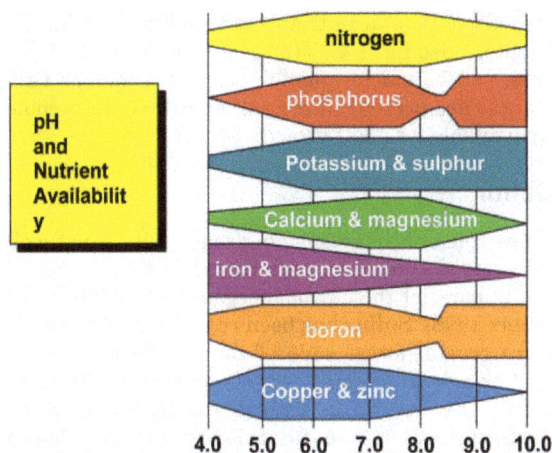

Figure 3.10—The availability of some common soil nutrients at different soil pH. (Figure courtesy of the USDA Forest Service, National Advanced Fire and Resource Institute, Tucson, AZ).

processes (also see chapter 4). Only those cycling processes affecting chemical changes in N are discussed in this chapter (that is, N volatilization). Biologically mediated processes affecting N are discussed in more detail as part of chapter 4. The changes in N availability produced during fire are discussed later in this chapter in a section describing the effect of fire on nutrient availability

Responses to Soil Heating—Volatilization is the chemically driven process most responsible for N losses during fire. There is a gradual increase in N loss by volatilization as temperature increases (Knight 1966, White and others 1973). The amount of loss at different temperatures has established the following sequence of N losses upon heating:

- Complete loss (100 percent) of N occurs at temperatures above 932 °F (500 °C).
- Between 75 and 100 percent of the N is lost at temperatures of 752 to 932 °F (400 to 500 °C).
- Between 50 and 75 percent of the N is lost at temperatures of 572 to 752 °F (300 to 400 °C).
- Between 25 and 50 percent of the N is lost at temperatures of 392 to 572 °F (200 to 300 °C).
- No N losses occur at temperatures below 392 °F (200 °C).

As a general rule the amount of total N that is volatilized during combustion is directly proportional to the amount of organic matter destroyed (Raison and others 1985a). It has been estimated that almost 99 percent of the volatilized N is converted to N_2 gas (DeBell and Ralston 1970). At lower temperatures, N_2 can be produced during organic matter decomposition without the volatilization of N compounds (Grier 1975). The N that is not completely volatilized either remains as part of the unburned fuels or it is converted to highly available NH_4-N that remains in the soil (DeBano and others 1979, Covington and Sackett 1986, Kutiel and Naveh 1987, DeBano 1991).

Estimates of the total N losses during prescribed fire must be based on both fire behavior and total fuel consumption because irregular burning patterns are common. As a result, combustion is not complete at all locations on the landscape (DeBano and others 1998). For example, during a prescribed burn in southern California, total N loss only amounted to 10 percent of the total N contained in the plant, litter, and upper soil layers before burning (DeBano and Conrad 1978). The greatest loss of N occurred in aboveground fuels and litter on the soil surface. In another study of N loss during a prescribed fire over dry and moist soils, about two-thirds of the total N was lost during burns over dry soils compared to only 25 percent when the litter and soil were moist (DeBano and others 1979). Although these losses were relatively small, it should be remembered that even small losses can adversely

USDA Forest Service Gen. Tech. Rep. RMRS-GTR-42-vol. 4. 2005

63

affect the long-term productivity of N- deficient ecosystems. The importance of N losses from ecosystems having different pools of N is considered in more detail below.

Monleon and others (1997) conducted understory burns on ponderosa pine sites burned 4 months, 5 years, and 12 years previously. The surface soils, 0 to 2 inches (0 to 5 cm), showed the only significant response. The 4-month sites had increased total C and inorganic N following burning and an increased C/N ratio. Burning the 5-year-old sites resulted in a decrease in total soil C and N and a decrease in the C/N ratio. Total soil C and N in the surface soils did not respond to burning on the 12-year-old site.

Nitrogen Losses—An Enigma— It has been conclusively established by numerous studies that total N is decreased as a result of combustion (DeBano and others 1998). The amount of N lost is generally proportional to the amount of organic matter combusted during the fire. The temperatures at which N is lost are discussed above. In contrast, available N is usually increased as a result of fire, particularly NH_4-N (Christensen 1973, DeBano and others 1979, Carballas and others 1993). This increased N availability enhances postfire plant growth, and gives the impression that more total N is present after fire. This increase in fertility, however, is misleading and can be short-lived. Any temporary increase in available N following fire is usually quickly utilized by plants within the first few years after burning.

Nitrogen Losses and Ecosystem Productivity— The consequences of N losses during fire on ecosystem productivity depend on the proportion of total N lost for a given ecosystem (Barnett 1989, DeBano and others 1998). In N-limited ecosystems even small losses of N by volatilization can impact long-term productivity (fig. 3.11).

The changes in site productivity are related to the proportion of total N in the system that is lost. For example, the left portion of figure 3.11 represents a situation where large quantities of N are presented on a site having high productivity. Moving to the right side of the graph, both total N capital and productivity decrease. This decrease is not linear because there are likely to be greater losses in productivity per unit loss of N capital on sites having lower productivity (right side of fig. 3.11) than on sites having higher site productivity (left side of fig. 3.11). As a result, the losses in site productivity per unit N loss (a_h to b_h) from sites of high productivity (l_h) are less than losses in site productivity per unit N loss (a_i to b_i) from sites having low productivity (l_i). This relationship points to the importance of somehow replenishing N lost during a fire on low productivity sites or when using prescribed fire in these situations, taking special care not to consume large amounts of the organic matter present.

Phosphorus

Phosphorus is probably the second most limited nutrient found in natural ecosystems. Deficiencies of P have been reported in P-fixing soils (Vlamis and others 1955) and as a result from N fertilization applications (Heilman and Gessel 1963). Phosphorus uptake and availability to plants is complicated by the relationship between mycorrhizae and organic matter and in most cases does not involve a simple absorption from the soil solution (Trappe and Bollen 1979). Phosphorus is lost at a higher temperature during soil heating than N, and only about 60 percent of the total P is lost by nonparticulate transfer when organic matter is totally combusted (Raison and others 1985a). The combustion of organic matter leaves a relatively large amount of highly available P in the surface ash found on the soil surface immediately following fire (see discussion on nutrient availability later in this chapter; also fig. 1.4). This highly available P, however, can be quickly immobilized if calcareous substances are present in the ash and thus can become unavailable for plant growth.

Sulfur

The role of S in ecosystem productivity is not well understood although its fluctuation in the soil is generally parallel to that of inorganic N (DeBano and others 1998). Sulfur has been reported as limiting in some coastal forest soils of the Pacific Northwest, particularly after forest stands have been fertilized with N (Barnett 1989). The loss of S by volatilization occurs at temperatures intermediate to that of N and P (Tiedemann 1987), and losses of 20 to 40 percent of the S in aboveground biomass have been reported during fires (Barnett 1989). Sulfur is similar to P (and unlike N) in that it cannot be fixed by biological processes, but instead is added primarily by burning

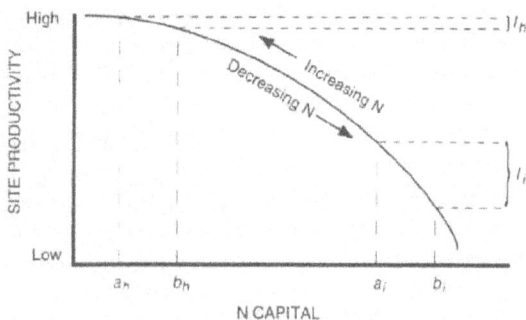

Figure 3.11—Relative importance of nitrogen low at different levels of site productivity. (After Barnett 1989).

fossil fuels (a source of acid rain), as fallout from volcanic eruptions, or by the weathering of rocks during soil development (DeBano and others 1998).

Soil Chemical Processes _____

Nutrient Cycling

Nutrients undergo a series of changes and transformations as they are cycled through wildland ecosystems. The sustained productivity of natural ecosystems depends on a regular and consistent cycling of nutrients that are essential for plant growth (DeBano and others 1998). Nutrient cycling in nonfire environments involves a number of complex pathways and includes both chemical and biological processes (fig. 3.3). Nutrients are added to the soil by precipitation, dry fall, N- fixation, and the geochemical weathering of rocks. Nutrients found in the soil organic matter are transformed by decomposition and mineralization into forms that are available to plants (see chapter 4). In nonfire environments, nutrient availability is regulated biologically by decomposition processes. As a result, the rate of decomposition varies widely depending on moisture, temperature, and type of organic matter. The decomposition process is sustained by litter fall (that is, leaf, wood, and other debris that falls to the forest floor). Through the process of decomposition, this material breaks down, releases nutrients, and moves into the soil as soil organic matter. Forest and other wildland soils, unlike agricultural soils where nutrients from external sources are applied as needed, rely on this internal cycling of nutrients to maintain plant growth (Perala and Alban 1982). As a result, nutrient losses from unburned ecosystems are usually low, although some losses can occur by volatilization, erosion, leaching, and denitrification. This pattern of tightly controlled nutrient cycling minimizes the loss of nutrients from these wildland systems in the absence of any major disturbance such as fire.

Fire, however, alters the nutrient cycling processes in wildland systems and dramatically replaces long-term biological decomposition rates with that of instantaneous thermal decomposition that occurs during the combustion of organic fuels (St John and Rundel 1976). The magnitude of these fire-related changes depends largely on fire severity (DeBano and others 1998). For example, high severity fires occurring during slash burning not only volatilize nutrients both in vegetation and from surface organic soil horizons, but heat is transferred into the soil, which further affects natural biological processes such as decomposition and mineralization (fig. 3.12; see also chapter 4). The effects of fire on soil have both short- and long-term consequences (that is, direct and indirect effects) on soil and site productivity because of the changes that occur in both the quantity and quality of organic matter.

In summary, many nutrients essential for plant growth including N, P, S, and some cations described earlier are all affected to some extent by fire. Nitrogen is likely the most limiting nutrient in natural systems (Maars and others 1983), followed by P and S. Cations released by burning may affect soil pH and result in the immobilization of P. The role of micronutrients in ecosystem productivity and their relationship to soil heating during fire is for the most part unclear. One study, however, did show that over half of the selenium in burned laboratory samples was recovered in the ash residue (King and others 1977).

Nutrient Loss Mechanisms

Nutrient losses during and following fire mainly involve chemical processes. The disposition of nutrients contained in plant biomass and soil organic matter during and following a fire generally occurs in one of the following ways:

- *Direct gaseous volatilization into the atmosphere takes place during fire.* Nitrogen can be transformed into N_2 along with other nitrogenous gases (DeBell and Ralston 1970).
- *Particulates are lost in smoke.* Phosphorus and cations are frequently lost into the atmosphere as particulate matter during combustion (Clayton 1976, Raison and others 1985a,b).
- *Nutrients remain in the ash deposited on the soil surface.* These highly available nutrients are vulnerable to postfire leaching into and through the soil, or they can also be lost during wind erosion (Christensen 1973, Grier 1975, Kauffman and others 1993).

Figure 3.12—Pinyon-juniper slash fire, Apache-Sitgreaves National Forest, Arizona. (Photo by Malchus Baker).

USDA Forest Service Gen. Tech. Rep. RMRS-GTR-42-vol. 4. 2005

65

- *Substantial losses of nutrients deposited in the surface ash layer can occur during surface runoff and erosion.* These losses are amplified by the creation of a water-repellent layer during the fire (see chapter 2; DeBano and Conrad 1976, and Raison and others 1993).
- *Some of the nutrients remain in a stable condition.* Nutrients can remain onsite as part of the incompletely combusted postfire vegetation and detritus (Boerner 1982).

Although the direct soil heating effect is probably limited to the surface (1 inch or 2.5 cm), the burning effect can be measured to a greater depth due to the leaching or movement of the highly mobile nutrients out of the surface layers. For example, leaching losses from the forest floor of a Southern pine forest understory burn increased from 2.3 times that of unburned litter for monovalent cations Na and K to 10 to 20 times for divalent cations Mg and Ca (Lewis 1974). Raison and others (1990) noted that while K, Na, and Mg are relatively soluble and can leach into and possibly through the soil, Ca is most likely retained on the cation exchange sites. Soil Ca levels may show a response in the surface soils for many years following burning. However, some cations more readily leached and as a result are easily lost from the site. For example, Prevost (1994) found that burning *Kalmia* spp. litter in the greenhouse increased the leaching of Mg but none of the other cations. Although ash and forest floor cations were released due to burning, there was no change in surface soil cation concentrations (0-2 inches or 0-5 cm). Soto and Diaz-Fierros (1993) measured changes in the pattern of cation leaching at differing temperatures for the six soils that represented six different parent materials. Leaching patterns were similar for all soil types. Leaching of divalent cations, Ca, and Mg, increased as the temperatures reached during heating increased, with a peak at 860 °F (460 °C). Monovalent cations, K, and Na, differed in that initially leaching decreased as temperature increased, reaching a minimum at 716 °F (380 °C). Then, leaching increased up to 1,292 °F (700 °C). The nutrients leached from the forest floor and the ash were adsorbed in the mineral soil. Surface soils were found to retain 89 to 98 percent of the nutrients leached from the plant ash (Soto and Diaz-Fierros 1993). As the leachates moved through the mineral soil, the pH of the solution decreased.

Nutrient Availability

The increased nutrient availability following fire results from the addition of ash, forest floor leachates, and soil organic matter oxidation products as the result of fire. The instantaneous combustion of organic matter described earlier directly changes the availability of all nutrients from that of being stored and slowly becoming available during the decomposition of the forest floor organic matter to that of being highly available as an inorganic form present in the ash layer after fire. Both short- and long-term availability of nutrients are affected by fire.

Extractable Ions—Chemical ions generally become more available in the surface soil as a result of fire. Grove and others (1986) found that immediately after fire, extractable nutrients increased in the 0 to 1.2 inch (0-3 cm) depth. Concentrations of S, NH_4, P, K, Na, zinc (Zn), Ca, and Mg increased. Everything except Zn and organic C increased in terms of total nutrients. At the lower depths sampled, 1.2 to 3.9 inches and 3.9 to 7.9 inches (3-10 and 10-20 cm), only extractable P and K were increased by burning. One year later nutrient levels were still greater than preburn concentrations, but had decreased. A study on an area of pine forest burned by a wildfire reported that in the soil, concentrations of P, Ca, and Mg, aluminum (Al), iron (Fe) had increased in response to different levels of fire severity (Groeschl and others 1993). In the areas exposed to a high-severity fire, C and N were significantly lower and soil pH was greater. In another study the soil and plant composition changes were studied in a jack pine (*P. banksiana*) stand whose understory had been burned 10 years earlier (Lynham and others 1998). In this study the soil pH increased in all soil layers following burn—O horizon, 0-2 inches (0-5 cm), and 2-3.9 inches (5-10 cm)—and remained 0.5 units greater than preburn 10 years later in the O horizon. Phosphorus, K, Ca, and Mg all increased in the mineral soil; P and K were still greater than preburn levels 10 years later. A stand replacement fire in the Southern Appalachians that resulted in a mosaic burn pattern affected exchangeable ions (Vose and others 1999). The midslope areas of this fire burned at a moderate severity, and a decrease in exchangeable K and Mg along with an increase in soil pH was measured. Soil Ca, total C and N did not respond in any of the burned areas. There are other studies that have also shown no effect or decreases of soil nutrients following burning (Sands 1983, Carreira and Niell 1992, Vose and others 1999).

Nitrogen—The two most abundant forms of available N in the soil are available NH_4- and NO_3-N. Both forms are affected by fire. Burning rapidly oxidizes the soil organic matter and volatilizes the organic N contained in the forest floor and soil organic matter, thereby releasing NH_4-N (Christensen 1973, Jurgensen and others 1981, Kovacic and others 1986, Kutiel and Naveh 1987, Marion and others 1991, Knoepp and Swank 1993a,b). The release of NH_4-N has been found by more detailed chemical analysis to involve the thermal decomposition of proteins and other nitrogen-rich organic matter. Specifically, the production of

NH_4-N is related to the decomposition of secondary amide groups and amino acids. These secondary amide groups are particularly sensitive to decomposition during heating and decompose when heated above 212 °F (100 °C) to yield NH_4-N (Russell and others 1974). The volatilization of more heat-resistant N compounds can occur up to 752 °F (400 °C). The temperatures required to volatilize nitrogen compounds are increased by the presence of clay particles in the soil (Juste and Dureau 1967).

Most NH_4-N that is volatilized is lost into the atmosphere, but significant amounts can move downward and condense in the mineral soil as exchange N. The ash produced by the fire can also contain substantial amounts of NH_4-N. As a result of these two processes, the inorganic N in the soil increases during fire (Kovacic and others 1986, Raison and others 1990, Knoepp and Swank 1993a,b). In contrast, NO_3-N is usually low immediately following fire and increases rapidly during the nitrification of NH_4 (see chapter 4). These NO_3-N concentrations may remain elevated for several years following fire (fig. 3.13).

The production of NO_3-and NH_4-N by fire depends on several factors. These include fire severity, forest type, and the use of fire in combination with other postharvesting activities.

Effect of fire severity: The amounts of NH_4-N that are produced as a result of fire generally increase with the severity and duration of the fire and the associated soil heating (fig. 3.13, 3.14, 3.15). Although large amounts of the total N in the aboveground fuels, litter, duff, and upper soil layers are lost into the atmosphere by volatilization, highly available NH_4-N still remains in the ash or in the upper mineral soil layers following fire (fig. 3.13). During both high and low severity fires, increase occurs in the amounts of NH_4-N that can be found both in the ash remaining on the soil surface following fire and in the upper mineral soil layers (Groeschl and others 1990, Covington and Sackett 1992). The

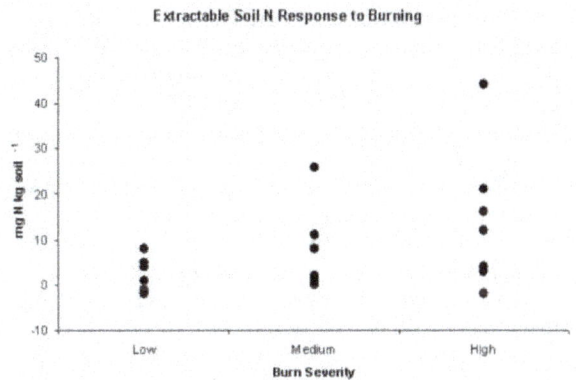

Figure 3.14—Extractable soil N in response to burn severity measured on eight studies.

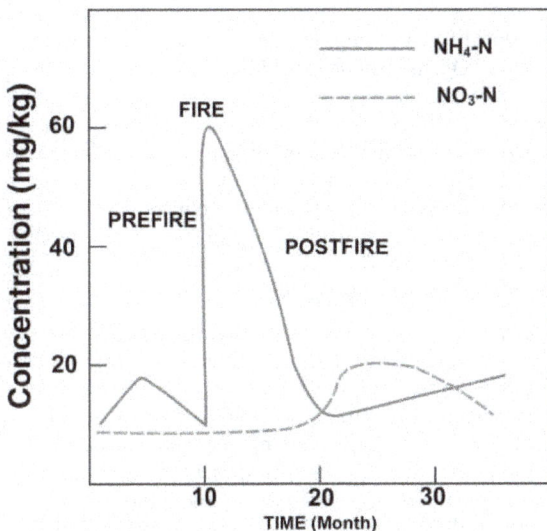

Figure 3.13—Soil NO_3-N and NH_4-N concentrations before, immediately following, and for several months following fire in Arizona chaparral. (Adapted from DeBano and others 1998. Fire's Effects on Ecosystems. Copyright © 1998 John Wiley & Sons, Inc. Reprinted with permission of John Wiley & Sons, Inc.).

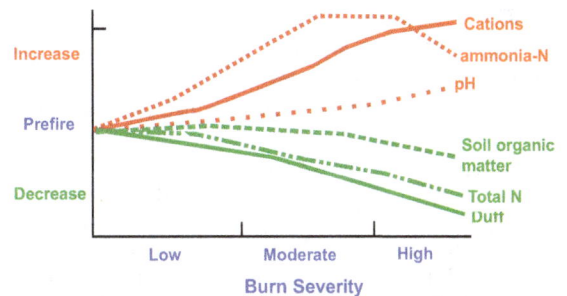

Figure 3.15—Generalized patterns of decreases in the forest floor (duff), total N, and organic matter, and increases in soil pH, cations, and NH_4 associated with increasing levels of fire severity. (Figure courtesy of the USDA Forest Service, National Advanced Fire and Resource Institute, Tucson, AZ).

USDA Forest Service Gen. Tech. Rep. RMRS-GTR-42-vol. 4. 2005

67

concentrations of NH_4-N found in the ash depend on the severity of the fire and amount of forest floor consumed, which in turn reflects stand age. During low severity fires, less of the forest floor is consumed and correspondingly less amounts of NH_4-N are produced (fig. 3.15). The amounts of NH_4-N produced during a fire may vary from increases of less than 10 ppm (10mg/kg) during low severity fires to 16 and 43 ppm (16 mg/kg and 43 mg/kg) during medium and high severity fires, respectively. The variability in the amounts of NH_4-N produced also increases with fire severity. The NH_4-N concentrations following burning are short lived (fig. 3.13), although Covington and Sackett (1992) reported that NH_4-N levels remained elevated for at least 1 year following a prescribed burn in ponderosa pine stands in northern Arizona.

The direct effects of fire on NO_3-N concentrations are less predictable. For example, the results of a study in the Shenandoah National Park showed that although total extractable inorganic N was elevated for 1 year following the fire, the NO_3-N on areas burned by either high or low severity fires increased only slightly (Groeschl and others 1990). A more common scenario for NO_3-N changes following fire results from the increased nitrification of the highly available N produced directly during the fire (for example, NH_4-N). The end result being that NH_4-N produced directly during fire is rapidly nitrified, and NO_3-N begins to increase following fire, depending on temperature and moisture conditions (fig. 3.13).

The effects of these increased levels of highly available N during and following fire are often beneficial to the recovering plants by providing a temporary increase in site fertility. However, these short-term benefits must be carefully weighed against the overall and long-term effect that the loss of total N during a fire has on the sustained productivity of the site (see the previous discussion on Nitrogen Losses –An Enigma).

Effect of forest type: Forest type can also affect the amount of NH_4-N produced by fire. This occurs mainly as a result of nature of the fire behavior and the amounts of litter that accumulate under different forest types. Such was the case in a study conducted in Idaho on pine/hemlock sites compared to sites occupied by a Douglas-fir/western larch forest (Mroz and others 1980). In this study the NH_4-N did not increase following burning of the pine/hemlock forest in contrast to the Douglas-fir/western larch forest site where it did increase. Within 3 to 7 days, however, mineralization and/or nitrification had begun on most sites.

Frequency of burning: Repeated burning and its frequency are often-asked questions by fire managers when conducting prescribed burns. In terms of N

availability, the effects of burning frequency depends largely upon the ecosystem, its inherent fertility (in terms of total and available N), total amounts of organic matter destroyed, and its ability to replenish the N lost by volatilization (DeBano and others 1998).

Several studies have focused on the effect of different frequencies of prescribed burning on changes in total and concentrations of total and available nitrogen. For example, a study on the effect of repeated burn low-intensity burning in Australian eucalypt forests (*Eucalyptus* spp.)showed that fire-free periods of about 10 or more years were required to allow natural processes time to replace the amount of N lost during the burning, assuming that about 50 percent of the total N in the fuel was volatilized (Raison and others 1993). In contrast, a study of repeated burning at 1-, 2-, and 5-year intervals in ponderosa pine forests in northern Arizona showed no significant differences in total N among the different burning frequencies, but available N (NH_4- and NO_3-N) was higher on the sites that repeatedly burned in comparison to the unburned controls (Covington and Sackett 1986). Researchers concluded from this Arizona study that frequent periodic burning can be used to enhance N availability in Southwestern ponderosa pine forests.

Studies on the effects of burning frequency on grasslands and shrubs have been reported to have less desirable outcomes. For example, the annual burning of tall grass prairies in the Great Plains of the Central United States resulted in greater inputs of lower quality plant residues, causing a significant reduction in soil organic N, lower microbial biomass, lower N availability, and higher C:N ratios in soil organic matter (Ojima and others 1994). Likewise, increases in available N may have adverse effects on some nutrient-deficient shrub ecosystems as has been reported by a study in a shrubland (fynbos) in South Africa. In a study of lowland fynbos, a twofold increase in soil nutrient concentrations produced by fire were detrimental to the survival of indigenous species that had evolved on these nutrient-impoverished landscapes (Musil and Midgley 1990).

Phosphorus—Responses of available soil P to burning are variable and more difficult to predict than those of other nutrients (Raison and others 1990). Phosphorus volatilizes at temperatures of about 1,418 °F (770 °C). The fate of this volatilized P is not well understood. One study indicated that the only response was on the surface soil, and P did not appear to move downward in the soil via volatilization and condensation, as N does (DeBano 1991). Grove and others (1986) found the opposite. They measured responses in all major cations, S, NH_4, and Zn following burning in the surface 0 to 1.2 inches (0-3 cm) of soil. In their study, only P and K concentrations also

68

USDA Forest Service Gen. Tech. Rep. RMRS-GTR-42-vol. 4. 2005

responded in the lower soil depths (1.2-3.9 inches and 3.9-7.9 inches; 3-10 cm and 10-20 cm).

As in the case of N, fire severity affects changes in extractable P. During high-severity fires, 50 to 60 percent of the total fuel P might be lost to volatilization (Raison and others 1990, DeBano 1991). Part of this volatilized P ends up as increased available P in both the soil and ash following burning. An extensive study of P responses to different burning severities was reported for eucalypt forests (Romanya and others 1994). The study sites included unburned, burned, and in an ash bed found under a burned slash pile. The greatest effects occurred in the surface soil (0-1 inch; 0-2.5 cm), and the response was dependent on fire severity. Extractable P concentrations increased with increasing fire severity, but the response decreased with depth. Organic P on the other hand reacted oppositely; concentrations were lower in the intensively burned areas and greater in the unburned and low-severity burned sites.

Fire affects the enzymatic activity and mineralization of P. One study compared these P responses in a controlled burn versus a wildfire (Saa and others 1993). When temperatures reached in the forest floor of the controlled burn were less than 329 °F (50 °C), extractable P concentrations (ortho-phosphate) showed no significant response. In contrast, a wildfire that produced higher soil temperatures reduced phosphatase activity and increased the mineralization of organic P, which increased ortho-phosphate P and decreased organic P. Laboratory experiments showed that phosphatase activity can be significantly reduced when heating dry soils but was absent in wet soils (DeBano and Klopatek 1988). In the pinyon-juniper soils being studied, bicarbonate extractable P was increased although the increases were short lived.

Ash-Bed Effect

Following fire, variable amounts of ash are left remaining on the soil surface until the ash is either blown away or is leached into the soil by precipitation (fig. 1.4). On severely burned sites, large layers of ash can be present (up to several centimeters thick). These thick accumulations of ash are conspicuously present after piling and burning (for example, burning slash piles). Ash deposits are usually greatest after the burning of concentrated fuels (piled slash and windrows) and least following low-severity fires.

The accumulation of thick layers of ashy residue remaining on the soil surface after a fire is referred to as the "ash bed effect" (Hatch 1960, Pryor 1963, Humphreys and Lambert 1965, Renbuss and others 1972). The severe burning conditions necessary to create these thick beds of ash affect most of the physical, chemical, and biological soil properties. Soil changes associated with ash beds can occur as a result of a fire itself (soil heating), the residual effect of the ash deposited on the soil surface (that is, the ash bed), or a combination of both (Raison 1979).

The amount and type of ash remaining after fire depend upon the characteristics of the fuels that are combusted, such as fuel densities (packing ratios), fuel moisture content, total amount of the fuel load consumed, and severity of the fire (Gillon and others 1995). As a result of the fire, the ash remaining after a fire can range from small amounts of charred dark-colored fuel residues to thick layers of white ash that are several centimeters thick (DeBano and others 1998). When densely packed fuels are completely combusted, large amounts of residual white ash are usually in one place on the soil surface following burning (such as after piling and burning slash, see Figure 1.4). The severe heating during the fire will change the color of the soil mineral particles to a reddish color, and where extreme soil heating has occurred, the mineral soil particles may be physically fused together. Silicon melts at temperatures of 2,577 °F (1,414 °C; see chapter 2).

Chemically, fire consumption of aboveground material determines the amount of ash produced. Ash consists mostly of carbonates and oxides of metals and silica along with small amounts of P, S, and N (Raison and others 1990). Calcium is usually the dominant cation found in these ash accumulations. Most of the cations are leached into the soil where they are retained on the cation exchange sites located on clay or humus particles and increase the mineral soil cation content (fig. 3.15). The pH may exceed 12. However, the composition of the preburn material and the temperature or severity of the fire determines the chemical properties of ash. Johnston and Elliott (1998) found that ash on uncut forest plots generally had the highest pH and the lowest P concentrations.

Physical changes associated with the ash bed effect mainly include changes in soil structure and permeability to water. The combustion of organic matter in the upper part of the soil profile can totally destroy soil structure, and the ashy material produced often seals the soil to water entry.

The biological impact of the ash bed effect is twofold. During the fire the severe soil heating can directly affect the long-term functioning of microbial populations because the high temperature essentially sterilizes the upper part of the soil. Plant roots and seeds are also destroyed so that the revegetation of these sites depends on long-term ecological succession to return to its former vegetative cover. Indirectly, the large amounts of ash can affect soil microbial populations. A study of the effects of ash, soil heating, and the ash-heat interaction on soil respiration in two Australian soils showed that large amounts of ash slightly

USDA Forest Service Gen. Tech. Rep. RMRS-GTR-42-vol. 4. 2005

69

decreased respiration, but small amounts had no effect (Raison and McGarity 1980). Additions of ash to sterilized soil produced no effect, indicating that ash acted via its influence on active soil biological populations. The chemical nature of ash was hypothesized to affect soil respiration by its effect on:

- Increasing pH.
- Changing the solubility of organic matter and associated minerals in water.
- Adding available nutrients for microbial populations.

Management Implications

Understanding the effects of fire on soil chemical properties is important when managing fire on all ecosystems, and particularly in fire-dependent systems. Fire and associated soil heating combusts organic matter and releases an abundant supply of highly soluble and available nutrients. The amount of change in the soil chemical properties is proportional to the amount of the organic matter combusted on the soil surface and in the underlying mineral soil. Not only are nutrients released from organic matter during combustion, but there can also be a corresponding loss of the cation exchange capacity of the organic humus materials. The loss of cation exchange capacity of the humus may be an important factor when burning over coarse-textured sandy soils because only a small exchange capacity of the remaining mineral particles is available to capture the highly mobile cations released during the fire. Excessive leaching and loss can thus result, which may be detrimental to maintaining site fertility on nutrient-limiting sandy soil.

An important chemical function of organic matter is its role in the cycling of nutrients, especially N. Nitrogen is most limiting in wildland ecosystems, and its losses by volatilization need to be evaluated before conducting prescribed burning programs. Nitrogen deficiencies often limit growth in some forest ecosystems. Xeric and pine dominated sites, which are typically prone to burning, often exhibit low N availability, with low inorganic N concentrations and low rates of potential mineralization measured on these sites (White 1996, Knoepp and Swank 1998). Forest disturbance, through natural or human-caused means, frequently results in an increase of both soil inorganic N concentrations and rates of potential N mineralization and nitrification. The N increases resulting from a combination of changes in soil moisture and temperature and the decreased plant uptake of N make more N available for sustaining microbial populations in the soil.

Historically, some wildland ecosystems have been exposed to frequent fire intervals. Many of these ecosystems are low in available N and other nutrients such as P and cations. The cycling of nutrients, especially N, may be slow, and the exclusion of fire from these systems often results in low N mineralization and nitrification rates. Frequent fire, however, can accelerate these biological rates of N mineralization because it destroys the inhibiting substances that hinder these processes. For example, in ponderosa pine forests in the Southwest, monoterpenes have been found to inhibit nitrification (White 1991). These monoterpenes are highly flammable and as a result are combusted during a fire. As a result, the removal of this inhibition by fire allows N mineralization and nitrification to proceed. It is hypothesized that these inhibitory compounds build up over time after a fire and decrease N mineralization. Significant differences in monoterpenes concentrations have been established between early and late successional stages, although specific changes over time have not been detectable because of the large variability between sites (White 1996).

Another study has shown that the xeric pine-hardwood sites in the Southern Appalachians are disappearing because of past land use, drought, insects, and the lack of regeneration by the fire-dependent pine species (Vose 2000). This information was used to develop an ecological model that could be used as a forest management tool to rejuvenate these *Pinus rigida* stands (fig. 3.16). This model specifies that a cycle of disturbance due to drought and insect

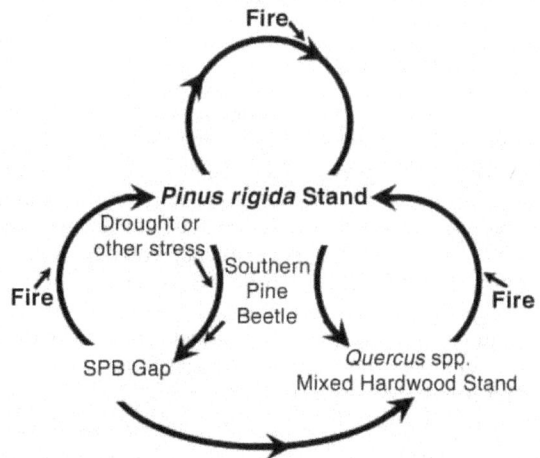

Figure 3.16—Effect of fire on nutrient cycling processes in *Pinus ridgida* stands. (Adapted from Vose 2000).

outbreaks followed by fire is necessary to maintain the pine component of these ecosystems. Without fire, mixed hardwood vegetation dominates the stand. Therefore, prescribed burning is proving to be an effective tool for enhancing ecosystem health and for sustaining, preserving, and restoring these unique habitats.

Fire severity is probably the one most important features of a fire that affects the chemical soil properties. Generalized relationships for several soil properties at different severities are presented in figure 3.15. Nitrogen, organic matter, and duff decrease as fire severity increases. Available NH_4-N and cations increase. The pH of the soil generally increases because of the loss of organic matter and its associated organic acids, which are replaced with an abundance of basic cations in the ash.

Summary

The most basic soil chemical property affected by soil heating during fires is organic matter. Organic matter not only plays a key role in the chemistry of the soil, but it also affects the physical properties (see chapter 2) and the biological properties (see chapter 4) of soils as well. Soil organic matter plays a key role in nutrient cycling, cation exchange, and water retention in soils. When organic matter is combusted, the stored nutrients are either volatilized or are changed into highly available forms that can be taken up readily by microbial organisms and vegetation. Those available nutrients not immobilized are easily lost by leaching or surface runoff and erosion. Nitrogen is the most important nutrient affected by fire, and it is easily volatilized and lost from the site at relatively low temperatures. The amount of change in organic matter and N is directly related to the magnitude of soil heating and the severity of the fire. High- and moderate-severity fires cause the greatest losses. Nitrogen loss by volatilization during fires is of particular concern on low-fertility sites because N can only be replaced by N-fixing organisms. Cations are not easily volatilized and usually remain on the site in a highly available form. An abundance of cations can be found in the thick ash layers (or ash-bed) remaining on the soil surface following high-severity fires.

USDA Forest Service Gen. Tech. Rep. RMRS-GTR-42-vol. 4. 2005

71

Notes

_____

_____

_____

_____

_____

_____

Matt D. Busse
Leonard F. DeBano

Chapter 4:
Soil Biology

Introduction

Soil biological properties involve a wide range of living organisms that inhabit the soil, along with the biologically mediated processes that they regulate. The welfare of these soil organisms directly affects the short- and long-term productivity and sustainability of wildland ecosystems (Borchers and Perry 1990). Soils are alive with large populations of microorganisms, roots and mycorrhizae, invertebrates, and burrowing animals that inhabit the upper part of the soil profile (Singer and Munns 1996). The biological component of soil also includes plant roots and their associated rhizosphere, vegetative reproductive structures, and seeds. These organisms proliferate in the soil matrix, particularly the upper layers that contain substantial amounts of organic matter. Collectively, these organisms contribute to soil productivity by enhancing decomposition, nitrogen (N) cycling, humus formation, soil physical and chemical properties, plant reproduction, disease incidence, and plant nutrition and stability.

Biological Components of Soils

The biological component of soils is made up of both living and dead biomass. Dead biomass consists of organic matter that is in various stages of decomposition, extending from undecomposed plant parts in the litter layer to highly decomposed humus materials that can be thoroughly mixed with the upper mineral layers of the soil profile. Both living and dead components are affected by fire. The effect of fire on organic matter is discussed in detail in chapter 3. This current chapter is devoted to the description of the important organisms that influence soil-litter systems and the effects that fire has upon them.

Living organisms can be classified several ways. One method of classification is whether they are flora or fauna. Soil flora includes algae, cyanobacteria, mycorrhiza, and plant roots. Soil fauna includes protozoa, earthworms, and insects. The category "soil fauna" has been further divided into micro-, meso-, and macrofauna based on body lengths of less than 0.2 mm, .20 to 10.4 mm, and greater than 10.4 mm,

USDA Forest Service Gen. Tech. Rep. RMRS-GTR-42-vol. 4. 2005

73

respectively (Wallwork 1970) (fig. 4.1). The term "*microorganisms*" encompasses a diverse group including bacteria, fungi, archaebacteria, protozoa, algae, and viruses. Some common representatives of the intermediate-sized soil organisms (that is, mesofauna) commonly found in soils are round worms, springtails, and mites, whereas the macrofauna are represented by a wide range of larger invertebrates such as many insects, scorpions, and earthworms.

Soil organisms ranging in size from microbes to megafauna are commonly concentrated in the surface horizons, because these soil layers contain large amounts of organic matter (see fig. A.1, part A), and are active sites for microbial processes, including decomposition and mineralization. Plant roots and invertebrates can occupy the forest floor (that is, L-, F-, and H- layer) or can be found in the uppermost layers of organic-rich mineral soil (that is, the upper part of the A-horizon). The H-layer represents the end product of the microbial decomposition activity that occurs in many soils.

Soil Microorganisms

Soil is teeming with life. Hundreds of millions of microorganisms are found in each handful of forest soil. No other living component in a forest (vegetation, wildlife, insects, and so forth) comes close to matching the sheer numbers and diversity of soil microorganisms. More important, microorganisms play a major role in nutrient cycling processes, decomposition of organic material, improvement of soil physical characteristics, and disease. They also play an important role in providing a labile pool of nutrients (Stevenson and Cole 1999). Hence, their influence on life in naturally occurring ecosystems is substantial. Also, some microorganisms form symbiotic relationships with plants, thereby creating a unique biological entity that can be easily affected by fire.

Free-Living Fungi and Bacteria—*Fungi* and *bacteria* are the workhorses of the forest soil organisms (fig. 4.2). Their functions in part include decomposition, nutrient turnover and acquisition (for

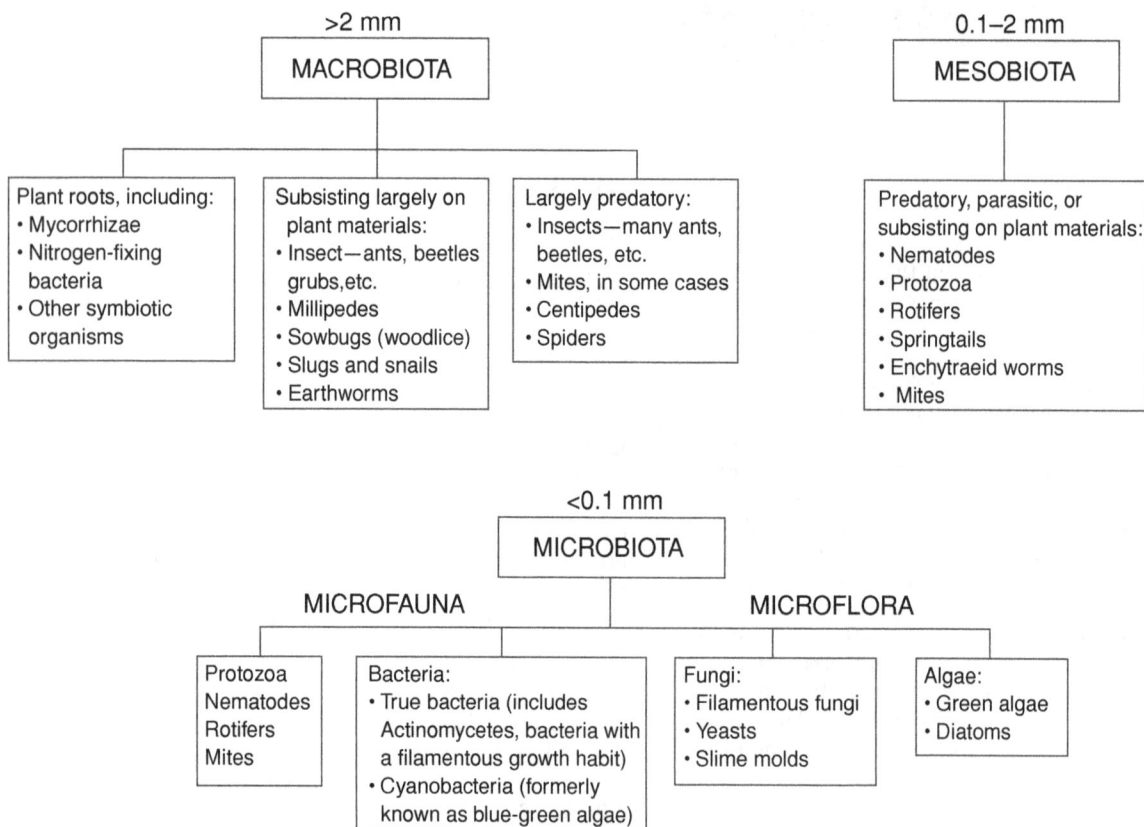

Figure 4.1—Types and sizes of soil organisms that can be affected by fire. (Adapted from Wallwork 1970).

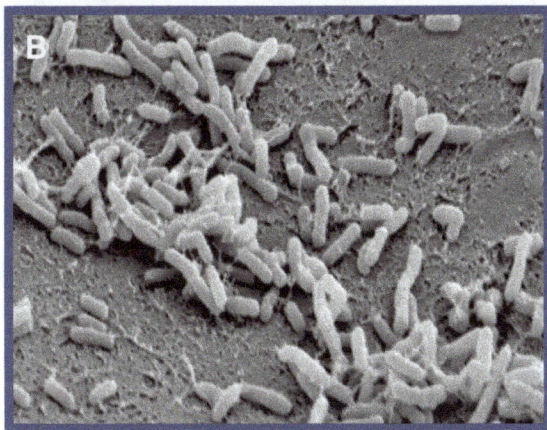

Figure 4.2—Fungi (A) and bacteria (B) are important forest soil organisms due to their roles in decomposition, nutrient turnover and acquisition, disease occurrence, and suppression and degradation of toxic materials. (Photos by Daniel Neary and Shirley Owens, Michigan State University).

example, mycorrhizal fungi and N-fixing bacteria), disease occurrence and suppression, and degradation of toxic materials. Bacteria (including actinomycetes and cyanobacteria) are the most numerous, commonly numbering about 100 million individuals in a gram of fertile agricultural topsoil (Paul and Clark 1989, Singer and Munns 1996). In forest soils, the numbers of bacteria are less, but fungal numbers can range from 5,000 to 900,000 individuals per gram of soil (Stevenson and Cole 1999). Bacteria and fungi have been estimated to contribute 2.2 tons/acre (5.0 Mg/ha) of biomass to some forest soils (Bollen 1974). Fungi, although less numerous than bacteria, often account for greater biomass due to their larger size and dominance in woody, organic material. The diversity of these two groups is almost unimaginable; thousands of species of each can be found in 0.002 pound (1 g) of soil (Torsvik

and others 1990, Molina and others 1999). And this diversity creates the collective ability of these microorganisms to adapt to fire or other environmental disturbances (Atlas and others 1991).

Other microbial groups also serve key roles in the biologically induced changes in forest floor litter and soil. Protozoa, by feeding on bacteria, release plant-available nutrients that were previously tied up in bacterial cells. The role of viruses and archaebacteria (formerly classified as bacteria) in forests, however, is not well understood. These viruses are protein-coated, acellular strands of DNA or RNA that can be predatory to many microorganisms, suggesting a contribution to nutrient turnover. Archaebacteria are common in extreme environments where drastic temperature, moisture, acidity, or nutrient conditions preclude other organisms. Their numbers in forest soils are assumed to be low.

A particularly important group of free-living microorganisms includes those concerned with the nutrient cycling processes described later in this chapter. Some of these biologically mediated processes include N fixation, mineralization, ammonification, nitrification, and the overall decomposition process.

Specialized Root-Microbial Associations—Some bacteria and fungi are in close association with plant roots and develop a symbiotic relationship with them. This suite of organisms is different than the free-living bacteria and fungi described above because they depend on a mutual relationship with plant roots. They are also distinct in their relationship to fire because both they and their host plants can be affected. The most common symbiotic relationships involve the root and:

- Mycorrhizal fungi, which enhance the plant's ability to obtain nutrients such as phosphorus (P) and zinc from the soil.
- Nitrogen-fixing bacteria, which convert N gas in the air to a form usable by plants. The effects of fire on N-fixation and other biological processes are discussed in more detail later in this chapter.

A useful concept used for describing the close relationship between roots and microorganisms is the *rhizosphere*. The rhizosphere is a cylindrical volume of the soil space that extends about 0.04 inch (1 mm) from the surface of roots (Singer and Munns 1996). The outer boundary of the rhizosphere is diffuse and inexactly defined because the effect of the root can extend variable distances into the soil. Simply, the rhizosphere may be thought of as including the root and its surrounding soil environment (fig. 4.3). Functionally, the rhizosphere is important because it contains a combination of the roots and associated microorganisms. The roots secrete products that stimulate

USDA Forest Service Gen. Tech. Rep. RMRS-GTR-42-vol. 4. 2005

75

Figure 4.3—The rhizosphere includes the root system (A), associated mycorrhizae, and varying volumes of the soil surrounding roots (B). (Photos courtesy of Nicholas Comerford, University of Florida).

the bacterial and fungal activity so that the fast-growing heterotrophic microbes are at least 10 times denser near the root than in the rest of the soil (Singer and Munns 1996). The heterotrophic microbial population plays an important role in the decomposition of organic matter and contributes to a desirable soil structure (granular) in the root zone.

Mycorrhizal fungi found in the rhizosphere depend on host plants for their well being (Borchers and Perry 1990). The development of mycorrhizae provides a way for the plant roots to extend farther into the soil. The thin fungal hyphae of the mycorrhizal fungi form a mutual relationship with the roots of some plants, thereby allowing the plant roots to proliferate a greater soil mass than by the roots alone. Root-fungal associations use about 5 to 30 percent of the total photosynthate that is translocated belowground by plants. Two types of mycorrhizae are found in soil, endo- (arbuscular) and ectomycorrhizae.

The endomycorrhizae produce structures within the plant roots (in deciduous trees, most annual crops, and other herbaceous species) that are called arbuscules (Coleman and Crossley 1996, Singer and Munz 1996). Arbuscular mycorrhiza fungi also send out hyphae, but only a few centimeters into the surrounding soil (Coleman and Crossley 1996).

In contrast to endomycorrhize fungi, the ectomycorrhizae, which are primarily basidiomycetes, grow between plant root cells (in many evergreen trees and shrubs), but not inside them as do endomycorrhizae (Coleman and Crossley 1996, Munns 1996). Ectomycorrhizae can send out hyphae for several meters into the surrounding soil to forage for nutrients and water that are essential for the host plant. Because of their ability to proliferate the soil, hyphae of ectomycorrhizae constitute a significant proportion of the C allocated to belowground net primary productivity in coniferous forests (Read 1991). The hyphae facilitate nutrient uptake, particularly of P, and are avid colonizers of organic matter where they enhance soil structure. Ectomycorrhizae are located at shallow depths in the soil profile and tend to be concentrated in the woody material during dry seasons and in the H-layer during moist conditions. Ectomycorrhizae are important decomposers and, as a result, obtain reduced C from the decomposing litter layer. Because the ectomycorrhizae are so near the soil surface, both the resting stages and the hyphae are easily damaged by soil heating during a fire.

Several groups of microorganisms form N-fixing symbiotic relationships with plants (Singer and Munns 1996). This type of symbiosis, commonly found in rhizosphere, involves a group of actinomycetes and rhizobia bacteria. The actinomycetes (*Frankia* spp.) infect the roots of many genera of shrubs and trees where they form N-fixing nodules. Rhizobial bacteria (*Rhizobium* and *Bradyrhizobium* spp.) also form nitrogen-fixing nodules with a large variety of plants belonging to the legume family, including alfalfa, clover, bean, pea, soybean, vetch, lupine, and lotus, among many others. Many of these are agricultural plants, although lupines, lotus, and clover are frequently found in abundance on freshly burned wildland areas.

Biological Crusts—*Biological crusts* (fig. 4.4) are found in hot, cool, and cold arid and semiarid regions throughout the world and frequently occupy the bare areas where vegetation cover is spare or totally absent (Belnap 1994, Belnap and others 2001). These surface communities are generically referred to as biological soil crusts, although they may specifically be called cryptogamic, cryptobiotic, microbiotic, or microphytic soil crusts. The biological soil crusts are made up of a complex community of cyanobacteria (blue-green algae), lichens, mosses, microfungi, and other bacteria

Figure 4.4—Biological crusts are found in hot, cool, and cold arid and semiarid regions through the world. (Photo courtesy of the U.S. Geological Survey).

(Isichei 1990, Johansen 1993, Loftin and White 1996, Belnap and others 2001). The algal component of this community has the ability to fix atmospheric C and is best recognized in forests as a component of lichens that colonize rocks, tree trunks and exposed soil surfaces. Lichens represent a symbiosis among fungi and blue-green algae or cyanobacteria. The blue-green algae photosynthesize and fix N, while the fungi provide water and mineral nutrients. They are typically found on erodible soils with some topographic relief, or in shallow soil pockets in slickrock habitats (Johansen 1993). In this environment, the filamentous growth of cyanobacteria and microfungi proliferates in the upper few millimeters of the soil, gluing loose particles together and forming a matrix that stabilizes and protects the soil surface from erosion (Belnap and others 2001). Cryptogamic crusts are also found in semiarid forest and shrublands, such as pinyon-juniper woodlands and sagebrush communities (Johansen 1993). These biological crusts can account for as much as 70 percent of all living ground cover in arid Western Hemisphere ecosystems (Belnap 1994).

The distribution of biological crusts is influenced by several environmental factors (Belnap and others 2001). The total amount of crust that develops is inversely related to vascular plant cover (in other words, the less plant cover, the more surface available for colonization and growth). Elevation also affects crust distribution and cover, and both are greatest at low elevation inland areas (less than 3,300 feet or 1,000 m) compared to mid-elevations 3,300 to 8,200 feet

(1,000 to 2,500 m). Stable soils and rocks near, or at, the soil surface enhance crust cover by collecting water and armoring the surface against physical disturbance (Belnap and others 2001). Stable, fine-textured soils support a greater percentage cover and a more diverse population of organisms making up these surface crusts. The season of precipitation also has a major influence on the dominance of biological crusts, and as a result, the regions that receive monsoons have the greatest diversity of cyanobacteria and the lowest abundance of lichens. Areas that are frequented by fog (such as portions of the California chaparral) frequently support lichens that intercept moisture from the air. Finally, but not least, surface disturbance has a strong influence on the welfare of biological crusts. The two most important historical impacts on crusts have been grazing and fire (Belnap and others 2001).

Cryptogamic crusts benefit soils in a number of ways. The ability of these crusts to fix N, accumulate C, and capture P enhances nutrient cycling in soils, especially during the early successional stages (DeBano and others 1998, Evans and Johansen 1999). Some specific benefits of cryptogamic crusts in soils arise from their ability to:

- Retard erosion on desert and steppe rangelands by binding individual soil particles (particularly in sandy soils). Soil particles are believed to be bound as a result of the production of extracellular polysaccharides (Lynch and Bragg 1985). The crusts can increase or decrease infiltration rates, although their effect on conserving soil moisture is not well-defined (Johansen 1993).
- Fix gaseous N. The cyanobacteria component of this symbiosis has been estimated to fix as much as 22 pounds/acre (25 kg/ha) of N annually (West and Skujins 1977). The amount of N fixed, however, is largely dependent upon the abundance and activity of the crusts and favorable climatic conditions (Loftin and White 1996).
- Enhance organic matter buildup. Biological crusts in semiarid areas of the Southwestern United States have been estimated to accumulate 5.3 to 20.5 pounds/acre (6 to 23 kg/ha) of C annually (Jeffries and others 1993).
- Increase P levels by retaining fine soil particles from loss by erosion (Kleiner and Harper 1977).
- Facilitate seedling establishment. Although germination can be inhibited by allelopathic substances produced by the crust, seedling establishment generally seems to benefit by their presence (St. Clair and others 1984).

USDA Forest Service Gen. Tech. Rep. RMRS-GTR-42-vol. 4. 2005

77

Soil Meso- and Macrofauna

The most common members of the mesofaunal group of soil organisms are the mites (members of the order Acari). The mites can be abundant in soils, particularly forest soils, where a 0.2 pound (100 g) sample of soil may contain as many as 500 mite species representing almost 100 genera (Coleman and Crossley 1996). Other mesofauna found in soil include *Rotifera*, *Nematoda* (round worms), and *Collembola* (springtails).

Macrofauna occupying the soil and litter in forest, shrubland, and grassland can be placed in two broad classes—those that spend all or most of their time in the litter and uppermost mineral soil layers, and those that inhabit these habitats only temporarily or not at all. Several of the more permanent faunal groups include those that reside in the soil and litter but also a group that dwells beneath stones, logs, under bark, or in similar protected habitats. These organisms are collectively called the cryptozoa. Three orders of higher insects—Isoptera (termites), Hymenoptera (ants, bees, wasps, and sawflies), and Coleoptera (beetles)—all play a major role in improving soil structure and enriching soil chemistry and associated food webs (Coleman and Crossley1996). The cryptozoa group includes millipedes, centipedes, and scorpions. These macrobiota enhance decomposition of organic matter, nutrient cycling, soil structure, and the long-term primary productivity of these ecosystems.

Earthworms (Oligochaeta) are a special class of the macrofauna that has long been recognized as an important component of healthy soil systems (Coleman and Crossley 1996, Lavelle 1988).Their large abundance and biomass in some soils make them a major factor in soil biology (fig. 4.5). Earthworms fall into three general groups: those that dwell in the surface litter, those that are active in the mineral soil layers, and those that move vertically between deeper soil layers and the soil surface. These organisms act as biological agents that decompose litter and mineralize C in both the litter and underlying soil (Zhang and Hendrix 1995). Earthworm digestion increases both the mineralization and humification of organic matter (Lavelle 1988). Earthworms are best known for their beneficial effects of building and conserving soil structure, which results from burrowing and soil ingestion activities (Lavelle 1988). Earthworm activities enhance soil structure by increasing the number of water-stable aggregates, and the creation of burrows and casts increases soil porosity, which improves both aeration and water movement through the soil.

Roots and Reproductive Structures

Many of the plant roots, vegetative reproductive structures, and seeds are found immediately on the surface of the soil or are distributed downward throughout the soil profile. These plant parts include tap roots, fibrous surface roots, rhizomes, stolons, root crowns, and bulbs. The welfare of the plant roots and reproductive structures is important to the sustainability of plant biomass and productivity of all terrestrial ecosystems found throughout the world, particularly in those ecosystems that experience repeated and regular fires. Closely associated with the roots are a suite of symbiotic microorganisms found in the soil (see the previous section Specialized Root-Microbial Associations).

Amphibians, Reptiles, and Small Mammals

Amphibians, reptiles, and small mammals inhabit holes and cavities in the upper part of the soil where they feed on invertebrates, plant parts (seeds), and other organic debris found on or near the soil surface. Burrowing activities and deposition of fecal material by these larger animals contribute to the aeration and fertility of wildland soils.

Biologically Mediated Processes in Soils

The living organisms described above are involved in numerous biological processes that regulate nutrient cycling and contribute to soil productivity and ecosystem health. The nutrient cycling processes for nonfire environments were presented in figure 3.3.

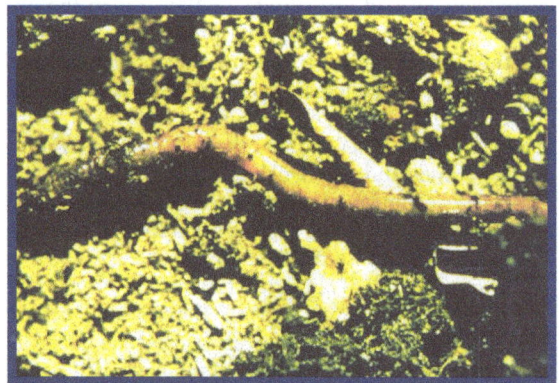

Figure 4.5—Earthworms have long been recognized as an important component of healthy soil systems that function as key biological agents in decomposing litter and mineralizing carbon in both the litter and underlying soil. (Photo courtesy of Earl Stone, University of Florida).

Nutrients enter the soil by precipitation, dry fall, N-fixation, and geochemical weathering of rocks. Nutrients in soil organic matter that accumulates on the soil surface are slowly changed to forms that are available to plants by the biological processes of decomposition and mineralization. The leakage of nutrients from these nonfire ecosystems is usually low, and only small losses occur via volatilization, erosion, leaching, and denitrification. Nutrient cycling in nonfire ecosystems represents a dynamic balance among the processes regulating decomposition, mineralization (ammonification and nitrification), N-fixation and denitrification, and nutrient immobilization and uptake by plants. The long-term sustainability of wildland ecosystems depends on this regular and consistent cycling of nutrients in order to sustain plant growth (DeBano and others 1998).

Decomposition and Mineralization

Organic matter that is deposited on a soil surface is decomposed by soil organisms and incorporated into the underlying soil (mainly as humus). During *decomposition,* C stored in the organic matter is recycled into the atmosphere as CO_2, while N is stored in microbial biomass until it is released during mineralization. Some of the C and N immobilized as microbial tissue can be microbially converted into resistant humic substances (via a process called humification). The humus fraction of soils in natural ecosystems undergoes both continuous decomposition and mineralization, so that the total soil organic matter and N content may remain in a steady state condition until disturbance, such as fire, occurs. Several factors that affect the decomposition of organic matter are the composition of the decomposing litter (substrate quality), environmental factors (moisture and temperature), types of microorganisms involved, and other soil factors including soil pH and deficiencies of inorganic nutrients (P, K, and micronutrients).

An important end product of decomposition is the release of nutrients that can be used by plants. These highly available nutrients are formed during mineralization, which is defined as the conversion of an element from an organic form to an inorganic state as the result of microbial activity (Soil Science Society of America 2001). The mineralization of organic matter involves the transformation of organic N compounds (such as proteins and amino acids) into NH_4 (ammonification) and, subsequently, into nitrite (NO_2) and NO_3 (nitrification); and the conversion of organic C into CO_2 (DeBano and others 1998). Both ammonification and nitrification are affected by fire. Nitrification is carried out mainly by autotrophic microorganisms, which derive their energy solely from the oxidation of NH_4 and NO_3 (Haynes 1986). Several genera of autotrophic bacteria that can oxidize NH_4-N to NO_2 are

Nitrosomonas, Nitrosolobus, and *Nitrospira.* The oxidation of NO_2 to NO_3, however, is almost exclusively done by *Nitrobacter* in natural systems (Haynes 1986). Approximately 30 percent of all the N that is nitrified in natural ecosystems is done by faunal populations (Verhoef and Brussaard 1990).

Nitrogen Cycling Processes

Nitrogen is unique among the soil nutrients because it is present in the soil almost entirely as part of organic compounds. No inorganic N reserve is normally available to replace the soil N lost by volatilization during a fire (Harvey and others 1976). Therefore, N is recovered from the atmosphere by N-fixation and atmospheric deposition (DeBano and others 1998).

Nitrogen-Fixation—*Nitrogen-fixation* is the conversion of molecular N (N_2) to ammonia and, subsequently, to organic combinations or forms utilizable in biological processes (Soil Science Society of America 2001). The atmosphere supplies N to soil in natural ecosystems mainly through organisms that fix inert N_2 into forms that can be used by plants. Nitrogen additions to the soil by N-fixing organisms (free-living and symbiotic) counterbalance the volatilized N that is lost during combustion and subsequent leaching of soluble N compounds into and through the soil following fire (DeBano and others 1998). Nitrogen-fixation is primarily by two groups of microorganisms found in the soil, namely those that fix N symbiotically and those that are free-living.

Symbiotic N-fixation: This form of N-fixation is carried out by symbiotic microorganisms that are associated with the roots of higher plants (symbiotic) and obtain the energy required for N-fixation from being the host plant. The most common symbiotic relationships found in wildland ecosystems are those formed by rhizobia or actinomycetes associated with plant roots. *Rhizobium* bacteria are found associated with the roots of leguminous plants that make up about 700 genera in the *Leguminosae* family (Haynes 1986). Locust trees are an example of legumes that enhance the N-status of forest soils (Klemmedson 1994). Nitrogen-fixation by actinomycetes is also widespread in wildland ecosystems, and an example of forest trees having this type of symbiosis is the genus *Alnus,* which has been reported to have enriched the soil with 18 to 129 pounds/acre (20 to156 kg/ha) of N per year depending on the local site conditions and stand density (Van Cleve and others 1971, Jurgensen and others 1979). Brush species having actinomycete-driven N-fixing capabilities are bitterbrush (*Purshia tridentata*) and mountain mahogany (*Cercocarpus* spp.).

Nonsymbiotic N-fixation: Over the past 3 to 4 decades, studies on N-fixation in soils have determined that free-living microbes are also able to fix N in

USDA Forest Service Gen. Tech. Rep. RMRS-GTR-42-vol. 4. 2005

79

substantial amounts, particularly in forest soils (Jurgensen and others 1997). These free-living microbes obtain the energy necessary for N-fixation during the decomposition of large woody debris (Harvey and others 1989). Coarse woody debris consists of tree limbs, boles, and roots that are greater than 3.0 inches (7.5 cm) in diameter and are in various stages of decay (Graham and others 1994). It is produced when trees die and their boles fall to the soil surface. The death of trees occurs because of old age, insect and disease attacks, devastating natural events, and human activities (timber harvesting debris, as one example). Much of the coarse woody debris found in forests is on the soil surface and is partially or totally covered by soil and humus layers. This coarse woody debris, and associated smaller organic matter, enhances the physical, chemical, and biological properties of the soil and thereby contributes directly to site productivity. In addition, it provides a favorable microenvironment for seedling establishment and growth.

A particularly important role of coarse woody debris is that it serves as a potentially valuable source of nonsymbiotic N-fixation. Extensive studies in the Inland Pacific Northwest show that up to 50 percent of the total N fixed on a site is contributed from the large woody debris component (Jurgensen and others 1997). Soil wood found in these forests is a product of brownrot decay and is made up of heartwood from pine and Douglas-fir trees. It is relatively resistant to decay and, as a result, can remain in the soil for hundreds of years. The accumulation of coarse woody debris and the fixation of N on warm, dry sites are lower than on the wetter, more productive sites.

Denitrification—*Denitrification* is the reverse process whereby NO_3 is reduced to N_2 and N_2O biologically. It is the major mechanism that returns N, which was originally fixed from the atmosphere, back into the atmosphere (Richardson and Vepraskas 2001). Denitrification typically occurs in saturated and water-logged soils found in wetlands (see chapter 8). Nitrogen losses are generally related to its availability, and significant amounts of N up to 54 pounds/acre/year (60 kg/ha/year) have been reported where NO_3 loadings occur as a result of nonpoint runoff or from other sources.

Fire Effects on Organisms and Biological Processes _____

Fire affects biological organisms either directly or indirectly. Direct effects are those short-term changes that result when any particular organism is exposed directly to the flames, glowing combustion, hot gases, or is trapped in the soil and other environments where enough heat is transferred into the organism's immediate surroundings to raise the temperature sufficiently to either kill or severely injure the organism. Indirect effects usually involve longer duration changes in the environment that impact the welfare of the biological organisms after the fire has occurred. These indirect effects can involve habitat, food supply, competition, and other more subtle changes that affect the reestablishment and succession of plants and animals following fire.

Soil Microorganisms

Environmental Constraints and Microbial Growth—In this section we provide a brief introduction on the effects of different environmental factors—temperature, moisture, and substrate availability—on microbial life so the reader will understand microbial reactions to fire. A subsequent section then describes the response of different microorganisms to fire.

Temperature: Microbial communities are well adapted to a wide range of prevailing temperature regimes. Drastic changes in temperature during soil heating can result in mortality and shifts in species composition of survivors (Baath and others 1995, Pietikainen and others 2000). Lethal temperatures are as low as 122 °F (50 °C) for some bacteria, particularly for important gram-negative organisms, such as nitrifiers, that have thin cell walls (Wells and others 1979). Above 392 °F (200 °C), virtually all bacteria are killed. Fungi are generally killed at lower temperatures than bacteria, and lethal temperatures range between 122 and 311 °F (50–155 °C) (Wells and others 1979). From a practical standpoint, the threshold temperatures for bacteria and fungi are usually reached to a depth of 2 inches (5 cm) or more in the mineral soil during medium- or high-severity fires (Theodorou and Bowen 1982, Shea 1993, Giardina and others 2000). As a result, decreases in microbial populations due to the direct effect of soil heating are common immediately following either wildfire (Acea and Carballas 1996, Hernandez and others 1997, Prietro-Fernandez and others 1998) or some higher severity prescribed burns (Ahlgren and Ahlgren 1965, Pietikainen and Fritze 1993). Indirect long-term change in soil temperature can also result from modifications to vegetation cover and microclimate following fire. However, compared to the immediate and detrimental effects of soil heating, these longer term temperature changes are subtle and can even produce a slight stimulation of microbial population size and activity (Bissett and Parkinson 1980).

Moisture and oxygen: Microorganisms thrive in moist soil. In fact, most can be considered aquatic organisms by nature of their requirement of aqueous solution for cell movement. Filamentous fungi and actinomycetes are an exception. Optimum moisture content for microbial activity is typically near 55 percent of soil water-holding capacity (Horwath and Paul 1994). A combination of water and oxygen (O_2) content in the soil affects the level of microbial activity. As soils dry, increased cellular energy is required to maintain turgor, resulting in reduced growth and activity. Microbial processes essentially cease when soil becomes air-dry, at which time cells become dormant. Conversely, excessive moisture displaces O_2 and inhibits the metabolic activity of the dominant aerobic microbial population.

Soil moisture is a crucial factor in determining microbial survival during fire. Water is capable of absorbing large amounts of heat energy, thereby resulting in less temperature rise and reduced fire severity for a given heat input. Frandsen and Ryan (1986) found a temperature reduction of more than 932 °F (500 °C) when soil was wetted prior to burning, thus providing a presumed advantage to microorganisms. Conversely, more biological damage can result in moist soil compared to dry soil at a given temperature maximum because water is a better conductor of heat than air, and microorganisms are more metabolically active in moist soil. For example, Dunn and others (1985) estimated that 95 percent of bacteria are killed in moist soil and only 25 percent are killed in dry soil at equivalent soil temperature (158 °F or 70 °C). Burning when soils are dry is recommended if severe soil heating is anticipated. An additional concern is the potential decrease in soil water availability following fire. Less soil water is available after forest fires as a result of decreased water infiltration and storage due to water repellency, and increased water loss by soil surface evaporation. Only one report was found where moisture changes following fire were suspected of altering microbial properties (Raison and others 1986). The reserchers found up to a 34 percent reduction in eucalyptus litter decay after moderate-severity prescribed fire and hypothesized that the change in decomposition resulted from moisture limitations created by increased surface evaporation. This response was suspected to be short-lived, however, because of the rapid reaccumulation of forest floor material in these systems and the anticipated reduction of surface evaporation.

Substrate availability: Most soil microorganisms are heterotrophic, meaning they require preformed, organic material in the forest floor or mineral soil for their source of energy. Therefore, any fire-induced changes in the quality or quantity of organic matter may have long-lasting implications for the biological activity of soil (Lucarotti and others 1978, Palmborg and Nordgren 1993, Pietikainen and others 2000). Surprisingly, most forests have insufficient reserves of degradable organic material to provide for optimal microbial growth. Forest soil microorganisms, in effect, are more likely to be C-limited than by water or other essential nutrients (that is, N and P). As a corollary, the removal of surface organic matter by high-severity fires can reduce microbial population size and activity (Fernandez and others 1997, Prietro-Fernandez and others 1998). Even with fires of low to medium severity, the consumption of downed woody material can influence an important reservoir of mycorrhizal fungi (Harvey and others 1976, 1980a).

In addition to heterotrophs, the microbial community includes a small yet biologically important group of bacteria that obtain their energy from the oxidation of inorganic compounds (autotrophs). Nitrifiers have received the most attention among autotrophic organisms because of their role in the N cycle and their sensitivity to soil heating (Dunn and others 1985). Nitrifiers obtain energy from the oxidation of NO_4 and NO_2, and release NO_3 as an end product.

Response of Soil Microorganisms to Fire—Fire affects most organisms that inhabit the belowground environment in both direct and indirect ways (Ahlgren and Ahlgren 1965, Borchers and Perry 1990). Fire impacts soil organisms directly by killing or injuring the organisms, and indirectly by its effect on plant succession, soil organic matter transformations, and microclimate. Heat penetration into the soil during a fire affects biological organisms located below the soil surface, depending on the heat transfer mechanism, soil moisture content, and duration of combustion.

Because many living organisms and the organic matter in soils are located on, or near, the soil surface, they are exposed to heat radiated by flaming surface fuels and smoldering forest floor fuels (fig. 4.6).

Figure 4.6—High-severity wildfires can remove nearly all the litter and duff and associated microbial populations present on the forest floor, leaving only gray ash. (Photo by Kevin Ryan).

USDA Forest Service Gen. Tech. Rep. RMRS-GTR-42-vol. 4. 2005

81

Depending on fire severity, organisms in the organic forest floor can be killed outright, although those in the deeper soil horizons, or in isolated unburned locations, survive. Even low-severity fires can damage organisms that are on or near the soil surface because most biota are damaged at lower temperatures than those that cause changes in physical and chemical soil properties during fire (table 4.1).

How do microorganisms respond to fire? Without question, fire is lethal. It also modifies the habitat of microorganisms by destroying organic matter, altering soil temperature and moisture regimes, and changing the postfire vegetation community and rates of organic matter accumulation. Consequently, changes in microbial population size and activity are common following wildfire and prescribed fire (see Ahlgren 1974, Raison 1979, Borchers and Perry 1990, Neary and others 1999 for reviews). Most reviews are quick to point out, however, that microbial responses are variable, or even unpredictable, depending on site conditions, fire intensity and severity, and sampling protocol. But if microbial responses are unpredictable, then no assistance can be offered to managers in developing fire prescriptions that meet operational objectives while minimizing risks to soil biota. The following discussion will challenge this concept in attempting to offer practical guidelines for fire managers.

If microbial life was only a function of temperature, moisture, and substrate availability, then the refinement of fire prescription guidelines based on soil biotic responses would be straightforward. Unfortunately, it is not that simple because the ability of most microorganisms to recover from disturbance is complex. Microbial communities are unmatched in physiological diversity and genetic malleability—properties that permit growth in any environment—and they have the ability to degrade nearly all known compounds. Resilience, therefore, is a trademark of the microbial community. As an example, rapid declines in soil microbial populations due to fire are usually transitory. Population sizes often match or surpass preburn levels within a growing season (Ahlgren and Ahlgren 1965, Renbuss and others 1973). Successful recolonization following fire is a function of several factors, including incomplete mortality of native populations, spore germination, influx of wind-blown organisms, and microbial growth stimulations from available nutrients. An important caveat is that not all microorganisms respond alike; differential responses by community members have been observed following fire. For example, Harvey and others (1980a) found poor recolonization of ectomycorrhizae, and Widden and Parkinson (1975) found a similar response for genera *Trichoderma*, an important antagonist to plant pathogens, following slash burning.

Some obvious observations can be made. Microbial responses to fire are easiest to predict at the opposing ends of the fire-severity continuum: (1) intense wildfire can have severe and sometimes long-lasting effects on microbial population size, diversity, and function; (2) low-severity underburning generally has an inconsequential effect on microorganisms. From there it becomes more difficult to predict microbial adaptation because results from medium- to high-severity slash fire and underburn studies vary widely among habitat types.

Wildfire: High-severity wildfires can remove nearly all the litter and duff and associated microbial populations present on the forest floor, leaving only gray or orange ash (fig. 4.6). Even mineral soil C is consumed during wildfires. Recent studies show a range of 5 to 60 percent loss of organic material in the surface mineral soil (Fernandez and others 1997, Hernandez and others 1997, Prietro-Fernandez and others 1998). Related declines in microbial biomass immediately following high-severity fires have been as high as

Table 4.1—Threshold temperatures for key biological organisms (Adapted from DeBano 1991, Neary and others 1999).

Biological component	Temperature		Reference
	°F	°C	
Plant roots	118	48	Hare 1961
Small mammals	120	49	Lyon and others 1978
Protein coagulation	140	60	Precht and others 1973
Fungi—wet soil	140	60	Dunn and others 1975
Seeds—wet soil	158	70	Martin and others 1975
Fungi—dry soil	176	80	Dunn and others 1975
Nitrosomonas spp. bacteria—wet soil	176	80	Dunn and DeBano 1977
Nitrosomonas spp. bacteria—dry soil	194	90	Dunn and DeBano 1977
Seeds—dry soil	194	90	Martin and others 1975
VA mycorrhizae	201	94	Klopatek and others 1988

96 percent (Hernandez and others 1997). Microbial respiration and extracellular enzyme production (Saa and others 1998)—strong indicators of microbial activity and viability—also decline dramatically following wildfire. Fungi appear more sensitive to wildfire than bacteria. For example, fungal propagules were undetectable 1 week after a stand-replacing fire in a central Oregon ponderosa pine (*Pinus ponderosa*) forest, while the viable bacteria population was only slightly reduced (D. Shields, personal communication). Similar responses were reported following wildfire in a pine forest of Spain (Vazquez and others 1993). Differences in response between bacteria and fungi may be attributable to greater heat sensitivity of fungi, soil pH increases after burning that favor bacteria, or the excessive loss of organic material.

No ecosystem remains sterile, even after severe disturbance. Most studies show stable recovery of microbial populations in the mineral soil to prefire levels within 1 to 4 years after wildfire (Vazquez and others 1993, Acea and Carballas 1996, Prietro-Fernandez and others 1998). However, reduced microbial biomass has been reported for as many as 11 years after fire (Dumontet and others 1996). Whether the microbial community will fully recolonize depends on the time required for recovering the forest floor layer. Not only are microbial populations and processes suppressed during this recovery period (Lucarotti and others 1978), but the potential erosive loss of soil is high. Or, as suggested by Giovannini and others (1987), burnt soils will slowly regenerate as long as erosive processes can be avoided.

Low-severity prescribed fire: Almost by definition, low-severity prescribed fire has a minimal effect on soil biota (fig. 4.7). The maximum temperatures are generally nonlethal, except for the upper litter layer (Shea 1993), and therefore the consumption of forest floor habitat is limited. Changes in microbial activity, in fact, often show a positive response to this type of fire, particularly with respect to N-fixation (Jorgensen and Wells 1971) and N availability (Schoch and Binkley 1986, White 1986, Knoepp and Swank 1993a,b). Rates of litter decay (White 1986, Monleon and Cromack 1996) and enzyme activity (Boerner and others 2000) are generally unaffected by low-severity underburning. Such results are not universal, however. Monleon and others (1997) found that N mineralization was reduced at sites burned either 5 or 12 years earlier by low- to medium-severity prescribed fire. They suggested that fire-induced changes in N mineralization possibly contributed to a decline in the long-term site productivity of ponderosa pine stands in central Oregon.

While single-entry underburning is generally considered harmless, repeated burning has been shown to substantially reduce microbial population size and activity (Jorgensen and Hodges 1970, Bell and Binkley 1989, Tongway and Hodgkinson 1992, Eivazi and Bayan 1996). This observation reflects a cumulative reduction in forest floor and total nutrients with frequent burning. Most studies have compared either annual burning or short-term repeated fires (2 to 4 years). The long-term impact of repeated burning every 7 to 20 or more years on soil organic matter, nutrient content, and microbial processes is not understood. As a consequence, Tiedemann and others (2000) urge caution in the use of frequent fire and suggest including partial harvesting as a complementary practice to reduce wildfire risk and extend the period between prescribed burning.

Slash burning: The effect of slash burning will depend on both the pattern and amounts of fuels burned. When slash is piled and burned (for example, pushing and burning operations used in some pinyon-juniper eradication programs) the burned areas are highly visible after the fire (fig. 4.8). On these burned areas, deep layers of white ash may accumulate, and the underlying soil is usually exposed to extended soil heating, which can sterilize to a depth of several centimeters. Although the severe heating under the piles of fuel are damaging to the soil, only a small percentage of the total area may be affected. To avoid this damage to the soil, the slash can be scattered and burned, thereby minimizing the severe soil heating that occurs under piled fuels.

The literature contains numerous references to site-specific responses of microorganisms to medium and high-severity slash burning. A common theme is that the response of microorganisms is dictated by their habitat: organisms in the forest floor struggle to survive and recolonize, while those in the mineral soil do

Figure 4.7—Low-severity prescribed fire has a minimal effect on soil biota because maximum temperatures are generally nonlethal, except for the upper litter layer, and consumption of forest floor habitat is limited. (Photo by Daniel Neary).

USDA Forest Service Gen. Tech. Rep. RMRS-GTR-42-vol. 4. 2005

83

Figure 4.8—Piling and burning slash after a fuel harvesting operation in pinyon-juniper woodlands can create thick layers of white ash and extensive soil heating. (Photo by Malchus Baker).

not. This comes as little surprise because the temperature in the forest floor can easily reach 1,110 °F (600 °C) or higher during burning (Renbuss and others 1973, Shea 1993), consuming both the microorganisms and their habitat. In fact, substantial changes in forest floor microbial biomass and community structure have been reported to occur during a soil heating experiment at a temperature of 445 °F (230 °C) (Pietikainen and others 2000). Soil heating was less severe in the mineral soil, resulting in a much shorter fluctuation period before the microbial community stabilizes. The following examples illustrate this theme.

Recent findings from slash burns in Finland show detrimental effects of slash burning on microbial biomass (Pietikainen and Fritze 1993), activity (Fritze and others 1994, 1998), and community structure (Baath and others 1995, Pietikainen and others 2000) in the forest floor. Failure of the microbial community to respond rapidly has been attributed to a decline in organic matter quantity (Baath and others 1995) and quality (Fritze and others 1993), and the pyrolytic production of toxic compounds (Fritze and others 1998). Also, microbial populations were unable to respond to the input of nutrient-rich ash. These studies were relatively short term, ranging from 1 month to 3 years postfire. As a result, the length of time required before the forest floor microbial community reaches preburn levels is unclear, yet it might take up to 12 years (Fritze and others 1993). Related declines in microbial function have been reported in other forest types. Staddon and others (1998) found microbial-mediated enzyme activity was suppressed 4 years after slash burning in jack pine (*P. banksiana*). Meanwhile,

Jurgensen and others (1992) found a 26 percent decline in N-fixation by free-living bacteria during the first 2 years following a relatively high-severity slash fire that consumed 61 percent of the forest floor in a cedar-hemlock forest.

In contrast with the forest floor, microbial recovery in mineral soil following intense slash burning is impressive. For example, Renbuss and others (1973) examined viable bacterial and fungal populations after a high-severity log pile burn that produced temperatures of 735 to 1,110 °F (400–600 °C) in the upper 2 inches (5 cm) of soil. Although the soil was initially sterilized by fire, bacteria had recolonized to preburn levels within 1 week after ignition. Their population size remained at or above the level of the control soil for the length of the study (1 year). Fungi and actinomycetes were slower to recolonize, yet their populations returned to prefire level by the end of the study. Chambers and Attiwill (1994) confirmed the "ash-bed" effect in a controlled soil heating experiment. No differences in microbial population sizes were found 133 days after heating soil to 1,110 °F (600 °C) when compared to unheated soil. Similar responses are common in field studies (Ahlgren and Ahlgren 1965, Theodorou and Bowen 1982, Deka and Mishra 1983, Van Reenen and others 1992, Staddon and others 1998), whereas some controlled soil heating experiments have found longer delays in recolonization (Dunn and others 1979, Diaz-Ravina and others 1996, Acea and Carballas 1999). Differences between field and controlled-environment studies suggest the importance of wind- or animal-transported inoculum for recolonization.

Specialized Root-Microbial Associations—Mycorrhizal fungi are easily affected by fire, and the extent of damage depends upon fire severity, the reproductive structures exposed to soil heating (such as spores or hyphae), and the type of fungi (such as endo- or ectomycorrhiaze). Mycorrhizae and roots frequently occupy the uppermost duff layers of soil and as a result are subjected to lethal soil temperatures during a fire because these layers are frequently combusted, particularly during medium- and high-severity fires. In general, vesicular arbuscular mycorrhizae are less affected by disturbances that destroy aerial biomass (including fire) than are ectomycorrhizal fungi because they form symbiotic relationships with a wider range of plant species (Puppi and Tartaglini 1991). Also, ectomycorrhizae are more abundant in the litter layer compared to vesicular arbuscular mycorrhizae, which tend to concentrate in the lower mineral soil horizons (Reddell and Malajczuk 1984). As a result, fire that destroys only the litter layer would favor vesicular arbuscular mycorrhizae.

The general relationships discussed above have been documented by several studies that show a decline in

84

USDA Forest Service Gen. Tech. Rep. RMRS-GTR-42-vol. 4. 2005

the formation of both ectomycorrhizae (Harvey and others 1980b, Schoenberger and Perry 1982, Parke and others 1984) and vesicular arbuscular mycorrhizae (Klopatek and others 1988, Vilariño and Arines 1991) within the first growing season following fire. Mycorrhizae were not per se eliminated by fire in these studies. Instead, the percentage of roots infected by mycorrhizal fungi was reduced. As an example, Klopatek and others (1988) found vesicular arbuscular mycorrhizae infection declined from 41 percent at preburn to 22 percent within 24 hours of burning the organic layer of a pinyon pine (*P. edulis* Engelm.) soil. Work by Harvey and others (1976, 1980a,b, 1981) has also clearly established a relationship between reduced ectomycorrhizal root tip formation and slash burning on difficult-to-regenerate sites in western Montana. They emphasize the importance of maintaining adequate soil humus and wood (up to 45 percent by volume) as refugia for mycorrhizae on these sites.

Not all studies, however, have reported a detrimental relationship between mycorrhizal formation and fire. Several studies have shown no effect of burning on mycorrhizae by the end of the first growing season (Pilz and Perry 1984, Deka and others 1990, Bellgard and others 1994, Miller and others 1998). In fact, Herr and others (1994) found a slight increase in ectomycorrhizal infection on eastern white pine (*P. strobus*) with increasing fire severity. Visser (1995) examined ectomycorrhizal development in an age sequence of jack pine stands regenerating following fire. More than 90 percent of the root tips were mycorrhizal regardless of time since burning (from 6 to 122 years). She suggested that successful recolonization of mycorrhizae on jack pine seedlings was a function of (1) avoidance of lethal temperatures by location in soil profile, (2) resistance of spores and resting structures to lethal temperatures, (3) wind or animal dispersal of spores, and (4) survival on alternative plant hosts such as manzanita (*Arctostaphylos* spp.).

Finally, there is no clear evidence that fire impairs the function of mycorrhizae in plant nutrition and growth. Studies showing a decline in mycorrhizal infection after fire have seen minor or no decline in seedling survival or short-term growth (Schoenberger and Perry 1982, Parke and others 1984). Two simplified explanations are plausible: (1) either the flush of nutrients after fire makes mycorrhizae temporarily superfluous, or (2) the decline in root infection (such as a decline from 40 to 20 percent root tip colonization following burning) has no relationship to function. The argument in the second explanation is that mycorrhizal function is still effective whether root systems are completely or partially infected. Intermediate- or long-term studies are needed to resolve this issue.

Biological Crusts—The patchy nature of native plant communities in arid and semiarid lands produces a discontinuous source of fuels and results in a mosaic of fire intensities (Whisenant 1990). The biological crusts themselves provide little fuel to carry fire and thereby provide a "refugia" that slows down the spread of fire and minimizes its severity (Rosentreter 1986).

However, once fire destroys cryptogamic crusts, it can take several years for their populations to redevelop to prefire levels (DeBano and others 1998). High-severity fires during the dry summer months cause the greatest damage to biological crusts. The recovery following fire can be fast or slow, depending upon the type of crust (algal or lichenous), soil conditions, and climate. For example, crusts can take much longer to develop in hotter and more arid shrublands such as occurred on a site in a blackbrush (*Coleogyne ramosissima*) community in southern Utah (Callison and others 1985). This site did not show any cryptogamic development following a severe range fire after 37 years. Likewise, annual fires for 7 years have been reported to destroy cryptogamic crusts on degraded semiarid woodlands in Australia (Greene and others 1990).

Low-severity fires, in contrast, may leave the structural matrix of the crust intact (Johansen and others 1993). For example, after a single fire in a semiarid shrub-steppe it only took 4 years for the crusts to reach prefire levels. This study showed that the components making up the cryptogamic crust (such as algae, cyanobacteria, or lichens) affect the rate of recovery following disturbance. Algae recovered from disturbance most rapidly and returned to prefire densities within 1 to 5 years. Historically, fires in the Southwestern deserts probably were exposed to small, low-intensity, and patchy fires because of the sparse and discontinuous vegetation (Allen 1998, Belnap and others 2001). This type of fire behavior most likely had a minimum effect on the biological crusts common to this area.

In general, algal cells of many species are usually able to survive even the most severe disturbance, so the dispersal into and recolonization of burned areas is faster (Johansen 1993). For example, the first organisms to recolonize the soil under burned English heaths were algae (Warcup 1981). Algae are also favored by the higher pH after fire, and species with windblown propagules are able to recolonize disturbed areas rapidly (DeBano and others 1998). In contrast, filamentous cyanophytes (blue-green bacteria) are less likely to recolonize by wind dispersal because of their size. Compared to algae, mosses and lichens are slower to reestablish themselves. Acrocarpic mosses were found to reoccupy burned sites in the Mediterranean area within 9 to 15 months after fire (De Las Heras and others 1993). Lichens that produce vegetative diaspores can move into disturbed areas more quickly than lichens that do not produce these propagules (Johansen

USDA Forest Service Gen. Tech. Rep. RMRS-GTR-42-vol. 4. 2005

85

and others 1984). Some lichens can take 10 to 20 years to develop diverse and abundant communities (Anderson and others 1982).

Soil Meso- and Macrofauna

Most research results on the effect of fire on meso- and macrofauna are reported in terms of general groups of soil invertebrates, including insects and other arthropod assemblages, and earthworms. Insects, for example, may have representatives in both the meso- and macrofauna groups (fig. 4.1; also see earlier discussion in this chapter on soil meso- and macrofauna). Another important ramification of fire and these organisms is the postfire infestation of forests and other ecosystems by insects as a result of the effect of fire on the health of the postfire vegetation. The response of all the above organisms depends to a large degree on the frequency and severity of fire.

The magnitude of short-term changes undergone by invertebrate populations in response to fire depends on both fire severity and frequency, the location of these organisms at the time of the fire, and the species subjected to fire. In the case of either an uncontrolled wildfire (high fire severity), and prescribed fire (low fire severity), the effect on invertebrates can be transitory or longer lasting (Lyon and others 1978). Both types of fires contain zones of high and low fire severity, but wildfires are more likely to burn larger, contiguous areas. The long-term abundance of arthropod populations, however, can remain high because of their resiliency to both intensity and frequency of burning (Andersen and Müller 2000).

Some studies show the effect of different severities and frequencies of fire on invertebrates. One study in *P. sylvestris* forests in Sweden showed that the overall mortality of invertebrates depended on the proportion of organic soil consumed by the fire and that the mortality ranged from 59 to 100 percent. Invertebrates that lived deeper in the soil had less mortality than those that colonized the vegetation and litter layers (Wikars and Schimmel 2001). Other characteristics that favored survival included greater mobility in the soil and thick protective cuticles (as is found in the taxa Oribatediae and Elateridae).

A study of the response of insects and other arthropods to prescribed burn frequency in prairie ecosystems showed that the changes in the physical environment and plant communities following prescribed fires can result in the development of distinctly different arthropod communities on the frequently burned sites compared to sites that were protected from burning (Reed 1997). Distinctive arthropod species and groups were supported by the changing succession stages following fire. In general, landscapes that have a range of sites representing different successional stages and sites that have different burn frequencies support the most species. However, on individual sites that are burned at intervals, a cycle of arthropod species richness, species composition, and numbers of individuals occur. The combined effect of fire frequency and time of burn on arthropod taxa were reported for tropical savannas found in Australia (Andersen and Muller 2000). A substantial resilience to fire of the arthropod assemblages was found. Only four of the 11 arthropod taxa were significantly affected by fire. Ants, crickets, and beetles declined in the absence of fire. Late season fires decreased spiders, homopterans, silverfish, and caterpillars.

Some invertebrates have traits that allow them to survive fire. These traits may arise in a variety of ways. Some may not have evolved specifically as an adaptation to fire, but rather more generally to hot and arid conditions. For example, some invertebrates have adaptations that enable them to conserve water and resist high ambient temperatures in seasonally dry habitats. Other groups of invertebrates possess traits, such as high mobility, that appear to be characteristic to particular taxonomic groups and not related to specific ecosystems or fire regimes. Still other adaptations appear to have evolved primarily in response to fire and can involve the complex long-term evolution of some rather esoteric anatomical features. For example, a recent study on detailed morphological and anatomical characteristics of a subfamily of beetles (Coleoptera: Clerinae) suggests that these invertebrates evolved thermoreceptor antenna, which enable the beetles to avoid death by fire in xeric environments (Opitz 2003). The evolution of this feature occurred over a span of tens of thousands of years.

The effects of fire on soil invertebrates have been reviewed by several authors (DeBano and others 1998, Lyon and others 2000a 2000b, Andersen and Müller 2000). Many of the reports cited describe invertebrates in savannas and other grasslands. These reports indicate that the effects of fire on soil invertebrates can occur via several mechanisms (Andersen and Müller 2000) that include direct mortality, through forced emigration, or through the immigration of pyrophilous species (such as wood-boring beetles that are attracted by heat and smoke to a burned area where they infest injured or dead trees). Short-term indirect effects include modification of the habitat and foraging sites, food supplies, microclimate, and rates of predation. Long-term indirect effects are manifest mainly in nutrient cycling and primary productivity.

Invertebrates residing more permanently in the upper soil layers are most likely affected when these soil layers are heated to lethal temperatures. Macroinvertebrates dwelling exclusively in litter were found to be particularly vulnerable to wildfire that destroys surface fuels and litter (Sgardelis and others 1995). However, invertebrates that permanently occupy deeper soil horizons are usually protected from even high-severity fires. Some macroinvetebrates have been found to move deeper into the soil during the summer and, as a result, they are insulated from lethal soil temperatures during fire. Most of the invertebrates in the top 1.0 inch (2.5 cm) of soil survive relatively cool burning wildfires or prescribed fires (Coults 1945). A reduction in litter quantity after fire can indirectly decrease both the number of invertebrate species and the species density of soil and litter invertebrates as unprotected mineral soil warms (Springett 1976).

The effect of soil heating on earthworms is not well understood. One study in tallgrass prairie, however, showed that the indirect effects of a fire are probably more important than direct heating of earthworm populations (James 1982). This study showed that fire increased earthworm activity because of differences in plant productivity following the fire. In general, the subsurface soil horizons are usually proliferated with roots and rhizomes, which in combination with more favorable soil moisture conditions create an ideal environment for earthworms at about 4 to 8 inches (10 to 20 cm) below the soil surface. A location this deep in the soil most likely protects earthworms from the direct effects of soil heating during fuel combustion, except in the case of severe long-duration fires that might occur under piles of slash and logs or in smoldering duff and roots. Other studies have shown that fire (in prairie grasslands and mixed forest types) frequently leads to an increase in exotic earthworm species at the expense of endemic species (Bhadauria and others 2000, Callaham and others 2003).

Roots and Reproductive Structures

Many of the plant roots, vegetative reproductive structures, and seeds are found immediately on the surface of the soil or distributed downward throughout the soil profile. These plant parts include tap roots, fibrous surface roots, rhizomes, stolons, root crowns, and bulbs. Many plant roots are found in the surface organic layers (L-, F-, and H-layers) and can be directly affected whenever these layers are heated or destroyed during a fire. Plant roots are sensitive to both duration of heating and the magnitude of the temperature reached. Temperatures of 140 °F (60 °C) for 1 minute are sufficient to coagulate protein (Precht and others 1973). Plant roots are sensitive to soil heating, and lethal temperatures can occur before proteins began to coagulate. The lethal temperature of plant tissue is highly dependent on the moisture content of the tissue. Those tissues containing higher moisture contents tend to be killed at lower temperatures and during a shorter interval of heating (Zwolinski 1990). Miller (2000) gives additional information on the effects of fire and soil heating on root mortality and the welfare of vegetative reproduction.

Plant roots that are insulated by the soil have a lower risk of being subjected to lethal temperatures during a fire (DeBano and others 1998). The two most important factors that insulate roots against soil heating are their depth in the soil and the soil water content. Generally, the deeper the plant roots are located in the soil, the greater will be the survival rate (Flinn and Wein 1977). Low-severity fire that destroys only the plant litter may kill only aboveground plant parts. In contrast, high-severity fires can consume all the surface organic matter and easily heat the mineral soil above the lethal temperature for roots.

Seed banks that are stored in the soil can be affected by fire. A majority of the seeds are stored in the litter and upper part of the soil beneath the vegetative canopy. Medium to high-severity fires heat the surface layers sufficiently to destroy any seeds that have been deposited. The lethal temperature for seeds is about 160 °F (70 °C) in wet soils and 190 °F (90 °C) in dry soils (Martin and others 1975). Although fire can destroy seeds, it also can enhance reproduction by seeds (Miller 2000). For example, fire can destroy allelopathic substances that inhibit seed production. Or, in the case of ponderosa pine regeneration, fire can provide a mineral seedbed required for germination and growth. The heating associated with fires may also stimulate the germination of seeds that lie dormant in the soil for years because of impermeable seed coats (such as seeds of chamise and hoaryleaf ceanothus).

Soil heating, heat transfer, and the effect of the lethal temperatures on the welfare of seeds and roots are more complicated in moist soils than when the soil is dry. Dry soil is a poor conductor of heat and, as a result, heat does not penetrate deeply in the soil, particularly if the residence time of the flaming front is short. The surface of dry soil can easily exceed the lethal temperature of living tissue of roots, while ambient daily soil temperatures can prevail 0.8 inch (2 cm) downward in the soil, with little damage occurring to the roots. Therefore, when the roots are in dry soil below 0.8 inch (2 cm), they are not likely to be damaged by soil heating unless the residence time of the flaming front is long. Conversely, those plant structures on or near the soil surface can easily be damaged. Also, the presence of moisture in the soil affects plant root and seed mortality (in other words, living biomass is killed at lower temperatures when the soil is wet compared to a dry soil).

Amphibians, Reptiles, and Small Mammals

The ability of amphibians, reptiles, and small mammals to survive wildland fires depends on their mobility and the uniformity, severity, size, and duration of any fire (Wright and Bailey 1982). Fire can cause direct injury and kill the animals themselves

USDA Forest Service Gen. Tech. Rep. RMRS-GTR-42-vol. 4. 2005

87

depending on how capable they are of avoiding and escaping the fire itself (Lyon and others 2000a). The effects of wildland fires on these animal populations can be found in a separate volume of this series (Smith 2000). Fire also affects the long-term welfare of these larger animals by changing their habitat (Lyon and others 2000b).

Biologically Mediated Processes

A wide range of microorganisms participate in cycling carbon and plant nutrients in a systematic and sustainable rate that is necessary for maintaining healthy ecosystems. Important biologically regulated processes carried out by these microbes that can be affected by fire include decomposition, mineralization, ammonification, nitrification, nitrogen-fixation, and denitrification.

Decomposition—Fire affects decomposition in two general ways (DeBano and others 1998). First, moderate- and high-severity fires kill the biological organisms that decompose organic matter (see the earlier discussions on microorganisms). The microorganisms most affected by fire include bacteria and fungi, which are numerically the most abundant organisms in terrestrial ecosystems and are the primary decomposers of organic matter in soil (Van Veen and Kuikman 1990). Second, a rapid, strictly chemical combustion process replaces the slower, biologically mediated decomposition processes that occur under nonfire conditions.

Nitrogen Mineralization—Two important mineralization process affected by fire are ammonification and nitrification. The sensitivity of both ammonifying and nitrifying bacteria to soil heating most likely plays an important role in the nutrition of plants because N is frequently limiting in wildland soils (DeBano and others 1998). This relationship has been demonstrated in unburned chaparral stands where high levels of total N can occur as organic N, but only relatively low levels of inorganic mineral N (NH_4- and NO_3-N) have been measured. It has been hypothesized that the low rate of mineralization in chaparral soils occurred because heterotrophic microorganisms responsible for mineralization were inhibited by allelopathic substances present, or because high lignin contents of chaparral plant leaves resist decomposition and subsequent mineralization of N (Christensen 1973). The hypothesis that higher concentrations of NH_4- and NO_3-N are generally present after a fire is based on the idea that NH_4- and NO_3-N are formed by different processes as a result of burning (Christensen and Mueller 1975, DeBano and others 1979). According to this hypothesis, relatively large amounts of NH_4-N are produced chemically by soil heating during a fire (see chapter 3) as well as being microbially produced

following a fire. Nitrogen in the form of NO_3, however, is not produced directly by heating during a fire, but instead is formed during subsequent nitrification of excess NH_4-N produced as a result of burning. This process is further complicated by the observation that postfire nitrification does not appear to be carried out by the classical nitrifying bacteria (for example, *Nitrosomonas* and *Nitrobacter*) because these bacteria are particularly sensitive to soil heating and other disturbances and, as a result, are absent (or at extremely low levels) for several months following burning (Dunn and others 1979). The absence of nitrifying bacteria after fire suggests that nitrification may be carried out by heterotrophic fungi. Dormant forms of heterotrophic fungi have been reported to be stimulated by mild heat treatments (Dunn and others 1985) and are thought to have contributed to the fungal growth that paralleled NO_3 production (Dunn and others 1979). Suppression of mineralization rates by allelopathic substances has been further substantiated by other studies in ponderosa pine forests in New Mexico (White 1991, 1986) and in pinyon-juniper woodlands (Everett and others. 1995).

Studies have shown that both *Nitrosomonas* and *Nitrobacter* are sensitive to soil heating during fire such as occurs with microorganisms described earlier (table 4.1). Studies in chaparral soils have shown that *Nitrosomonas* bacteria are killed in dry soil at temperatures of 250 to 280 °F (120 to 140 °C) as contrasted to a moist soil where the lethal temperature is between 165 and 175 °F (75 and 80 °C) (Dunn and DeBano 1977, Dunn and others 1985). *Nitrobacter* bacteria are even more sensitive and are killed at 212 °F (100 °C) in dry soil and at 120 °F (50 °C) in a moist soil. Unlike heterotrophs, which must adapt to decreased organic matter availability, nitrifiers are provided with a sharp increase in available substrate (NH_4) after fire (Raison 1979). Consequently, the initial nitrifying bacterial population decline due to soil heating typically is reversed within the first year in response to the "flush" of available substrate (Jurgensen and others 1981, Acea and Carballas 1996).

Unfortunately, the increases in NH_4- and NO_3-N following a fire are relatively short-lived and can return to prefire levels within 2 years after burning (DeBano and others 1998). Studies conducted after burning tropical forests in Costa Rica showed that the increase in available N returned to background levels in 6 months (Matson and others 1987). In an Arizona study the NH_4-N was immediately increased during a fire and then slowly decreased the following 10 months, at which time NO_3-N began increasing (see fig. 3.13). Within a year, both NH_4- and NO_3-N levels returned to prefire levels. In another study, N mineralization increased and remained elevated for 1 year in *Agropyron spicatum* and *Stipa comata* grasslands for 2 years following prescribed burning in mountain shrublands (Hobbs and Schimel 1984).

Nitrogen-Fixation—The effect of fire on N-fixation involves the effect of heating on the living protoplasm present in symbiotic and nonsymbiotic microorganisms that fix N. Symbiotic N-fixing microorganisms can be affected in at least two ways. First, the destruction of the host plant during combustion affects the symbiotic relationship by removing the source of energy. Second, the symbiotic microorganisms present in the roots may be killed if the upper organic layers are consumed during a high-severity fire. Conversely, little or no direct damage would be expected in the case of deeper roots, which are far removed from the soil surface, or during a cooler burning prescribed fire (DeBano and others 1998). Nonsymbiotic bacteria respond similarly to other microbes discussed earlier.

On the other hand, nonsymbiotic processes, which receive energy from the biological oxidation of organic matter, also have been reported to fix substantial amounts of N. For example, more than one-third of the N-fixing capacity of forest soils has been reported to be provided by microorganisms responsible for decaying wood on the surface and in the soil profile (Harvey and others 1989). The management of woody residues (coarse woody debris) within a fire prescription thus becomes an important consideration in forest management. Therefore, it is important to retain a substantial amount of large woody debris on forest sites after timber harvesting or when using prescribed fire. For example, the amounts of residual woody debris recommended in figure 4.9 are considered necessary for maintaining the productivity of forests in Arizona, Idaho, and Montana.

The indirect effect of fire on N-fixation focuses on the role that both symbiotic and nonsymbiotic organisms play in N-fixation and replenishment in wildland ecosystems (DeBano and others 1998). Currently this role is still under debate. For example, the results reported from a study in forests of the Northwestern United States showed that soil N-additions by symbiotic N-fixation were not as large as previously assumed for these ecosystems (Harvey and others 1989). However, some vegetation types, such as alder trees in riparian ecosystems and dense stands of snowbrush (*Ceanothus velutinus*), have been reported to fix substantial amounts of N (Jurgensen and others 1979). The fixation of N in these forest and brushland areas is by symbiosis.

Burning may create a favorable environment for the establishment of N-fixing plants in some plant communities that are subjected to frequent fire. For example, fire exclusion in ponderosa pine-Douglas fir forests in the Northwest has been reported to lead to such widespread changes in forest structure, composition, and functioning that N-fixing plants species have been reduced (Newland and DeLuca 2000).

Denitrification—Little research has been done on the biological losses of N in relation to fire, partly because of the overwhelming losses that occur chemically by volatilization (DeBano and others 1998). Denitrification, however, can be an important factor when using fire in wetlands (see chapter 8).

Management Implications

Rarely are microorganisms and their processes more than a passing thought in forest management plans. For example, fuel-reduction programs rarely consider the potential effects of surface fire on soil organisms such as mycorrhizal fungi or autotrophic bacteria (W. Johnson, personal communication). A recent survey and management policy developed for protecting a small percentage of the fungi and lichen within the critical range of the northern spotted owl is the exception to the rule (Molina and others 1999). This situation is not surprising for at least two reasons. First, most microorganisms are invisible to the naked eye and are thus "out of sight" and, as a result, "out of mind." Second, no simple, inexpensive field test is available to measure microbial populations or their processes. Thus, managers have no practical means of determining microbial responses to operational prescribed burns. In addition, the results published from fire effects studies have not always presented a clear picture of how microorganisms respond to fire, thus leaving managers guessing at how responsive microorganisms are (or wondering whether they should care) for given forest types and anticipated fire severities.

However, the following general concepts of microbial responsiveness to fire can be gleaned from past studies:

- Microorganisms are skilled at recolonizing disturbed forests. Their resiliency is a function of unsurpassed physiological and genetic diversity.
- Fire effects are greatest in the forest floor and decline rapidly with mineral soil depth. Recovery of microbial populations in the forest floor is not guaranteed, particularly in dry systems with slow reaccumulation of organic material.
- Severe wildfire and prescribed fire reduce organic matter content and increase the potential for loss of soil by erosion.
- Prescriptions that avoid drastic changes in the environmental factors controlling microbial life—soil temperature, moisture, and substrate availability—will be the most successful at meeting operational objectives while ensuring a functioning soil biotic community.

USDA Forest Service Gen. Tech. Rep. RMRS-GTR-42-vol. 4. 2005

89

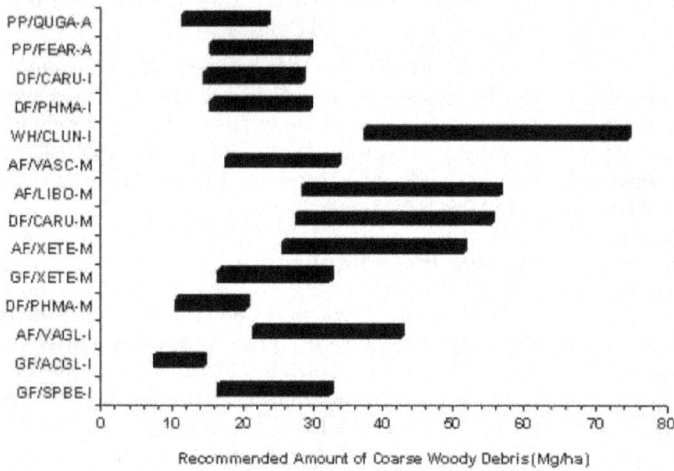

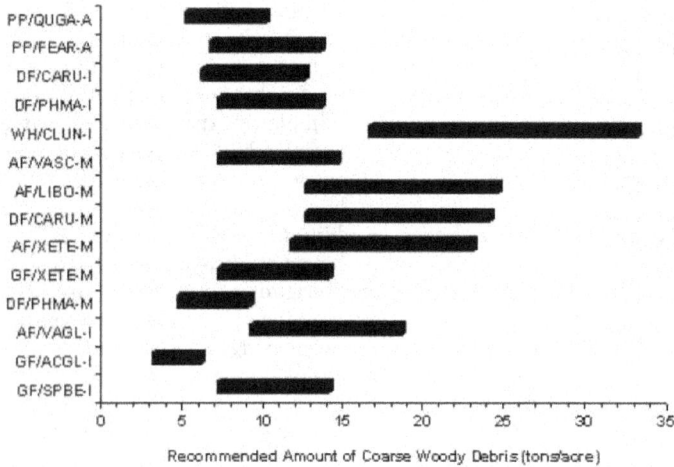

Figure 4.9—Amounts of residual coarse woody debris to leave after timber harvesting necessary to maintain site productivity in Arizona, Idaho, and Montana forests. Accompanying chart lists habitat type acronyms. (Adapted from Graham and others 1994).

State	Acronym	Habitat type
Idaho	GF/SPBE-I	Grand fir / snowberry
	GF/ACGL-I	Grand fir / mountain maple
	AF/VAGL-I	Subalpine fir / huckleberry
	WH/CLUN-I	Western hemlock / queencup beadlily
	DF/PHMA-I	Douglas-fir / ninebark
	DF/CARU-I	Douglas-fir / pinegrass
Montana	DF/PHMA-M	Douglas-fir / ninebark
	GF/XETE-M	Grand fir / bear grass
	AF/XETE-M	Subalpine fir / beargrass
	DF/CARU-M	Douglas-fir / pinegrass
	AF/LIBO-M	Subalpine fir / twintower
	AF/VASC-M	Subalpine fir / whortleberry
Arizona	PP/FEAR-A	Ponderosa pine / Arizona fescue
	PP/QUGA-A	Ponderosa pine / Gambel oak

90

USDA Forest Service Gen. Tech. Rep. RMRS-GTR-42-vol. 4. 2005

- Knowledge gaps persist, particularly regarding repeated fire and its effect on microorganisms.

Based on the above concepts, the following recommendations are offered:

- Minimize loss of forest floor (litter and duff). Microorganisms are most vulnerable to heat damage and habitat changes in this layer. This presents a quandary for prescribed fire practioners: How much organic material (fuel) should be removed to reduce wildfire danger while still maintaining an adequate supply for forest function? Tiedemann and others (2000) recommend burning when the upper layer of the forest floor is dry enough to carry fire and the lower layers are wet enough to avoid consumption. Further, they recommend extending the recovery time between repeated fires if these conditions are not achieved or if exposure of mineral soil is desired for tree regeneration.
- Avoid burning when soil is moist *if* the anticipated fire severity is high. Mortality of microorganisms is greater in moist soil than in dry soil at high temperatures.
- Provide adequate inoculum for microbial recolonization by burning with mosaic patterns (there is no assurance that indigenous populations will survive soil heating).
- Supplement burning with other silvicultural practices (partial harvest, crushing, mulching) to reduce fuel buildup. Repeated burning of the forest floor can result in detrimental effects to microbial biomass and activity.

Summary

Soil microorganisms are complex. Community members range in activity from those merely trying to survive to others responsible for biochemical reactions that are among the most elegant and intricate known. How microorganisms respond to fire will depend on numerous factors, including fire intensity and severity, site characteristics, and preburn community composition. Some generalities can be made, however. First, most studies have shown strong resilience by microbial communities to fire. Recolonization to preburn levels is common, with the amount of time required for recovery generally varying in proportion to fire severity. Second, the effect of fire is greatest in the forest floor (litter and duff). We recommend prescriptions that consume major fuels but protect forest floor, humus layers, and soil humus.

USDA Forest Service Gen. Tech. Rep. RMRS-GTR-42-vol. 4. 2005

91

Notes

_____

_____

_____

_____

_____

_____

Part B

Effects of
Fire on Water

Daniel G. Neary
Peter F. Ffolliott

Part B—The Water Resource: Its Importance, Characteristics, and General Responses to Fire

Introduction

Effects of fire on the hydrologic cycle are determined largely by the severity of the fire, decisions made relative to any suppression activities, and the immediate postfire precipitation regime. Because information is typically scarce for portions of the spectrum of conditions in which a fire might occur, it is not possible to adequately describe the possible impacts of fire in all conceivable situations. But by understanding the nature of the hydrologic processes impacted, we can interpret the impact of fire on these processes at least to the degree needed to make adequate management decisions.

This chapter covers the hydrologic processes represented by the components of the hydrologic cycle (Brooks and others 2003). We review how changes in hydrologic processes that are brought about by, or attributed to, the occurrence of fire, can translate into changes in streamflow regimes.

Hydrologic Cycle

The *hydrologic cycle* represents the processes and pathways by which water is circulated from land and water bodies to the atmosphere and back again. While the hydrologic cycle is complex in nature and dynamic in its functioning, it can be simplified as a system of water-storage components and the solid, liquid, or gaseous flows of water within and between storage points (fig. B.1). Precipitation inputs (rain, snow, sleet, and so forth) to a watershed are affected little by burning. However, interception, infiltration, evapotranspiration, soil moisture storage, and the overland flow of water can be significantly affected by fire. It must be kept in mind that these components of the hydrologic cycle are closely interrelated, and therefore, it is difficult in practice to isolate the impacts of fire on one component alone.

A generalized percentage breakdown of water inputs, fluxes, and outputs in undisturbed forested watersheds is shown in figure B.2 (Hewlett 1982). These

USDA Forest Service Gen. Tech. Rep. RMRS-GTR-42-vol. 4. 2005

95

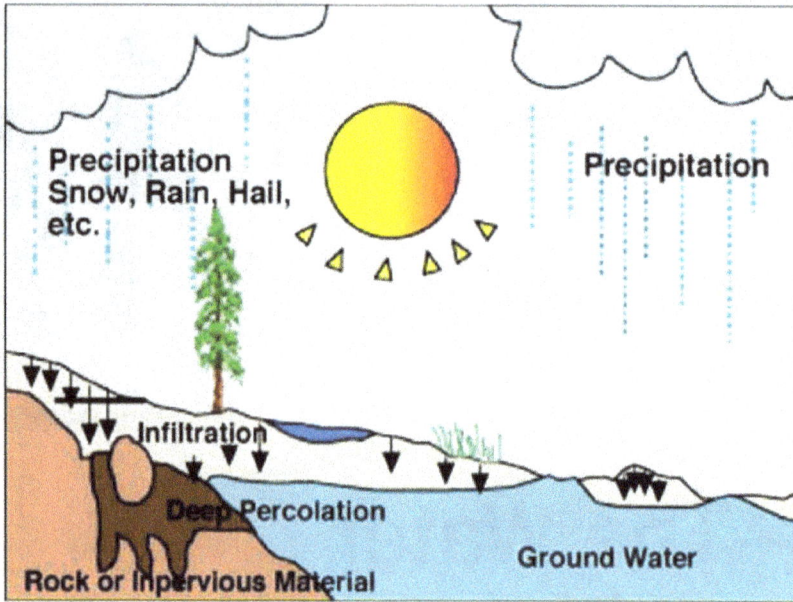

Figure B.1—Generalized diagram of the hydrologic cycle. (Figure courtesy of the USDA Forest Service, National Advanced Fire and Resource Institute, Tucson, AZ, Tucson, AZ).

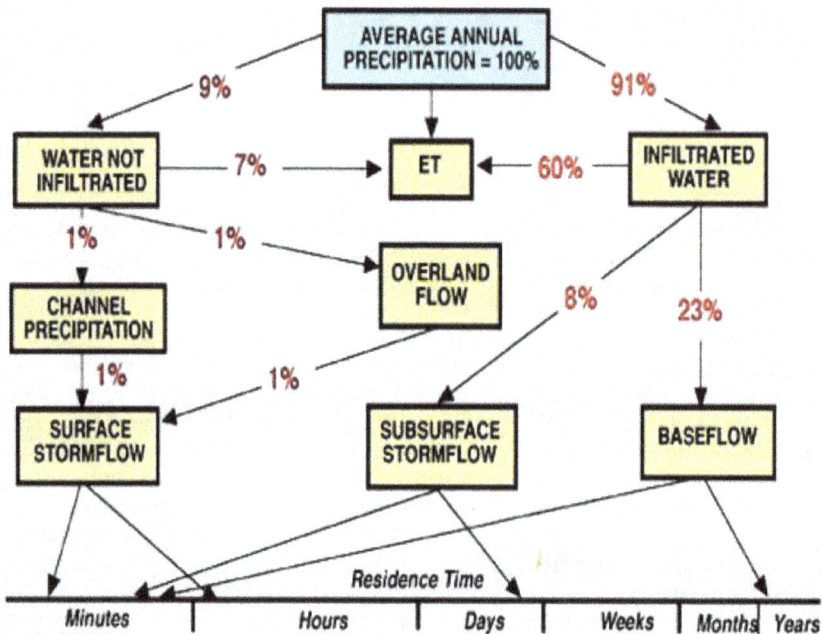

Figure B.2—Inputs, outputs, and fluxes of water in forested watersheds. (Adapted from Hewlett 1982, Principles of Hydrology, Copyright © University of Georgia Press, Athens, GA; Figure courtesy of the USDA Forest Service, National Advanced Fire and Resource Institute, Tucson, AZ).

movements of water can change somewhat in shrub and grassland ecosystems and are altered in watersheds disturbed by harvesting (fig. B.3), burning, insect defoliation, windthrow, land-use conversions, mining, agriculture, and so forth. Precipitation inputs consist of rain, snow, sleet, and so forth. Fluxes, or movement pathways within a watershed, consist of interception, stemflow, throughfall, infiltration, surface runoff, interflow, baseflow, and stormflow.

Interception

Interception is the hydrologic process by which vegetative canopies and accumulations of litter and other decomposed organic matter on the soil surface interrupt the fall of precipitation from the atmosphere to the soil surface. Interception plays a hydrologic role of protecting the soil surface from the energy of falling raindrops. Without this dissipation of energy, the mineral soil surface can become compacted or dislodged by raindrop splash, which then impacts the infiltration characteristics of the soil surface and the pathways of water to stream systems within a watershed.

Much of the precipitation that is intercepted returns to the atmosphere by evaporation and, therefore, becomes a loss of water from the soil surface. As a consequence, interception is a storage term that is subtracted from the gross precipitation input to a watershed in water budget studies. However, not all of the precipitation intercepted by a vegetative canopy or litter layer is returned to the atmosphere. Some of the water intercepted by a vegetative canopy drips off the foliage (*throughfall*) or flows down the stems of trees to the soil surface (*stemflow*). This is especially the case with the occurrence of large storms of long

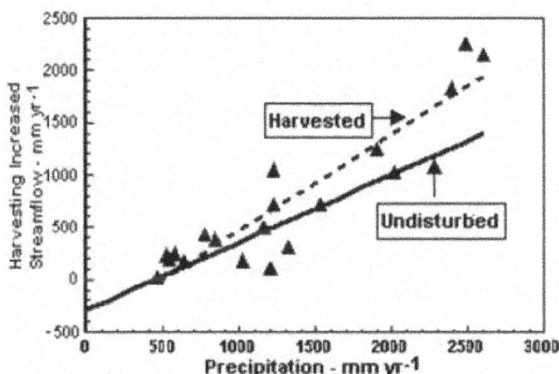

Figure B.3—Annual streamflow response to timber harvesting and precipitation. (From Neary, D.G. 2002. Chapter 5.3: Hydrologic values Page 200, Figure 5.3-6. In: Bioenergy from Sustainable Forestry, All Rights Reserved Copyright © Kluwer Academic Publishers).

duration (Brooks and others 2003). A portion of the water that is intercepted by litter layers also drains to the soil surface.

There is considerable variability in the magnitude of rainfall interception by vegetative canopies. Interception losses in temperate forests of North America range from 0.05 to 0.26 inch (13 to 66 mm) in individual storm events (Helvey 1971, Luce 1995). This amounts to less than 5 to more than 35 percent of the annual rainfall input to a watershed (Aldon 1960, Rothacher 1963, Helvey and Patric 1965, Fuhrer 1981, Roth and Chang 1981, Plamondon and others 1984). Interception losses in the sparsely stocked woodlands, shrublands, and grasslands of arid and semiarid regions are typically less than 10 percent of the annual rainfall (Skau 1964a, Tromble 1983, Haworth and McPherson 1991). Brooks and others (2003) discuss formulas used to calculate rainfall interception. Regardless of the region, however, interception of rainfall represents a transient form of water storage in the vegetative cover of a watershed.

Interception of snowfall is more difficult to quantify than the interception of rainfall, largely because neither the initial amount of snowfall nor the amount of water in the snow that accumulates on foliage of vegetative canopies can be measured adequately. In many situations, much of the intercepted snow is ultimately deposited in the *snowpack* accumulating on the ground through wind erosion and snowmelt, with subsequent dripping and freezing in the snowpack on the ground (Miller 1966, Hoover and Leaf 1967, Satterlund and Haupt 1970, Tennyson and others 1974).

Interception of either rain or snow by vegetative canopies is largely a function of:

- The form (rain or snow), intensity, and duration of the precipitation event.
- The wind velocity, water vapor gradient away from intercepting surfaces, and other storm characteristics.
- The type of vegetation (broad- or needle-leaved), number of vegetative layers in the canopy, and amount of surface leaf area.

Throughfall is precipitation that falls through a plant canopy and lands on bare soil or litter. Interception of throughfall precipitation by litter and other decomposed organic matter on the soil surface ranges from 5 to 35 percent of the gross precipitation input to a watershed (Brooks and others 2003). Storage of intercepted water in litter layers can represent a relatively large proportion of small 1 inch (25 mm) and less rainfall events. It can amount to 0.08 inch (2 mm) on trees, 0.04 inch (1 mm) on shrubs, and 0.12 inch (3 mm) on litter. Interception and the storage of water in litter on the soil surface are related to the depth,

USDA Forest Service Gen. Tech. Rep. RMRS-GTR-42-vol. 4. 2005

97

density, and relative stage of the development of the layers.

One obvious hydrologic consequence of fire destroying vegetative canopies, reducing litter accumulations, or both is its consequent effects on interception losses. It is one of the largest changes in hydrologic response to short-duration, high-intensity summer rain storms brought about by fire. Most of the vegetative canopy and litter is completely lost in severe wildfires, and as a result, comparatively little postfire interception of precipitation occurs (Bond and van Wilgen 1996, Pyne and others 1996, DeBano and others 1998). The effect of fire on interception in this case is a likely increase in the amount of net precipitation reaching the soil surface—that is, the amount of throughfall. When only small quantities of a vegetative cover or litter are consumed in a fire of low severity, the effect of fire on the interception process is less pronounced. Persistence of prefire levels of litter and other decomposed organic matter is important in protecting the soil surface in those situations where vegetation is destroyed by fire. Increased soil loss through erosive processes is often a consequence when large quantities of both protective layers (vegetation and litter) are lost to fire.

Infiltration

Precipitation that reaches the soil surface moves into the soil mantle, forms puddles of water on the soil surface, or flows over the soil surface. The process of water entering the soil is *infiltration*. The maximum rate at which water can enter the soil is the infiltration capacity. Water that infiltrates into the soil either moves slowly downward and laterally to a stream channel by *interflow* or downward still farther to a groundwater aquifer. When more water is supplied to a site than can infiltrate, the excess water flows off the surface, by the process of *overland flow* or *surface runoff*, to a stream channel.

The relative proportion of the net precipitation that infiltrates into the soil and moves to a stream channel or percolates through the soil to the groundwater aquifer largely determines the amount and timing of streamflow that ultimately occurs on a watershed. Infiltrometer measurements indicate that undisturbed forest soils have high infiltration capacities compared to other types of soil (Meeuwig 1971, Johnson 1978, Johnson and Beschta 1980, Sidle and Drlica 1981). This high rate of infiltration is a major factor contributing to the popularly held idea that forests have a moderating effect on streamflow regimes.

Forest soils are generally porous and open on the surface because of the accumulations of organic matter on the soil surface, the relatively high organic content of forest soils, and the large number of macropores that typically occur as a result of earthworm, insect, and other burrowing animal activities. Infiltration into forest soils in more humid temperate regions is generally higher than that observed in the soils of arid and semiarid regions because of their more permeable structure and the greater stability of aggregates (Hewlett and Troendle 1975). Infiltration capacities of the soils in arid and semiarid regions are often higher on sites dominated by tree species than on shrub or grass-dominated sites (table B.1). Regardless of region, however, variables affecting the infiltration component include:

- The soil texture, structure, porosity, and so forth.
- The accumulations of litter and other decomposed organic matter on the soil surface.
- The composition and structure of the vegetative cover.
- The land use and resultant vegetative changes, which influence infiltration capacities primarily by altering soil water storage.
- Precipitation rate versus infiltration rate.

The latter factor is an important one to consider in the interior Western United States and parts of the South. Typical summer severe thunderstorms often have high, short-duration rainfall bursts (for example, 10 to 15 minute downpours at the rate of 2 inches/hour

Table B.1—Infiltration rates under various surface conditions. (Adapted from Hewlett 1982, copyright University of Georgia Press, Athens, GA).

Surface conditions	Infiltration		
	Rate		Description
	in/hr	*mm/hr*	
1. Intact forest floor	>6.3	>160	Very rapid
2. Vegetation	0.2 – 2.0	5 – 50	Slow to moderate
3. Bare soil	0.0 – 1.0	0 – 25	Very to moderately slow
4. Water repellent soil	0.0 – 0.04	0 – 10	Very slow to none

(50 mm/hr). These rainfalls are often confined to 250 to 500 acres (1 to 2 km^2) (Neary 2002).

Infiltration properties of soils are often altered when fire destroys vegetation and litter covers on a watershed (Pyne and others 1996, DeBano and others 1998, Brooks and others 2003). When the burning has been severe enough to exposes bare soil, infiltration can be reduced due to:

- A collapse of the soil structure and a subsequent increase in bulk density of the soil because of the removal of organic matter, which serves as a binding material.
- The consequent reduction is soil porosity.
- Impacts of raindrops on the soil surface causing compaction and a further loss of soil porosity.
- The kinetic forces of raindrop impact displacing surface soil particles and causing a sealing of surface pores.
- Ash and charcoal residues clogging soil pores.

Variables that affect both the infiltration capacity and cumulative infiltration into the soil can be affected by fire to varying degrees, often resulting in decreased infiltration (Zwolinski 1971, Biswell 1973, MaNabb and others 1989), increased overland flow (DeBano and others 1998, Brooks and others 2003), and, ultimately, increased streamflow discharge. Rates of infiltration are a function of a number of factors such as soil texture, vegetation and litter cover, and soil porosity. Infiltration rates with litter present can often exceed rainfall intensities greater than 6.3 inches/hour (greater than 160 mm/hr). Infiltration decreases when soil particle sizes and cover are reduced. Rainfall infiltration rates less than 1 inch/hour (less than 25 mm/hr) in bare sands become less than 0.2 inch/hour (less than 5 mm/hr) in clay-textured soils. The variables that influence infiltration include:

- The vegetative cover type.
- The portion of soil surface covered by litter accumulations and other decomposed organic matter.
- The weight (depth) of the litter and other organic material.
- The soil texture, structure, porosity, bulk density, and so forth.

Another soil property that influences the infiltration process is the *wettability* of the soil (see chapter 2). Soils in some vegetative types and regions can develop a characteristic of water repellency following the occurrence of a fire, which (in turn) can reduce infiltration capacities. Although the presence of these *hydrophobic soils* is frequent in these situations, the causes of this condition are not always well known (DeBano 1981 2000a,b, DeBano and others 1998). Most hydrophobic soils repel water as a result of organic, long-chained hydrocarbon substances coating the soil particles. As a consequence, water "beads up" on the soil surface and will not readily penetrate the surface (see fig. 2.8 and 2.9), resulting in a change in infiltration. With this condition, accelerated overland flow and increased surface erosion can occur, especially on steeper slopes.

Hydrophobic soils are typically found in the chaparral shrublands (comprising *Quercus turbinella* and other sclerophyllous species) of southern California. However, hydrophobic soils can be found after fires in other vegetation types. Fires that occur frequently in the chaparral region intensify the hydrophobic condition and, apparently, volatilize organic substances that accumulate in the litter layer in the interval between fires (DeBano 1981, Dunn and others 1988, DeBano and others 1998). The resulting water repellent layer is then driven deeper into the soil profile. This layering arrangement allows rainfall to infiltrate to only a limited depth before the wetting front reaches the water repellent layer, often causing concurrent increases in the amount of overland waterflow. The soil layer above this water repellent layer is also easily eroded and, therefore, affects sedimentation and debris flow production after fire.

A fire can also influence the microclimate of a site by causing greater air and soil temperature extremes (Fowler and Helvey 1978, Pyne and others 1996, DeBano and others 1998, Brooks and others 2003). In cooler temperate regions, these temperature changes can increase potentials for concrete-type soil frost to form, which can then cause a reduction in infiltration capacity that, therefore, is indirectly related to burning (Bullard 1954).

Evapotranspiration

Evaporation from soils, plant surfaces, and water bodies, and water losses from transpiring plants, are collectively the *evapotranspiration* component of the hydrologic cycle. Part of the evapotranspiration component is when vegetation canopies intercept precipitation that is evaporating from plant foliage. Evapotranspiration is often a high percentage of the precipitation in a water budget, approaching 100 percent on some forested watersheds.

The evapotranspiration component of the hydrologic cycle interests hydrologists and watershed managers because its magnitude largely determines the proportion of the total precipitation input to a watershed that is likely to eventually become streamflow or result in groundwater recharge. Evapotranspiration also represents the component of the hydrologic cycle that is influenced the most by vegetative changes on a watershed that are brought about by planned and

USDA Forest Service Gen. Tech. Rep. RMRS-GTR-42-vol. 4. 2005

99

unplanned land management activities. The evapotranspiration process largely controls the hydrologic response of a watershed to rainfall and snowmelt events; nevertheless, hydrologists and watershed managers still understand little about the process itself or the feedback mechanisms that control the evapotranspiration process in natural environments (Morton 1990, Ffolliott and Brooks 1996, Brooks and others 2003). It is known, however, that the composition, density, and structure of vegetation influence transpiration losses through time. Differences in the transpiration rates among plant communities and individual plant species on a watershed are attributed largely to:

- Differences in rooting characteristics
- Stomatal response
- Albedo of leaf surfaces
- The length of the growing season

Estimated evapotranspiration values in the temperate forests of North America range from 40 to over 85 percent of the annual precipitation. However, these estimates of evapotranspiration vary greatly with different compositions and structures of forest overstories (Croft and Monninger 1953, Brown and Thompson 1965, Johnson 1970). On a watershed-scale, it has been estimated that 80 to 95 percent of the annual precipitation is evaporated from land surfaces or transpired by plants on the forested watersheds in the Southwestern United States, leaving only 5 to 20 percent available for runoff (Ffolliott and Thorud 1977). By contrast, runoff approaches 50 percent of the rainfall, and there are larger snowmelt inputs on the higher mountain watersheds of the Western United States; nevertheless, the evapotranspiration component is still large and potentially subject to modification.

Evapotranspiration represents the largest loss of water in terms of the components of the hydrologic cycle. This is a problem in arid and semiarid regions because of low precipitation (Pillsbury and others 1963, Skau 1964b, Branson and others 1976). In tropical areas, evapotranspiration is high but so is rainfall. In some situations, soil water storage following the end of the growing season in these harsh environments is nil, regardless of the type of vegetative cover, indicating that large quantities of precipitation are lost through the evapotranspiration process.

Watershed management studies throughout the world have demonstrated that streamflow can increase following vegetative changes that reduce evapotranspiration losses (Bosch and Hewlett 1982, Troendle and King 1985, Hornbeck and others 1993, Whitehead and Robinson 1993). That is, following a vegetative change, less precipitation is converted into vapor through the evapotranspiration process, and as a

consequence, more water is available for streamflow. Vegetation-modifying or vegetation-replacing fire, therefore, can change evapotranspiration (Bond and van Wilgen 1996, Pyne and others 1996, DeBano and others 1998). Fire that modifies the composition and structure of the vegetation by removing foliar volume will result in less evapotranspiration losses from a watershed. Fire that causes a replacement of deep-rooted, high profile trees or shrubs by shallow-rooted, low profile grasses and forbs is also likely to reduce evapotranspiration losses. In either instance, less evapotranspiration loss following a fire often translates into increased streamflow.

Soil Water Storage

The maximum amount of water that a soil body retains against the force of gravity is the *field capacity* of the soil. When water is added to a soil that is already charged to field capacity, the excess water either flows overland to a stream channel or drains from the soil. Soil is normally charged to, or near to, field capacity in periods of high precipitation events and at the start of the plant growing season. However, much of the water that is stored in the soil is consumed by plants in periods of sparse precipitation, and by the evapotranspiration process as the growing season progresses. The soil water deficit occurring at the end of the growing season is satisfied when high precipitation amounts occur once again.

The amount of stored soil water that is lost to evapotranspiration is largely a function of the vegetative type occupying the watershed site. Trees and shrubs have roots that can penetrate deep into the soil and, as a consequence, are able to extract water throughout much of the soil body. On the other hand, grasses, grasslike plants, and forbs have relatively shallow root systems and are only able to use water in the upper foot or so of the soil mantle. Water that infiltrates into the soil surface is stored in the upper layers of the soil profile, percolates through the soil body, or both. Vegetative change has a lesser effect on subsoil properties that influence soil water storage than on those properties impacting on infiltration. It follows, therefore, that effects of vegetative change on the subsurface soil properties that influence soil water storage are not likely to be controlling factors in the hydrologic cycle of a watershed (Brooks and others 2003). However, a vegetative change that affects both the evapotranspiration and infiltration processes can influence soil water storage.

The effects of fire on soil water storage result mostly from the loss of vegetation by the burn, which lowers the evapotranspiration losses (DeBano and others 1998, Brooks and others 2003). Lower evapotranspiration losses (in turn) leave more water in the soil at the

100

USDA Forest Service Gen. Tech. Rep. RMRS-GTR-42-vol. 4. 2005

end of the growing season than would be present if the vegetation had not been burned (Tiedemann and others 1979, Wells and others 1979, DeBano and others 1998). Overland flows of water and, ultimately, streamflow regimes become more responsive to subsequent precipitation events as a consequence of this increased soil water storage. It is often likely that soil water deficits on the burned sites at the end of the growing season will return to prefire levels in time if the vegetative cover also recovers to conditions that characterized the watershed before burning.

Effects of fire on the water storage of rangeland soils are more variable than those in forested soils. Some investigators have reported that the soil water storage is higher on burned sites, others have found lower soil water storage on these sites, and still others observed no change (Wells and others 1979). Varying severities of fire are often cited as the reason for these differences. Increases in soil water content and pore pressures can be similar to those observed after forest harvesting (Sidle 1985).

Snow Accumulation and Melt Patterns

Much of this introduction to the water resource section of this publication has focused on rainfall as the form of precipitation. However, snowfall is also an important form of precipitation input to watershed lands in many regions. The snowpack melts that accumulate at higher latitudes and higher elevations are often a primary source of water to downstream users. It is not surprising, therefore, that hydrologists and watershed managers can be interested in snow accumulation and melt patterns and the effects of vegetative change on these patterns.

The total snow on a watershed at any point in time throughout the winter is largely a function of the total snowfall (Baker 1990, Satterlund and Adams 1992, Brooks and others 2003). However, greater snow accumulations tend to be found at higher elevations on a watershed than at lower elevations because of the generally greater snowfall and lower temperatures at the higher elevations (Anderson and others 1976, Harr 1976; Ffolliott and others 1989, Ffolliott and Baker 2000). More snow accumulates and is retained longer into the winter season on "cooler" than on "warmer" sites because of lower solar radiation levels impinging on the former sites. More snow also accumulates in sparsely stocked forests than in more dense forests, and additional snowfall is deposited in small openings in a forest canopy because of increased turbulence (Troendle 1983, Ffolliott and others 1989, Brooks and others 2003) or through the reduction in the amount of snow intercepted by the forest canopy (Troendle and Meiman 1984, Satterlund and Adams 1992, Brooks and others 2003). Once snowmelt is

initiated in the spring, the rate of melt becomes more rapid in the forest openings than under dense vegetative canopies because of greater levels of solar radiation impinging on the open site. The main effect is due to the reduction in snow pack interception in crowns and subsequent sublimation rather than any "redistribution" effect (Troendle and King 1985).

Fire affects snow accumulation and melt patterns when the burn creates openings in formerly dense vegetative canopies. Not only is the amount of snowfall interception decreased after a fire has destroyed the canopies, but additional snowfall is frequently deposited into the created openings due to the disruptions in wind turbulence over the canopy surface (Satterlund and Haupt 1970). The characteristics of these openings are dependent on the severity of the fire. A wildfire of high severity can destroy much of the forest cover on a watershed, creating many large openings in the process. However, only a few relatively small openings are likely to be created by a low severity fire that consumes only the surface fuels.

Charred trees and other black bodies protruding from a snowpack after a fire can change the reflectivity of the ground surface, inducing more surface heating and earlier and more rapid rates of snowmelt. Earlier snowmelt in the spring can also be attributed to a reduced soil water deficit on a burned site. In other words, not as much water is needed to satisfy a deficit, and as a consequence, overland flow originating from snowmelt starts earlier in the season. Such changes in the timing of snowmelt can also alter the timing, magnitude, and duration of the streamflow regimes.

Overland Flow

That portion of the net precipitation that flows off the soil surface is *overland flow*, also called *surface runoff*, which is a major contributor to many streamflow systems and the main contributor to most intermittent streams. This hydrologic process occurs when the rainfall intensity or the rate of snowmelt exceeds the infiltration capacity of a site. Overland flow is the pathway that moves net precipitation most directly to a stream channel and, in doing so, quickly produces streamflow (the following section on Streamflow Regimes discusses the pathways and processes by which excess precipitation becomes streamflow).

The relative contribution of overland flow to the streamflow from a watershed is variable, depending largely on how impervious the soil surface is. Overland flow generally occurs on sites that are impervious, locally saturated, or where the infiltration capacity has been exceeded by the net precipitation or rate of snowmelt (Satterlund and Adams 1992, Brooks and others 2003). Some overland flow can be detained enroute to a stream channel by the roughness of the

USDA Forest Service Gen. Tech. Rep. RMRS-GTR-42-vol. 4. 2005

101

soil surface and, therefore, slowed its movement to the channel. Influences that vegetation and the soil exert on interception, evapotranspiration, infiltration rates, and the soil moisture content ultimately affect the magnitude of overland flow.

Overland flow is a comparatively large component to streamflow hydrographs for highly impervious areas such as urban landscapes, is typically insignificant for forested watersheds with well-drained and deep soils, but is a problem where soils are shallow, rocky, or fine textured (such as high clay or silt content).

An increase in overland flow often results when a fire decreases interception and infiltration rates. This is a major factor in the observed increases in streamflow and flood peakflows, particularly after high severity wildfires. A high severity wildfire can consume all or nearly all of the protective vegetative cover and litter layer over extensive watershed areas, producing a significant effect on the magnitude of overland flow and, as discussed below, on streamflow from a watershed (Tiedemann and others 1979, Baker 1990, DeBano and others 1998). Formation of hydrophobic soils following fire also reduces infiltration, increases overland flow, and speeds delivery of the overland flow to stream channels (Hibbert and others 1974, Rice 1974, Scott and Van Wyk 1990). Persistence of the increased overland flow following fire relates to the rate at which burned sites become revegetated. Prescribed burning often has its greatest hydrologic influence on the infiltration processes and, as a consequence, on the potential for increased overland flow.

Baseflows and Springs

Baseflow is the streamflow between storm events and originates from infiltrated rainfall or groundwater flow. It can increase when the watershed condition is maintained and deep-rooted vegetation on the watershed is harvested or otherwise cut, removed in converting from one vegetation type to another, or killed by fire, insects, or disease. However, baseflow is likely to decrease when the watershed condition deteriorates as a consequence of the disturbance, and more excess precipitation leaves a watershed as overland flow. In extreme situations, perennial streams that are sustained by baseflow become ephemeral.

Baseflows are important in maintaining perennial flow through the year. They are critical for aquatic species habitat and survival. Baseflows can increase if watershed condition remains good (infiltration remains adequate) and deep-rooted vegetation is cut (harvesting), removed (species conversion), or killed by fire, insects, disease, herbicides, and so forth. If watershed condition deteriorates and more precipitation leaves as surface runoff, baseflows will decrease. In extreme conditions, perennial streams become

ephemeral. The effect on biota in aquatic ecosystems then becomes devastating. Even in subtropical or tropical areas, deterioration of watershed condition can result in the loss of perennial baseflow and ultimately in desertification.

Crouse (1961) reported increased baseflows from burned watersheds on the San Dimas Experimental Forest in southern California. While these watersheds had been cleared of their chaparral shrubs and associated vegetation by burning, seeded to grass, and maintained in a grass cover to induce higher streamflow discharges, the author and others (Dunn and others 1988, DeBano and others 1998) felt that the wildfire had made a significant contribution to the increased baseflow.

Berndt (1971) observed immediate increases in baseflow following a wildfire on a 1,410 acre (564 ha) watershed in eastern Washington. While the causative hydrologic mechanisms involved were unknown, the removal of riparian (streambank) vegetation by the fire also eliminated diurnal fluctuations of flow. The increased baseflow persisted above prefire levels for 3 years after the fire.

Pathways and Processes _____

Before considering how fire affects streamflow regimes, however, it is useful to review the pathways and processes of waterflow from a watershed's hillslope to a stream channel. *Excess water* represents that portion of total precipitation that flows off the land surface plus that which drains from the soil and, therefore, is neither consumed by evapotranspiration nor leaked into deep groundwater aquifers (table B.2). Various pathways by which excess water eventually becomes streamflow (Brooks and others 2003) include:

- Interception of precipitation that falls directly into a stream channel, a streamflow pathway referred to as *channel interception*.
- Overland flow (see previous discussion).
- *Subsurface flow* that represents the part of precipitation that infiltrates into the soil and arrives at a stream channel in a short enough period to be considered part of the stormflow hydrograph. The stormflow components of a hygrograph are the sum of channel interception, overland flow, and subsurface flow.
- A perennial stream that is fed by baseflow that sustains streamflow between precipitation or snowmelt events.

It is almost impossible to separate pathways of waterflow in most investigations of streamflow responses to the effects of fire or other watershed disturbances (Dunne and Leopold 1978, Satterlund and Adams 1992, DeBano and others 1998, Brooks and

102

USDA Forest Service Gen. Tech. Rep. RMRS-GTR-42-vol. 4. 2005

Table B.2—Annual precipitation inputs and resultant annual streamflow totals for different vegetative types. (Adapted from Dortignac 1956).

Vegetation zone	Precipitation		Runoff		Streamflow as a percent of precipitation	Altitude	
	in	*mm*	*in*	*mm*	*Percent*	*ft*	*m*
Spruce fir-aspen	30.39	772	9.49	241	29.0	8,990	2,740
Mountain grassland	22.99	584	5.98	152	26.0	8,000	2,440
Ponderosa pine	22.99	584	3.82	97	17.0	7,510	2,290
Sagebrush	15.20	386	0.79	20	5.3	4,990	1,520
Pinyon-juniper	14.29	363	0.39	10	2.8	4,990	1,520
Semiarid grassland	12.20	310	0.12	3	0.8	4,990	1,520
Greasewood/saltbush	10.39	264	0.16	4	1.4	4,495	1,370
Creosote bush	8.39	213	0.04	1	0.6	4,495	1,370

others 2003). Responses of streamflow to fire are more generally evaluated when possible and where appropriate by separating the stormflow component of the hydrograph from the baseflow (when a baseflow is present) and then studying the two flow components separately (fig. B.4).

The Variable Source Area Concept (Hewlett and Hibbert 1967) describes how the perennial channel system expands during precipitation as areas at the head and adjacent to perennial channels become saturated during the event. Figure B.5 depicts the concept of how a stream channel system expands during precipitation. It is important to understand this concept in order to understand the hydrologic response of a watershed and its component areas, particularly after disturbances such as fire. Important hydrologic and geomorphic processes (sediment transport, channel scour and fill, streambank erosion, fish habitat damage, riparian vegetation damage, nutrient transport, woody debris transport and deposition, and so forth.) occur during stormflows when water volumes and velocities are at their highest. Low severity fires usually do not affect the flow pathways shown in figure B.2, but severe fires shift more of the movement of water to the "water not infiltrated" side of flow diagram. The magnitude and duration of stormflow is a function of the intensity and duration of precipitation as well as a factor called watershed condition.

Watershed Condition

The timing, magnitude, and duration of a stormflow response to fire are largely a function of the hydrologic condition of the watershed. *Watershed condition* is a term that describes the ability of a watershed system to receive and process precipitation without ecosystem or hydrologic degradation (Brooks and others 2003).

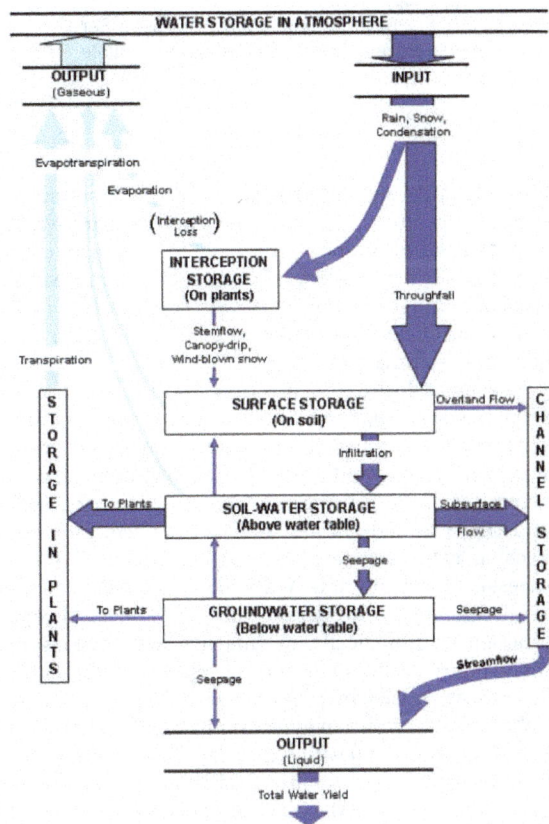

Figure B.4—Relationship between the pathways of flow from a watershed and the resultant streamflow hydrograph. (Adapted from Anderson and others 1976).

USDA Forest Service Gen. Tech. Rep. RMRS-GTR-42-vol. 4. 2005

103

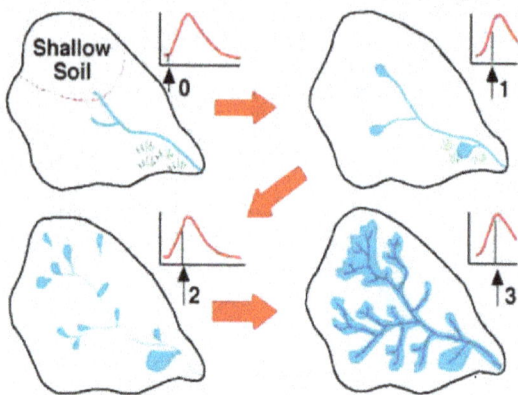

Figure B.5—Variable area source concept of streamflow network expansion, from time = 0 to time = 3, during storm events. (Adapted from Hewlett 1982, Principles of Hydrology, Copyright © University of Georgia Press, Athens, GA; Figure courtesy of the USDA Forest Service, National Advanced Fire and Resource Institute, Tucson, AZ).

Rainfall infiltrates into the soil, and the baseflow of perennial streams is sustained between storms when a watershed is in good condition. In this situation, rainfall does not contribute significantly to increased erosion because most of the excess precipitation does not flow over the soil surface where it can detach and transport sediments.

A severe wildfire can alter the condition of a watershed, however, reducing it to a generally poorer state. With poor watershed condition, the amount of infiltrated rainfall is reduced significantly, more excess precipitation then flows over the surface of the soil, and there is little or no baseflow between storms. Erosion rates are relatively higher because of the excessive overland flow. Watersheds in arid or semi-arid regions with rocky and thin soils almost always function hydrologically in this manner because of their inherent climate, soils, and vegetative features. These watersheds are prone to damaging flash floods.

Watershed condition at a point in time is controlled largely by the composition and density of the vegetative cover, the accumulations of litter and other organic material, and the amount of exposed rocks and bare soils that characterizes the watershed. Because a wildfire of high severity can destroy the vegetation and litter layer on a watershed and alter the physical properties of the soil, the infiltration and percolation capacities of the soil are detrimentally impacted (see chapter 2). In turn, these cumulative fire effects can change the watershed condition from good to poor,

resulting in ever-increasing overland flow, erosion, and soil loss. An analogy to this situation is a loss of function in the human skin with increasing severity of burning. As watershed condition deteriorates, the inherent hydrologic processes become altered, increasing the likelihood of adverse hydrologic responses to fire. The change in infiltration rates going from "good" (intact forest floor) to "poor" (bare soil plus water repellency) is one example of this response.

Streamflow Discharge

The *streamflow discharge* for a specified time interval (year, month, season, and so forth) is reflected by stormflow, baseflow, or combinations of the two pathways. Stormflow results directly from a precipitation or snowmelt event. When vegetation and organic matter on the soil surface are destroyed by fire, interception and evapotranspiration are reduced, infiltration is decreased, and overland flow and subsurface flow can increase. In turn, increases in overland flow and subsurface flow often translate into:

- Increases in stormflow from the burned watershed.
- Increases in baseflow if the stream is perennial in its flow.
- Increases in streamflow discharge as a consequence of the increases in stormflow, baseflow, or both.

Onsite fire effects must be determined initially, and these effects can then be evaluated within the context of the entire watershed to determine the responses of streamflow discharge and, more generally, the other changes in a streamflow regime due to a fire. While determining the onsite effects of a fire can be relatively straightforward through appropriate measurements and evaluations, determining effects of fire on a watershed-scale is more difficult (Pyne and others 1996, DeBano and others 1998). With the latter, combining all processes and pathways of waterflow to the outlet of the watershed, and routing this flow to downstream points of interest or use, are necessary (Brooks and others 2003). Concerning the streamflow discharge at the outlet of a watershed, the effects of fire on the timing can be diminished and the magnitude can be lengthened as the postfire streamflow event moves downstream.

Hydrologists and watershed managers, when studying the effects of fire on the streamflow regime of a watershed, are also concerned with the likelihood of changes in timing of flow. Information on this topic is limited, but some researchers note that streamflow from burned watersheds often responds to rainfall inputs faster than watersheds supporting a protective vegetative cover, producing streamflow events where time-to-peak is earlier (Campbell and others 1977, DeBano and others 1998, Brooks and others 2003). Earlier time-to-peak, coupled with higher peakflow, can increase the frequency of flooding. Timing of snowmelt in the spring can also be advanced by fire in some instances. Early snowmelt can be initiated by lower snow reflectivity (albedo) caused by blackened trees and increased surface exposure where vegetative cover has been eliminated (Helvey 1973).

There is little doubt that wildfire often has an influence on streamflow discharge, especially a wildfire of high severity. The combined effects of a loss of vegetative cover, a decrease in the accumulations of litter and other decomposed organic matter on the soil surface, and the possible formation of water repellent soils are among the causative mechanisms for the increase in streamflow discharge (Tiedemann and others 1979, Baker 1990, Pyne and others 1996, DeBano and others 1998, Brooks and others 2003). While the increases in streamflow discharge are highly variable, they are generally greater in regions with higher precipitation as illustrated by studies in the United States and elsewhere (fig. B.3).

Water Quality _____

Increases in streamflow following a fire can result in little to substantial impacts on the physical, chemical, and biological quality of water in streams, rivers, and lakes. The magnitude of these effects is largely dependent on the size, intensity, and severity of the fire, the condition of the watershed when rainfall starts, and the intensity, duration, and total amount of rainfall. Postfire streamflow can transport solid and dissolved materials that adversely affect the quality of water for human, agricultural, or industrial purposes. The most obvious effects are produced by sediments. See chapter 2 for more discussion on these components of water quality, which in the following chapters on the water resource are relative to information on municipal water supply quality.

Water quality refers to the physical, chemical, and biological characteristics of water relative to a particular use. Important characteristics of interest to hydrologists and watershed managers include sediment, water temperature, and dissolved chemical constituents such as nitrogen, phosphorus, calcium, magnesium, and potassium. Bacteriological quality is also important if water is used for human consumption or recreation; this is the case with many waters that are both within, and that drain from, forested lands.

A *water quality standard* refers to the physical, chemical, or biological characteristics of water in relation to a specified use. Changes in water quality due to a watershed management practice or natural and human-caused disturbances can make the water

flowing from the watershed unsuitable for drinking. However, it might still be acceptable for other uses. In some instances, laws or regulations prevent water quality characteristics from becoming degraded to the point where a water quality standard is jeopardized (DeBano and others 1998, Landsberg and Tiedemann 2000, Brooks and others 2003). The main purpose of these laws and regulations is maintaining the quality of water for a possible and maybe unforeseen future use.

Hydrologists and watershed managers often confront the issue of whether forest or rangeland fires will create conditions in natural or impounded waters that are outside of the established water quality standards. Water quality standards and criteria established by the Environmental Protection Agency (EPA) (1999) are the "benchmarks" for water quality throughout the United States. Water quality criteria are the number or narrative benchmarks used to assess the quality of water. Standards include criteria, beneficial uses, and an antidegradation policy. The most adverse effects from wildfires on water quality standards come from physical effects of the sediment and ash that are deposited into streams.

Several chemical constituents that are regulated by water quality standards are likely to be impacted by burning. *Primary standards* in the EPA regulations cover nitrate-nitrogen, nitrite-nitrogen, and other substances that are not immediately associated with fire. *Secondary standards* apply to pH, sulfate, total dissolved solids, chloride, iron, turbidity, and several other constituents. Secondary standards are also set for color and odor. Chapter 6 examines the water quality standards that can be impacted, and in some cases exceeded, by fire.

Changes in the hydrologic cycle caused by fires can affect the rate of soil erosion, and the subsequent transport and deposition of eroded soil as sediment into streams, lakes, and reservoirs (DeBano and others 1998, Brooks and others 2003). Chapter 5 looks at fire-produced alterations in the hydrologic cycle that in turn affect soil erosion, sedimentation, and water quality.

Maintaining a vegetative cover or a cover of litter and other organic material on the soil surface is the best means of preventing excessive soil erosion rates. However, fire can remove these protective covers and accelerate soil erosion (Dunne and Leopold 1978, Satterlund and Adams 1992, Brooks and others 2003). Increased soil erosion is often the most visible effect of a fire other than the loss of vegetation by burning.

Aquatic Biology _____

Prior to the 1990s, little information existed on the effects of wildfire on fishes, other aquatic organisms such as macroinvertebrates, and their habitats.

USDA Forest Service Gen. Tech. Rep. RMRS-GTR-42-vol. 4. 2005

105

Severson and Rinne (1988) reported that most of the focus of postwildfire effects on riparian-stream ecosystems has traditionally been on hydrological and erosional responses. The Yellowstone Complex Fires in 1988 ushered in an extensive effort to examine both the direct and indirect effects of wildfire on aquatic ecosystems (Minshall and others 1989a, Minshall and Brock 1991). Most of the information available on fire effects on fishes and their habitats was generated in the 1990s and on a regional basis. By the late 1990s, state-of-knowledge papers on the topic of fire, aquatic ecosystem, and fishes were drafted by Rieman and Clayton (1997) and Gresswell (1999). These two papers suggest future research and management direction for both corroborating aquatics-fisheries and fire management and conservation of native, sensitive species. These papers are have become the base for 21st century fisheries and aquatic management relative to both wild and prescription fires. Chapter 7 addresses both the direct and indirect effects of wildland fires and associated suppression activities on aquatic ecosystems.

Daniel G. Neary
Peter F. Ffolliott
Johanna D. Landsberg

Chapter 5:
Fire and Streamflow Regimes

Introduction

Forested watersheds are some of the most important sources of water supply in the world. Maintenance of good hydrologic condition is crucial to protecting the quantity and quality of streamflow on these important lands (fig. 5.1). The effects of all types of forest disturbance on storm peak flood flows are highly variable and complex, producing some of the most profound hydrologic impacts that forest managers have to consider (Anderson and others 1976). Wildfire is the forest disturbance that has the greatest potential to change watershed condition (DeBano and others 1998).

Wildfires exert a tremendous influence on the hydrologic conditions of watersheds in many forest ecosystems in the world depending on a fire's severity, duration, and frequency. Fire in these forested areas is an important natural disturbance mechanism that plays a role of variable significance depending on climate, fire frequency, and geomorphic conditions. This is particularly true in regions where frequent fires, steep terrain, vegetation, and postfire seasonal

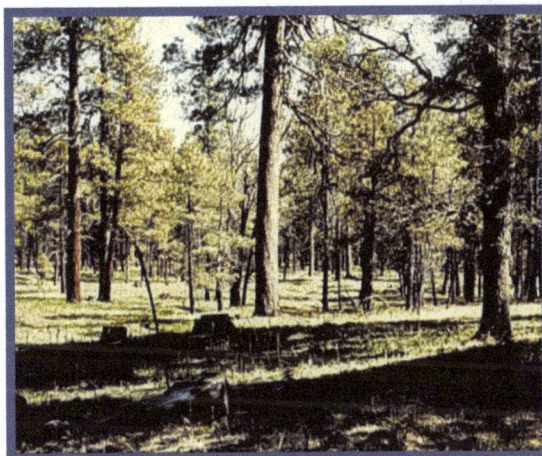

Figure 5.1—Ponderosa pine watershed, Coconino National Forest, with a good watershed condition. (Photo by Malchus Baker, Jr.).

USDA Forest Service Gen. Tech. Rep. RMRS-GTR-42-vol. 4. 2005

107

precipitation interact to produce dramatic impacts (Swanson 1981, DeBano and others 1998, Neary and others 1999).

Watershed condition, or the ability of a catchment system to receive and process precipitation without ecosystem degradation, is a good predictor of the potential impacts of fire on water and other resources (such as roads, recreation facilities, riparian vegetation, and so forth). The surface cover of a watershed consists of the organic forest floor, vegetation, bare soil, and rock. Disruption of the organic surface cover and alteration of the mineral soil by wildfire can produce changes in the hydrology of a watershed well beyond the range of historic variability (DeBano and others 1998). Low severity fires rarely produce adverse effects on watershed condition. High severity fires usually do (fig. 5.2). Most wildfires are a chaotic mix of severities, but in parts of the world, high severity is becoming a dominant feature of fires since about 1990 (Neary and others 1999, Robichaud and others 2000). Successful management of watersheds in a postwildfire environment requires an understanding of the changes in watershed condition and hydrologic responses induced by fire. Flood flows are the largest hydrologic response and most damaging to many resources (fig. 5.3, Neary 1995).

The objective of this chapter is to examine some of the effects of fire on watershed hydrology.

Soil Water Storage

The effects of fire on soil water storage can be illustrated by a wildfire that occurred on a 1,410 acre (564 ha) watershed in eastern Washington that effectively changed the magnitude of the autumnal soil

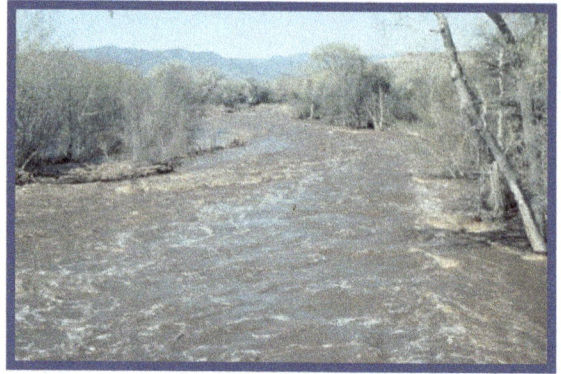

Figure 5.3—Postwildfire flood flows are the largest hydrologic response, and most damaging to many resources after the fire itself. These floods can be 100-fold greater than prefire flood flows. (Photo by John Rinne).

water deficit on the watershed (Klock and Helvey 1976). The mixed conifer forest vegetation on this watershed apparently depleted all of the available soil water in the upper 48 inches (120 cm) of the soil profile immediately before the August 1970 fire (fig. 5.4). The difference between the soil water deficits from 1970 to 1971 was about 4.6 inches (116 mm), which (researchers concluded) contributed a significant part to the increased streamflow discharge reported by Helvey and others (1976). The transpiration draft of large conifer trees had been removed by 1971, and the

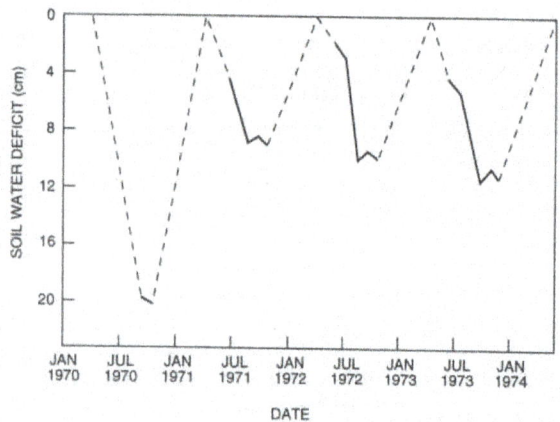

Figure 5.4—Trends in the autumnal soil water deficit in the upper 48 inches (120 cm) of the soil profile for 3 years following the August 1970 wildfire Entiat Experimental Forest, Washington. (Adapted from Klock and Helvey 1976. Copyright © 1976 Tall Timbers Research Station).

Figure 5.2—Flare-up on the Rodeo-Chediski Fire, 2002, Apache-Sitgreaves National Forest. (Photo by USDA Forest Service).

108

USDA Forest Service Gen. Tech. Rep. RMRS-GTR-42-vol. 4. 2005

observed soil water deficit was an apparent result of surface evaporation and transpiration by the newly established postfire vegetation. The increased autumnal soil water deficit in 1972 and 1974 was caused by the greater evapotranspiration demand by increased vegetative regrowth (Tiedemann and Klock 1976). The trends for postfire years 1971 to 1974 suggest that the minimum soil contents might reach prefire levels in about 5 years after the wildfire.

This trend does not necessarily hold true in all instances, however. For example, a reduction in water storage was observed in the upper 12 inches (30 cm) of the soil on a watershed in northern Arizona where the ponderosa pine (*Pinus ponderosa*) forest had been severely burned, in comparison to the soil storage of an adjacent unburned watershed (Campbell and others 1977). Greater overland flow from the burned watershed was the factor underlying this difference. Water repellency of the soil and the increased drying of the more exposed soil surface contributed to this increased flow.

Effects of fire on the water storage of rangeland soils are more variable than those in forested soils. Some investigators have reported that the soil water storage is higher on burned sites, others have found lower soil water storage on these sites, and still others observed no change (Wells and others 1979). Varying severities of fire are often cited as the reason for these differences.

Baseflows and Springs

Wildfires in 1996 and 2000 resulted in a number of anecdotal reports of springs beginning to flow after years of being dry. This sort of response is common in the Southwest and regions such as central Texas when an area is cleared or burned (Thurow, personal communication). Often trees such as juniper or live oak have increased in density and size along with the onset of effective fire control in the early 1900s. As trees begin to dominate ecosystems such as fire-climax grassland savannas, the trees alter the water balance because they have substantially greater interception loss and transpiration capacity. The local soils and geology determine whether water yield occurs as spring flow or groundwater recharge. The seasonal patterns of the amount and timing precipitation and potential evapotranspiration determine whether there is any excess water to contribute to water yield. For example, on arid sites all precipitation would be lost to evapotranspiration and thus essentially nothing would percolate fast enough beyond the root system, or evaporate from the soil surface, to recharge springs or aquifers. Some shrub sites may yield substantial amounts of water, others may yield nothing (Wu and others 2001).

On the Three Bar chaparral watersheds of the Tonto National Forest in Arizona, watersheds that had no flow or were intermittent responded to a wildfire (Hibbert and others 1974). Watershed B yielded no flow prior to the fire. It flowed continuously for 18 months, then was intermittent until treated with herbicides when it returned to perennial flow for 10 plus years. Watershed D was dry 67 percent of the time prior to the wildfire. Following the fire, it then resumed continuous flow until brush regrew, when it returned to intermittent flow. On the Whitespar Watersheds, Watershed B resumed perennial flow after 38 acres (15 ha) adjacent to the channel were treated with herbicide. Similar responses have been found in California chaparral (Crouse 1961). Treatment of the Natural Drainages, Sierra Ancha, with herbicides did not produce perennial flow but did increase the duration of intermittent flows. The soils at this site are more shallow than those at Three Bar or Whitespar. The key here is control of vegetation in deep-soil, arid systems where transpiration from deep-rooted plants can consume water that would otherwise become perennial streamflow. The response is quickly terminated as deep-rooted, brush-chaparral trees regrow. Perennial flow can only be maintained by converting to (or back to) shallow-rooted herbaceous species. Increases in any single year are affected by rainfall. Often, 80 percent of the increased water yield over a 10-year period occurs in just a few wetter-than-average years, not every year.

Streamflow Regimes

Annual streamflow totals (annual water yields) generally increase as precipitation inputs to a watershed increase. Streamflows originating on forest watersheds, therefore, are generally greater than those originating on grassland watersheds, and streamflows from grasslands are greater than those originating on desert watersheds. Furthermore, annual streamflow totals frequently increase when mature forests are harvested or otherwise cut, attacked by insects, or burned (Bosch and Hewlett 1982, Troendle and King 1985, Hornbeck and others 1993, Whitehead and Robinson 1993). The observed increases in streamflow following these disturbances often diminish with decreasing precipitation inputs to a watershed (see fig. B.3).

Effects of Wildfires

Annual streamflow discharge from a 1,410 acre (564 ha) watershed in the Cascade Range of eastern Washington, on which a wildfire killed nearly 100 percent of the mixed conifer forest vegetation, increased dramatically relative to a prefire streamflow

USDA Forest Service Gen. Tech. Rep. RMRS-GTR-42-vol. 4. 2005

109

relationship between the watershed that was burned and an unburned control (fig. 5.5). Differences between the measured and predicted streamflow discharge varied from nearly 4.3 inches (107 mm) in a dry year (1977) to about 19.1 inches (477 mm) in a wet year (1972). Soil water storage remained high for the period of record largely because of abnormally high precipitation (rain and snow) inputs (Helvey 1980). As a consequence, the burned and control watersheds became more sensitive to precipitation.

Campbell and others (1977) observed a 3.5 times increase of 0.8 inch (20 mm) in average annual stormflow discharge from a small 20.2 acre (8.1 ha) severely burned watershed following the occurrence of a wildfire in a Southwestern ponderosa pine forest. Average annual stormflow discharge from a smaller 10 acre (4 ha) moderately burned watershed increased 2.3 times to almost 0.6 inch (15 mm) in relation to an unburned (control) watershed. Average runoff efficiency (i.e., a term that refers to the percentage of runoff to precipitation) increased from 0.8 percent on the unburned watershed to 3.6 and 2.8 percent on the severely burned and moderately burned watersheds, respectively. In comparison to the moderately burned watershed, the average runoff efficiency on the severely burned watershed was 357 percent greater when the precipitation input was rain and 51 percent less in snowmelt periods. The researchers speculated that the observed differences during rainfall events

were largely due to the lower tree density, a greater reduction in litter cover, and a more extensive formations hydrophobic soil, resulting in lower evapotranspiration losses and more stormflow on the severely burned watershed than on the moderately burned watershed. In the spring snowmelt period, the lower tree density of the severely burned watershed allowed more of the snowpack to be lost to evaporation. As a result, less stormflow occurred than on the more shaded, moderately burned watershed.

In the fire-prone interior chaparral shrublands of the Southwestern United States, annual streamflow discharge from their watersheds can increase by varying magnitudes, at least temporarily, as a result of wildfires of high severity (Davis 1984, Hibbert 1984, Baker and others 1998). The combined effects of loss of vegetative cover, decreased litter accumulations, and formation of water repellent soils following the burning are the presumed reasons for these streamflow increases (Rundel 1977, Baker 1999).

In the first year after a 365 acre (146 ha) watershed in southern France, near the Mediterranean Sea, was burned over by a wildfire, streamflow discharge increased 30 percent to nearly 2.4 inches (60 mm) (Lavabre and others 1993). The prefire vegetation on the watershed was primarily a mixture of maquis, cork oak, and chestnut trees. The researchers attributed the increase in annual streamflow discharge to the reduction in evapotranspiration due to the destruction of this vegetation by the fire.

Average annual streamflow discharge increased by about 10 percent to 4.8 inches (120 mm) on a 612 acre (245 ha) watershed in the Cape Region of South Africa following a wildfire that consumed most of the indigenous fynbos (sclerophyllous) shrubs (Scott 1993). This increase was related mostly to the reductions in interception and evapotranspiration losses.

Effects of Prescribed Burning

Streamflow responses to prescribed fire (fig. 5.6) are smaller in magnitude in contrast to the responses to wildfire. It is generally not the purpose of prescribed burning to completely consume extensive areas of litter and other decomposed organic matter on the soil surface (Ffolliott and others 1996, DeBano and others 1998) and, therefore, the drastic alterations in streamflow discharges that are common after severe wildfires do not normally occur. To illustrate this point, the average annual streamflow discharge was not changed relative to prefire levels in 6 years following a prescribed fire on 43 percent of a 1,178 acre (471 ha) watershed in northern Arizona supporting a ponderosa pine forest (Gottfried and DeBano 1990). The

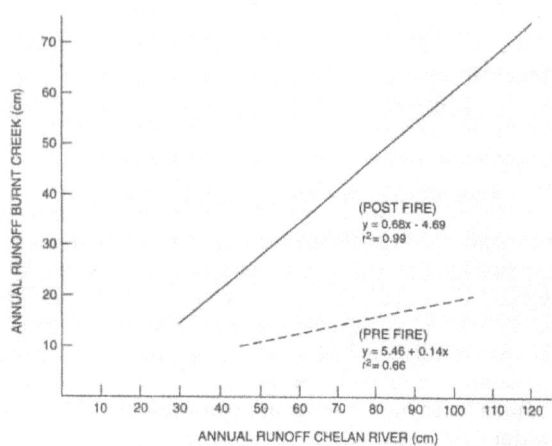

Figure 5.5—Annual streamflow from a burned watershed before and after the fire in relation to annual streamflow from a control. (Adapted from Helvey 1980. Copyright © 1980 American Water Resources Association.)

Figure 5.6—Grass prescribed fire in interior Alaska. (Photo by Karen Wattenmaker).

prescribed burning plan specifying a 70 percent reduction in fine fuels and a 40 percent reduction in heavy fuels on the burned area was satisfactorily met, with minimal damage to the residual stand of trees.

A burn that was prescribed to reduce the accumulated fuel loads on a 450 acre (180 ha) watershed in the Cape Region of South Africa resulted in a 15 percent increase to 3.2 inches (80 mm) in average annual streamflow discharge (Scott 1993). Most of the fynbos shrubs that vegetated the watershed were not damaged by the prescribed fire. The effectiveness of the prescribed fire was less than anticipated because of the unseasonably high rainfall amounts at the time of burning.

A prescribed fire in a Texas grassland community (table 5.4) resulted in a large increase (1,150 percent) in streamflow discharge in comparison to an unburned watershed in the first year after burning (Wright and others 1982). The increased postfire streamflow discharge was short lived, however, with streamflows returning to prefire levels shortly after the burning.

Burning of logging residues (slash) in timber harvesting operations, burning of competing vegetation to prepare a site for planting, and burning of forests and woodlands in the process of clearing land for agricultural production are common practices in many parts of the United States and the world. Depending on their intensity and extent, the burnings prescribed for these purposes might cause changes in streamflow discharge from the watersheds on which these treatments are conducted. But, in analyzing the responses of streamflow discharge to prescribed fire, it is difficult to isolate effects of these burning treatments from the accompanying hydrological impacts of timber harvesting operations, site preparation, and clearing of forest vegetation.

Results of other studies on the changes in streamflow discharge following either a wildfire or prescribed burning are presented in tables 5.1 through 5.5. These results are summarized on the basis of Bailey's (1995) ecoregion classifications for comparisons purposes.

Peakflows

Peakflows are a special subset of streamflow regimes that deserve considerable attention. The effects of forest disturbance on storm peakflows are highly variable and complex. They can produce some of the most profound impacts that forest managers have to

Table 5.1—Increased water yield from prescribed fire (Rx) burned watersheds, Eastern United States ecoregions.

State	Treatment	Area		Precipitation		1st year runoff		Increase	Recovery
		acre	ha	inch	mm	inch	mm	%	years
M221 CENTRAL APPALACHIAN BROADLEAF-CONIFER FOREST PROVINCE[1]									
NC	Hardwoods			67.91	1,725				
	Control	40	16			29.09	739	—	—
	Cut, Rx burn	40	16			35.00	889	20	5
	Swank and Miner 1968								
231 SOUTHEASTERN MIXED FOREST PROVINCE									
SC	Loblolly pine			54.72	1,390				
	Control	5	2			4.88	124	—	—
	Under burn	5	2			7.09	180	45	2
	Rx burn, cut	5	2			8.54	217	75	2
	Van Lear and others 1985								

[1]Bailey's Ecoregions (1995).

USDA Forest Service Gen. Tech. Rep. RMRS-GTR-42-vol. 4. 2005

111

Bailey's Ecoregions (1995).

Table 5.2—Increased water yield from prescribed fire (Rx) and wildfire burned watersheds, Cascade Mountains ecoregions, U.S.A.

State	Treatment	Area		Precipitation		1st year runoff		Increase	Recovery
		acre	ha	inch	mm	inch	mm	%	years
M242 CASCADE MIXED-CONIFER-MEADOW FOREST PROVINCE[1]									
WA	Ponderosa pine			22.83	580				
	Prefire basis	1,277	517			8.70	221	—	—
	Wildfire	1,277	517			12.36	314	42	Unknown
	Helvey 1980								
OR	Douglas-fir			97.76	2,483				
	Control	175	71			74.21	1,885	—	—
	Cut, Rx burn	175	71			87.60	2,225	18	>5
	Bosch and Hewlett 1982								
OR	Douglas-fir			94.02	2,388				
	Control	237	96			54.17	1,376	—	—
	Cut, Rx fire	237	96			72.36	1,838	34	>5
	Bosch and Hewlett 1982								

[1]Bailey's Ecoregions (1995).

Table 5.3—Increased water yield from prescribed fire (Rx) and wildfire burned watersheds, Colorado Plateau and Arizona-New Mexico mountains semidesert ecoregion, U.S.A.

Ecoregion/State	Treatment	Area		Precipitation		1st year runoff		Increase	Recovery
		acre	ha	inch	mm	inch	mm	%	years
313 COLORADO PLATEAU SEMIDESERT PROVINCE[1]									
AZ	Chaparral			29.13	740				
	Control	6	2			2.5			
		9	8			2	64	—	—
	Rx burn	8	3			6.1	15	144	>11
		2	3			4	6		
	Davis 1984								
AZ	Chaparral			23.03	585				
	Control	9	3			3.2			
		6	9			3	82	—	—
	Wildfire	9	3			5.1	30	59	Unknown
		6	9			2			
	Hibbert and others 1982								
AZ	Chaparral			25.79	655				
	Control	4	1			0			
		7	9				0	—	
	Wildfire	4	1			4.8	12	>9,999+	>9
		7	9			8	4		
	Control	4	3			0.7			
		7	9			5	19	—	
	Wildfire	4	3			11.		1,421	>9
		7	9			38	89		
	Hibbert 1971								

[1]Bailey's Ecoregions (1995).

Table 5.4—Increased water yield from prescribed fire (Rx) and wildfire burned watersheds, Arizona-New Mexico mountains semidesert, and Southwest plateau and Plains dry steppe ecoregions, U.S.A.

State	Treatment	Area		Precipitation		1st year runoff		Increase	Recovery
		acre	ha	inch	mm	inch	mm	%	years
M313 AZ-NM MOUNTAINS SEMIDESERT-WOODLAND-CONIFER PROVINCE[1]									
AZ	Pinyon-juniper			18.90	480				
	Control	12	5			1.34	34	—	—
	Rx burn	12	5			1.54	39	15	5
	Control	12	5			1.69	43	—	
	Rx burn	12	5			2.20	56	30	>5
	Hibbert and others 1982								
AZ	Pinyon-juniper			18.98	482				
	Control	331	134			0.79	20	—	—
	Slash Rx burn	331	134			0.43	11	−45	4
	Control	363	147			0.71	18	—	
	Herbicide	363	147			1.10	28	56	>4
	Clary and others 1974								
AZ	Ponderosa pine			29.02	737				
	Control	44	18			0.24	6	—	
	Wildfire, low	25	10			0.35	9	50	2
	Wildfire, mod.	10	4			0.79	20	233	7
	Wildfire, high	20	8			1.06	27	350	15
	DeBano and others 1996								
315 SOUTHWEST PLATEAU AND PLAINS DRY STEPPE AND SHRUB PROVINCE[1]									
TX	Juniper/grass	2	<1	25.98	660				
	Control	2	<1			0.08	2	—	
	Rx fire	2	<1			0.98	25	1,150	5
	Rx fire, seeded	2	<1			0.43	10	400	2
	Wright and others 1982								

[1] Bailey's Ecoregions (1995).

Table 5.5—Increased water yield from prescribed burned watersheds, Southern Rocky Mountains ecoregion, U.S.A.

State	Treatment	Area		Precipitation		1st year runoff		Increase	Recovery
		acre	ha	inch	mm	inch	mm	%	years
M331 SOUTHERN ROCKY MOUNTAIN STEPPE-WOODLAND-CONIFER PROVINCE[1]									
CO	Aspen, mixed conifer			21.10	536				
	Control	4.88	24			6.18	157	—	—
	Clearcut, Rx burn	4.88	24			7.52	191	22	5
	Bosch and Hewlett 1982								

[1] Bailey's Ecoregions (1995).

USDA Forest Service Gen. Tech. Rep. RMRS-GTR-42-vol. 4. 2005

113

consider. The magnitude of increased peakflow following fire (table 5.6) is more variable than streamflow discharges (tables 5.1 to 5.5) and is usually well out of the range of responses produced by forest harvesting. Increases in peakflow as a result of a high severity wildfire are generally related to a variety of processes including the occurrence of intense and short duration rainfall events, slope steepness on burned watersheds, and the formation of soil water repellency after burning (DeBano and others 1998, Brooks and others 2003). Postfire streamflow events with excessively high peakflows are often characteristic of flooding regimes

Peakflows are important events in channel formation, sediment transport, and sediment redistribution in riverine systems (Rosgen 1996, Brooks and others 2003). These extreme events often lead to significant changes in the hydrologic functioning of the stream system and, at times, a devastating loss of cultural resources. Peakflows are important considerations in the design of structures (such as bridges, roads, dams, levees, commercial and residential buildings, and so forth). Fire has the potential to increase peakflows well beyond the normal range of variability observed in watersheds under fully vegetated conditions (table 5.6). For this reason, understanding of peakflow response to fire is one of the most important aspects of understanding the effects of fire on water resources.

Peakflow Mechanisms

A number of mechanisms occur singly or in combination to produce increased postfire peakflows (fig. 5.7). These include obvious mechanisms such as unusual

Table 5.6—Effects of harvesting and fire on peakflows in different habitat types.

Location	Treatment	Peakflow increase factor	Reference
M212 ADIRONDACK-NEW ENGLAND MIXED FOREST PROVINCE			
Hardwoods, NH	Clearcut	+2.0	Hornbeck 1973
M221 CENTRAL APPALACHIAN BROADLEAF-CONIFER FOREST PROVINCE			
Hardwoods, NC	Clearcut	+1.1	Hewlett and Helvey 1970
Hardwoods, WV	Clearcut	+1.2	Reinhart and others 1963
232 COASTAL PLAIN MIXED FOREST PROVINCE			
Loblolly Pine, NC	Rx Fire	0.0	Anderson and others 1976
M242 CASCADE MIXED-CONIFER-MEADOW FOREST PROVINCE			
Douglas-fir, OR	Cut 50%, burn	+1.1	Anderson 1974
	Clearcut, burn	+1.3	
	Wildfire	+1.4	
M262 CALIFORNIA COASTAL RANGE WOODLAND-SHRUB-CONIFER PROVINCE			
Chaparral, CA	Wildfire	+20.0	Sinclair and Hamilton 1955
		+870.0	Krammes and Rice 1963
		+6.5	Hoyt and Troxell 1934
313 COLORADO PLATEAU SEMI-DESERT PROVINCE			
Chaparral, AZ	Wildfire	+5.0 (Sum)	Rich 1962
		+150.0 (Sum)	
		+5.8 (Fall)	
		+0.0 (Winter)	
M 313 AZ-NM MOUNTAINS SEMIDESERT-WOODLAND-CONIFER PROVINCE			
Ponderosa pine, AZ	Wildfire	+96.1	Campbell & others 1977
	Wildfire, Mod.	+23.0	
	Wildfire, Severe	+406.6	
	Wildfire, Severe	+2,232.0	Ffolliott and Neary 2003
M331 SOUTHERN ROCKY MOUNTAINS STEPPE-WOODLAND-CONIFER PROVINCE			
Aspen-conifer, CO	Clearcut, Rx burn	−1.50	Bailey 1948
Ponderosa pine, NM	Wildfire	+100.00	Bolin and Ward 1987

Figure 5.7—Flood flow at Heber, AZ, after the Rodeo-Chediski Fire, 2002, Apache-Sitgreaves National Forest. (Photo by Dave Maurer).

rainfall intensities, destruction of vegetation, reductions in litter accumulations and other decomposed organic matter, alteration of soil physical properties, and development of soil hydrophobicity.

A special circumstance sometimes occurs with postwildfire peakflows that can contribute to the large responses (up to three orders of magnitude increase). Cascading debris dam failures have the potential to produce much higher peakflow levels than would be expected from given rainfall events on bare or water repellent soils. This process consists of the establishment of a series of debris dams from large woody debris in and adjacent to stream channels, buildup of water behind the dams, and sequential failure of the first and subsequent downstream debris dams (fig. 5.8).

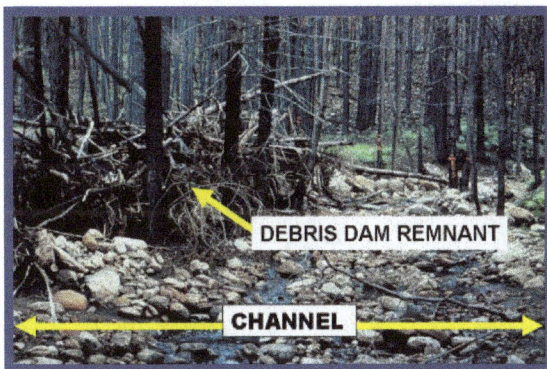

Figure 5.8—Remnant of a debris dam in the channel of Dude Creek after the Dude Fire, Tonto National Forest, 1991. (Photo by John Rinne).

Concern about this process has led to the use of one type of BAER channel treatment, debris removal. Channels particularly prone to this process would include those with large amounts of woody debris and a high density of riparian trees or boulders, which could act as the dam formation mechanism. After the 1991 Dude Fire in Arizona, Rinne (1994) reported that little of the tagged prefire woody debris moved after a significant postfire flood event. On the other hand, some unusually high flood flows after the 2000 Cerro Grande Fire in New Mexico left distinct evidence of woody debris dam formation and failure (Kuyumjian, G.A., USDA Forest Service, personal communication).

Fire Effects

Anderson and others (1976) provided a good review of peakflow response to disturbance. These responses are influenced by fire severity. Low severity prescribed burning has little or no effect on peakflow because it does not generally alter watershed condition.

Intense short duration storms that are characterized by high rainfall intensity and low volume have been associated with high stream peakflows and significant erosion events after fires (Neary and others 1999). In the Intermountain West, high intensity, short duration rainfall is relatively common (Farmer and Fletcher 1972). Five-minute rainfall rates of 8.38 to 9.25 inches/hour (213 and 235 mm/hour) have been associated with peakflows from recently burned areas that were increased five times that for adjacent, unburned areas (Croft and Marston 1950). A 15-minute rainfall burst at a rate of 2.64 inches/hour (67 mm/hour) after the 2000 Coon Creek Fire in Arizona produced a peakflow that was in excess of sevenfold greater than the previous peakflow during 40 years of streamflow gauging. Moody and Martin (2001) reported on a threshold for rainfall intensity (0.39 inch/hour or 10 mm/hour in 30 minutes) above which flood peakflows increase rapidly in the Rocky Mountains. Robichaud (2002) collected rainfall intensity on 12 areas burned by wildfire in the Bitterroot Valley of Montana. He measured precipitation intensities that ranged from 3 to 15 mm in 10 minutes. The high end of the range was an equivalent to 75 mm/hour (greater than a 100-year return interval). It is these types of extreme rainfall events, in association with altered watershed condition, that produce large increases in stream peakflows and erosion.

Peakflows after forest cutting can increase or decrease depending on location, the percentage of the watershed cut, precipitation regime, and season (table 5.6, fig. B.3). Most studies show increases in peakflows of 9 to 100 percent. The concern with increases in annual flood peakflows is that the increases could lead

USDA Forest Service Gen. Tech. Rep. RMRS-GTR-42-vol. 4. 2005

115

to channel instability and degradation, and to increased property damage in flood-prone urban areas.

Fire has a range of effects on stream peakflows. Low severity, prescribed fires have little or no effect because they do not substantially alter watershed condition (table 5.6). Severe wildfire has much larger effects on peakflows. The Tillamook Burn in 1933 in Oregon increased the total annual flow of two watersheds by 1.09-fold and increased the annual peakflow by 1.45-fold (Anderson and others 1976). A 127 ha wildfire in Arizona increased summer peakflows by 5- to 150-fold, but had no effect on winter peakflows. Another wildfire in Arizona produced a peakflow 58-fold greater than an unburned watershed during record autumn rainfalls. Campbell and others (1977) documented the effects of fire severity on peakflows. A moderate severity wildfire increased peakflow by 23-fold, but high severity wildfire increased peakflow response three orders of magnitude to 406.6-fold greater than undisturbed conditions. Krammes and Rice (1963) measured an 870-fold increase in peakflow in California chaparral. In New Mexico, Bolin and Ward (1987) reported a 100-fold increase in peakflow after wildfire in a ponderosa pine and pinyon-juniper forest. Watersheds in the Southwest are much more prone to these enormous peakflow responses due to interactions of fire regimes, soils, geology, slope, and climate (Swanson 1981).

Following the Rodeo-Chediski Fire of 2002, peakflows were orders of magnitude larger than earlier recorded. The estimated peakflow on a gauged watershed that experienced high severity stand-replacing fire was almost 8.9 ft^3/sec (0.25 m^3/sec) or nearly 900 times that measured prefire (Ffolliott and Neary 2003). The peakflow on a watershed subjected to low-to-medium

severity fire was estimated to be about one-half less, but still far in excess of the previous observations. A subsequent and higher peakflow on the severely burned watershed was estimated to be 232 ft^3/sec (6.57 m^3/sec) or about 2,232 times that measured in snowmelt runoff prior to the wildfire. This latter peakflow increase represents the highest known relative postfire peak flow increase that has been measured in the ponderosa pine forest ecosystems of Arizona or, more generally, the Southwestern United States. However, the specific discharge (94.2 ft^3/sec/mile^2 or 1.02 m^3/sec/ km^2) was on the lower end of range of discharges measured by Biggio and Cannon (2001).

Another concern is the timing of stormflows or response time. Burned watersheds respond to rainfall faster, producing more "flash floods." They also may increase the number of runoff events. Campbell and others (1977) measured six events on an unburned watershed after the Rattle Burn and 25 on a high-severity burned watershed. Hydrophobic conditions, bare soils, and litter and plant cover loss will cause flood peaks to arrive faster and at higher levels. Flood warning times are reduced by "flashy" flow, and higher flood levels can be devastating to property and human life. Recovery times after fires can range from years to many decades.

Still another aspect of the postfire peakflow issue is the fact that the largest discharges often occur in small areas. Biggio and Cannon (2001) examined runoff after wildfires in the Western United States. They found that specific discharges were greatest from relatively small areas (less than 0.4 mile^2 or 1 km^2, fig. 5.9). The smaller watersheds in their study had specific discharges averaging 17,664.3 ft^3/sec/mile^2 (193.0 m^3/sec/km^2), while those in the next higher sized watershed category

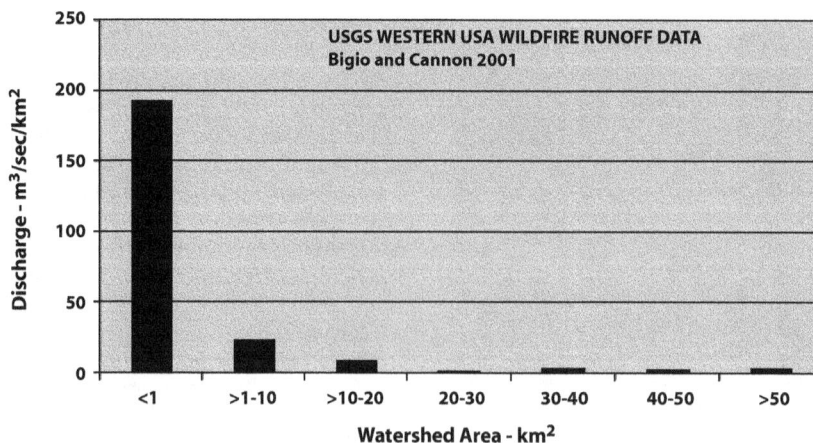

Figure 5.9—Post-wildfire flood specific discharge and watershed area. (Adapted from Biggio and Cannon 2001).

(up to 4 mile^2 or 10 km^2) averaged 2,077.6 ft^3/sec/mile^2 (22.7 m^3/sec/km^2).

So, the net effect on watershed systems and aquatic habitat of increased peakflows is a function of the area burned, watershed characteristics, and the severity of the fire. Small areas in flat terrain subjected to prescribed fires will have little if any effect on water resources, especially if Best Management Practices are used. Peakflows after wildfires that burn large areas in steep terrain can produce significant impacts, but peakflows are probably greatest out of smaller sized watersheds less than 0.4 mile^2 (1 km^2). Burned area emergency rehabilitation (BAER) techniques may be able to mitigate some of the impacts of wildfire (see chapter 10). However, the ability of these techniques to moderate the impacts of rainfalls that produce extreme peakflow events is not well documented (Robichaud and others 2000).

Management Implications _____

Fires affect watersheds, resulting in changes that affect many resource values, including municipal water, visual aspects, recreation, floral and fauna existence and welfare, transportation, and human activity and well being. The vast extent of possible water and watershed changes makes management of these lands crucial.

Prescribed fires with low to moderate burn severity rarely produce adverse hydrologic effects that land managers need to be concerned about. Postwildfire floods are the main concern, particularly the timing of storm flows (response time) and magnitudes of flood peaks. Because burned watersheds respond to rainfall faster, producing more "flash floods," they also may increase the number of runoff events. Flood warning times are reduced by these "flashy" flows, and higher flood levels can be devastating to property and human life. Another aspect of this is the fact that recovery times can range from years to many decades.

So, the net effect on watershed systems and aquatic habitat of increased peak flows is a function of the area burned, watershed characteristics, and the severity of the fire. Small areas in flat terrain subjected to prescribed fires will have little if any effect on water resources, especially if Best Management Practices are utilized. Peak flows after wildfires that burn large areas in steep terrain can produce significant impacts. Burned area emergency rehabilitation watershed techniques may be able to mitigate some of the impacts of wildfire.

However, the ability of these techniques to moderate the impacts of rainfalls that produce extreme peakflows is poorly documented (Robichaud and others 2000). In some circumstances, a lack of willingness has existed to implement effective runoff control measures in a timely and thorough way because of

visual and environmental concerns as well as hasty, improper installation techniques. In some instances, BAER techniques such as contour trenching have been able to reduce peakflow events where short duration rainfall intensities are high but total storm volumes are low (Robichaud and others 2000).

Increased flood flow peaks have the potential to damage both natural and cultural resources. Large floods often change stream geomorphology. Aquatic biota and riparian ecosystems can be severely impacted by unusual flood flows. Culverts, bridges, roads, dams, and irrigation structures are at risk by flood flows outside of their design parameters. Recreation facilities, houses, businesses, and community structures can also be affected. Most important, human health and safety can be at serious risk.

Forest managers need to understand these risks in order to design adequate rehabilitation projects, order and enforce recreation area closures, establish flood-warning protocols, and conduct appropriate land management activities.

Summary _____

Fires affect water cycle processes to a greater or lesser extent depending on severity. Table 5.7 contains a general summary of the effects.

Fires can produce some substantial effects on the streamflow regime of both small streams and rivers. Tables 5.1 to 5.6 shows that fires can affect annual and seasonal water yield, peakflows and floods, baseflows, and timing of flows. Adequate baseflows are necessary to support the continued existence of many wildlife populations. Water yields are important because many forest, scrubland, and grassland watersheds function as municipal water supplies. Peakflows and floods are of great concern because of their potential impacts on human safety and property. Next to the physical destruction of a fire itself, postfire floods are the most damaging aspect of fire in the wildland environment. It is important that resource specialists and managers become aware of the potential of fires to increase peakflows.

Following wildfires, flood peak flows can increase dramatically, severely affecting stream physical conditions, aquatic habitat, aquatic biota, cultural resources, and human health and safety. Often, increased flood peak flows of up to 100 times those previously recorded, well beyond observed ranges of variability in managed watersheds, have been measured after wildfires. Potentials exist for peak flood flows to jump to 2,300 times prewildfire levels. Managers must be aware of these potential watershed responses in order to adequately and safely manage their lands and other resources in the postwildfire environment.

USDA Forest Service Gen. Tech. Rep. RMRS-GTR-42-vol. 4. 2005

117

Table 5.7—A summary of the changes in hydrologic processes produced by wildland fires.

Hydrologic process	Type of change	Specific effect
1. Interception	Reduced	Moisture storage smaller Greater runoff in small storms Increased water yield
2. Litter storage of water	Reduced	Less water stored (0.05 in/in or 0.5 mm/cm litter) Overland flow increased
3. Transpiration	Temporary elimination	Streamflow increased Soil moisture increased
4. Infiltration	Reduced	Overland flow increased Stormflow increased
5. Streamflow	Changed	Increased in most ecosystems Decreased in snow systems Decreased in fog-drip systems
6. Baseflow	Changed	Decreased (less infiltration) Increased (less evapotranspiration) Summer low flows (+ and -)
7. Stormflow	Increased	Volume greater Peakflows larger Time to peakflow shorter Flash flood frequency greater Flood levels higher Stream erosive power increased
8. Snow accumulation	Changed	Fires <10 ac (<4 ha), increased snowpack Fires >10 ac (> 4 ha), decreased snowpack Snowmelt rate increased Evaporation and sublimation greater

USDA Forest Service Gen. Tech. Rep. RMRS-GTR-42-vol. 4. 2005

Daniel G. Neary
Johanna D. Landsberg
Arthur R. Tiedemann
Peter F. Ffolliott

Chapter 6:
Water Quality

Introduction

Increases in streamflow discharges following a fire can result in little to substantial effects on the physical, chemical, and biological quality of the water in streams, rivers, and lakes. The magnitude of these effects is largely dependent on the size, intensity, and severity of the fire, and on the condition of the watershed at the time of burning (fig. 6.1). Higher postfire streamflow discharges can result in an additional transport to stream channels or other water bodies of solid and dissolved materials that adversely affect the quality of water for human, agricultural, or industrial purposes. The most obvious effects are produced by suspended and bedload sediments. These components of water quality were introduced and discussed in chapter 2, and in this chapter they are referred to in the discussion on water quality relative to municipal water supply quality.

Fire affects water quality characteristics through the changes that the burning causes in the hydrologic cycle and streamflow regimes (see chapter 5). The effect of fire on water quality is the topic of this chapter.

Water Quality Characteristics and Standards

Water quality refers to the physical, chemical, and biological characteristics of water in reference to a particular use. Among the physical characteristics of interest to hydrologists and watershed managers are sediment concentrations, turbidity, and water temperature. Dissolved chemical constituents of importance include nitrogen (N), phosphorus (P), calcium (Ca), magnesium (Mg), and potassium (K). Some of these nutrients are adsorbed on organic and inorganic sediment particles. Bacteriological quality is also important if water is used for human consumption or recreation. The processes in the hydrologic cycle directly or indirectly affect the magnitude of soil erosion and, as a consequence, the transport and deposition of sediment in water and other physical, chemical, and biological quality characteristics that collectively determine the quality of water.

A *water quality standard* refers to the physical, chemical, or biological criteria or characteristics of water in relation to a specified use. It also includes the beneficial uses and antidegradation policy. For example, a water

USDA Forest Service Gen. Tech. Rep. RMRS-GTR-42-vol. 4. 2005

119

Figure 6.1—Schoonover Fire just prior to crowning, 2002, Montana. (Photo by USDA Forest Service).

quality standard for irrigation is not necessarily acceptable for drinking water. Changes in water quality due to a watershed management practice could make the water flowing from the area unsuitable for drinking, but at the same time it could be acceptable for irrigation, fisheries, and other uses. In some instances there are laws or regulations to prevent water quality characteristics from becoming degraded to the point where a water quality standard is jeopardized (DeBano and others 1998, Landsberg and Tiedemann 2000, Brooks and others 2003). The main purpose of these laws and regulations is maintaining the quality of water for a possible, unforeseen, future use.

A major issue that hydrologist and watershed managers often confront is whether forest or rangeland fires will create conditions in stream, river, or lake waters that are outside of the established water quality standards and, as a consequence, will require remedial actions to bring the water within the standards. Water quality standards that have been established by the U.S. Environmental Protection Agency (1999) are the "benchmarks" for water quality throughout the United States. The most adverse effects from wildfires on water quality standards come from physical effects of the sediment and ash that are deposited into streams. However, stream chemistry parameters can be exceeded.

Several chemical constituents that are regulated by water quality standards are likely to be impacted by burning. Primary standards in the regulations of the U.S. Environmental Protection Agency cover nitrate-nitrogen (NO_3-N), nitrite-nitrogen (NO_2-N), and other substances that are not immediately associated with fire. Secondary standards apply to pH, sulfate (SO_4-S),

total dissolved solids (TDS), chloride (Cl), iron (Fe), turbidity, and several other constituents. Secondary standards are also set for color and odor. Phosphate phosphorus (PO_4-P) can affect water quality because of its ability to affect the color and odor of water by accelerating the eutrophication process. That water quality standards can be impacted and exceeded by fire in some cases is also examined in this chapter.

Soil Erosion and Sedimentation Processes

Changes in the hydrologic cycle caused by fire can also affect the rate of soil erosion, the subsequent transport and deposition of the eroded soil as sediment, and the chemical characteristics that collectively determine the quality of water (DeBano and others 1998, Brooks and others 2003). Alterations that burning can cause in the hydrologic cycle, that in turn affect soil erosion, sedimentation, and water quality, are considered in chapter 5.

Soil erosion is the physical process of the force of raindrops or eddies in overland flow (surface runoff) dislodging soil particles, which are then transported by water or wind or the force of gravity (Dunne and Leopold 1978, Satterlund and Adams 1992, Brooks and others 2003). *Sedimentation* is the process of deposition of sediment in stream channels or downstream reservoirs or other point of use.

Increased soil erosion, or sediment, is often the most visible effect of a fire other than the loss of vegetation by burning. Maintaining a vegetative cover or a cover of litter and other organic material on the soil surface of a watershed is the best means of preventing excessive soil erosion rates. However, fire can cause the loss of these protective covers and in turn cause excessive soil erosion and soil lost from the burned site (Dunne and Leopold 1978, Satterlund and Adams 1992, Brooks and others 2003).

Natural rates of sedimentation are generally lower in high rainfall regions than in arid and semiarid regions (Brooks and others 2003). As a result of the infrequent "big storms" that are characteristic of arid and semiarid environments, sedimentation is often viewed as a discontinuous (unsteady) process where sediment runs from its source through a stream channel system with intermittent periods of storage (Wolman 1977, Baker 1990, Baker and others 1998). The disproportionate amount of sediment transported by these big storms makes it difficult to determine a "normal rate" of sedimentation on either undisturbed or burned watersheds in arid and semiarid regions (DeBano and others 1998).

Only a portion of the sediment is passed through and out of a watershed with a single storm event. Most of the sediment that is generated by a storm is deposited at the base of hillslopes, in floodplains following high overland flows, and within stream or river channels (DeBano and others 1998, Brooks and others 2003). The relationships between hillslope soil erosion and downslope and downstream sedimentation involve a complexity of channel processes and their dynamics, both of which are poorly understood. Nevertheless, the sediment that is eventually deposited into the channels following a fire changes the physical characteristics of the water flowing from the burned watershed.

Physical Characteristics of Water

Among the more important physical characteristics of postfire streamflow regimes of main interest to hydrologists and watershed managers are suspended sediment concentrations and turbidity and elevated streamflow temperatures (thermal pollution). Suspended sediment consisting of silts and colloids of soil materials has impacts on water quality in terms of human, agricultural, and industrial uses of the water and aquatic organisms and their environments (see chapter 7). Elevated streamflow temperatures can also impact water quality characteristics and aquatic organisms and environments.

Sediment

Watersheds that have been severely denuded by a wildfire are often vulnerable to accelerated rates of soil erosion and, therefore, can yield large (but often variable) amounts of postfire sediment (fig. 6.2).

Figure 6.2—Stermer Ridge watershed burned at high severity during the Rodeo-Chediski Fire, Arizona, 2002. Note recently deposited sediment in the lower portion of the photo. (Photo by Daniel Neary).

Wildfires generally produce more sediment than prescribed burning. The large inputs of sediment into a stream following a wildfire can tax the transport capacity of the stream and lead to channel deposition (aggradation). However, prescribed burns by their design do not normally consume extensive layers of litter or accumulations of other organic materials. Hence, sedimentation is generally less than that resulting from a wildfire.

Suspended Sediment Concentrations and Turbidity—Suspended sediment concentrations and turbidity are often the most dramatic of water quality responses to fire. *Turbidity* is an expression of the optical property of water that scatters light (Dunne and Leopold 1978, Satterlund and Adams 1992, Brooks and others 2003). Turbidity reduces the depth to which sunlight can penetrate into water and, therefore, influences the rate of photosynthesis. Sediment concentrations are commonly expressed in parts per million (ppm) or milligrams per liter (mg/L) (International System of Units), while turbidity is measured in nephelometeric units. Postfire increases in suspended sediment concentrations and turbidity can result from erosion and overland flow, channel scouring because of the increased streamflow discharge, creep accumulations in stream channels, or combinations of all three actions after a fire.

Less is known about the effect of fire on turbidity than on the sedimentation processes. One problem contributing to this lack of information is that turbidity has been historically difficult to measure because it is highly transient, variable, inconsistent, and varies by instrument used. With the development of continuous turbidimeters or nephelometers, some suspended sediment estimates are now based on continuous turbidity measurements. These turbidity estimates must be translated to suspended sediment using turbidity-to-suspended sediment rating curves that are time consuming, carefully calibrated, site specific, and instrument specific.

Nevertheless, it has been observed that postfire turbidity levels in stream water are affected by the steepness of the burned watershed (table 6.1). The turbidity of overland flow from burned steep slopes in central Texas that had been converted from woodland to an herbaceous cover was higher than that of overland flow from burned slopes of lesser steepness (Wright and others 1976, 1982). Turbidity increases after fires are generally a result of the postfire suspension of ash and silt-to-clay sized soil particles in the water (fig. 6.3).

The primary standards for suspended sediment concentration and turbidity with respect to drinking water are written in terms of turbidity (U.S. Environmental Protection Agency 1999). However, only two of the studies reviewed by Landsberg and Tiedemann

USDA Forest Service Gen. Tech. Rep. RMRS-GTR-42-vol. 4. 2005

121

Table 6.1—Water turbidity after fire or in combination with other treatments (Adapted from Landsberg and Tiedemann 2000).

Treatment	Habitat	Location	Treatment Pre	Treatment Post	Reference
			Jackson Turbidity Units		
Rx fire, pile, and burn	Juniper	Central Texas			Wright and
		3-4% slope	12	12	others 1976
		8-20% slope	20	53	
		37-61% slope	12	132	
Pile and burn	Juniper	Central Texas	12	162	Wright and
Pile, burn, seed			12	72	others 1982

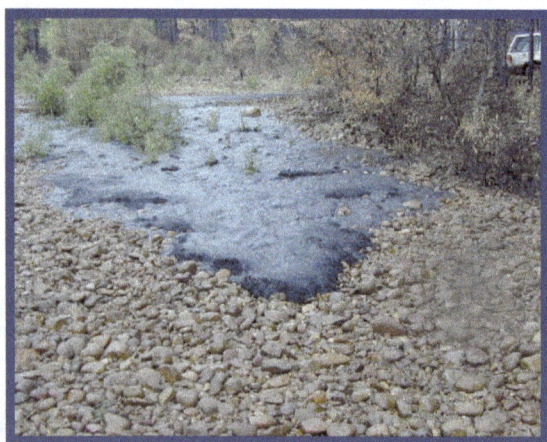

Figure 6.3—Ash slurry flow in a ephemeral drainage after the Rodeo-Chediski Fire, Arizona, 2002, Apache-Sitgreaves National Forest. (Photo by Daniel Neary).

(2000) in their synthesis of the scientific literature on the effects of fire on drinking water used turbidity measured in nephelometric units as a measure of the suspended sediment concentrations of water (table 6.1). Other studies reported concentrations of suspended sediment in ppm or mg/L as the measure (table 6.2). While suspended sediment concentrations in ppm (mg/L) have been converted to turbidity in nephelometric units in studies of nutrient losses by soil erosion after wildfire, the relationship is site-specific. Beschta (1980) found that a relationship between suspended sediment concentrations and turbidity can be established for a specified watershed in some instances, but that the relationship differs significantly among watersheds. He suggested, therefore, that this relationship be established on a watershed-by-watershed basis.

The few postfire values for turbidity found in the literature (table 6.1) exceed the allowable water quality standard for turbidty (U.S. Environmental Protection

Table 6.2—Suspended sediment concentrations in streamflow after fire alone or fire in combination with other treatments (Adapted from Landsberg and Tiedemann 2000).

Treatment	Habitat	Location	Treatment Pre	Treatment Post	Reference
			ppm (mg/L)		
Wildfire	Taiga	Interior Alaska	10.6	6.0	Lotspeich and others 1970
Clearcut, slash burn	Douglas-fir	W. Oregon	2.0	150.0	Fredriksen 1971
Wildfire	Ponderosa pine	E. Washington	—	1,200.0	Helvey 1980
Pile and burn	Juniper	Texas	1.1	3.7	Wright and others 1982
Pile, burn, and seed	Juniper	Texas	1.0	3.7	Wright and others 1982
Prescribed fire	Loblolly pine	South Carolina	26.0	33.0	Douglass and Van Lear 1983
Wildfire	Mixed conifer	Montana	<3.0	32.0	Hauser and Spence 1998

122

USDA Forest Service Gen. Tech. Rep. RMRS-GTR-42-vol. 4. 2005

Agency 1999). Therefore, Landsberg and Tiedemann (2000) recommended that the effect of fire on turbidity per se needs further investigation to better understand the processes involved. Because of the elevated suspended sediment concentrations after fire and fire-related treatments, it is difficult to imagine that some of these concentrations would not have produced turbidity above the permitted level for water supplies.

Sediment Yields—While the level of suspended sediment concentrations is a primary factor of environmental concern, sediment yield is important in estimating the sediment buildup in reservoirs or other impoundments (Wetzel 1983). The *sediment yield* of a watershed is the total sediment outflow from the watershed for a specified period of time and a defined point in the stream channel. Sediment yields are dependent mostly on the physical characteristics of the sediment, the supply of soil particles to a stream channel, the magnitude and rate of streamflow discharge, and the condition of the watershed. The values presented in tables 2.8 to 2.10 are indicative of the level of variability that can be expected in the changes in sediment yields following a fire. This variability generally reflects the interacting factors of geology, soil, topography, vegetation, fire characteristics, weather patterns, and land use practices on the impacted watershed. The higher values resulted from fires on steep slopes and areas of decomposing granite that readily erode. These higher sediment yields are sufficient in their magnitude to generate concern about soil impoverishment and water turbidity. The lower values were generally associated with flat sites and lower severity fires. Although the magnitudes shown are site specific, the changes in sediment transport presented reflect responses that can be expected from wildfire and prescribed burning.

Postfire sediment yields are generally the highest in the first year or so after burning, especially when the burned watershed has been exposed to large, high-intensity rainfall events immediately after the fire has exposed the soil surface. These sediment yields are indicative of the partial or complete consumption of litter and other decomposed organic matter on the soil surface, a reduction in infiltration, and a consequent increase in overland flow (DeBano and others 1998, Brooks and others 2003). Sediment yields typically decline in subsequent years as the protective vegetation becomes reestablished on the burned watershed.

Water Temperature

Water temperature is a critical water quality characteristic of many streams and aquatic habitats. Temperature controls the survival of certain flora and fauna in the water that are sensitive to water temperature. The removal of streambank vegetation by burning can cause water temperature to rise, causing thermal pollution to occur, which in turn can increase biological activity in a stream (DeBano and others 1998, Brooks and others 2003). Increases in biological activity place a greater demand on the dissolved oxygen content of the water, one of the more important water quality characteristics from a biological perspective.

There are no established national standards for the temperature of drinking water. However, under the Clean Water Act, States are required to develop water quality standards to protect beneficial uses such as fish habitat and water quality restoration. The U.S. Environmental Protection Agency provides oversight and approval of these State standards. Currently, about 86 percent of the national listings of waterbodies with temperature-impaired water quality are in the Pacific Northwest (Ice and others, in press). One of the problems with these standards is identifying natural temperature patterns caused by vegetation, geology, geomorphology, climate, season, and natural disturbance history. Also, increases in stream water temperatures can have important and often detrimental effects on stream eutrophication. Acceleration of stream eutrophication can adversely affect the color, taste, and smell of drinking water.

Severe wildfires can function like streamside timber clearcuts in raising the temperature of streams due to direct heating of the water surface. Increases up to 62 °F (16.7 °C) have been measured in streamflows following fire, and following timber harvesting and fire in combination (table 6.3). When riparian (streamside) vegetation is removed by fire or other means, the stream surface is exposed to direct solar radiation, and stream temperatures increase (Levno and Rothacher 1969, Brown 1970, Swift and Messner 1971, Gibbons and Salo 1973, Brooks and others 2003).

Another important aspect of the temperature issue is the increase in fish mortality posed by stream temperature increases. The main concerns relative to aquatic biota are the reduction in the concentrations of dissolved oxygen (O_2) that occurs with rising temperatures, fish pathogen activity, and elevated metabolic activity. All of these can impair the survivability and sustainability of aquatic populations and communities. Dissolved O_2 contents are affected by temperature, altitude, water turbulence, aquatic organism respiration, aquatic plant photosynthesis, inorganic reactions, and tributary inflow. When O_2 concentrations become less than 10 ppm (less than 10 mg/L), they create problems for salmonid fishes. Increases of 2 to 9 °F (1-5 °C), which are not a problem at sea level, become problematic for salmonids at high altitude. Warm water fishes can tolerate stream temperatures below 10 ppm (10 mg/L) and are not as easily impacted by O_2 concentration declines.

USDA Forest Service Gen. Tech. Rep. RMRS-GTR-42-vol. 4. 2005

123

Table 6.3—Temperature increases in streamflow after fire alone and in combination with other treatments.

Location	Treatment	Buffer	Temperature increase		Reference
			°F	(°C)	
M221 CENTRAL APPALACHIAN BROADLEAF-CONIFER FOREST PROVINCE					
Pennsylvania	Clearcut	Yes	3.1	(1.7) mn[1]	Lynch and Corbett 1990
North Carolina	Clearcut, farm	No	20.9	(11.6) mx	Swift and Messner 1971
	Clearcut	No	5.9	(3.3) mx	
	Understory cut	No	2.0	(1.1) mx	
M242 CASCADE MIXED-CONIFER-MEADOW FOREST PROVINCE					
Oregon	Clearcut	Yes	14.0	(7.8) mn	Brown and Krygier 1970
Oregon	Patch cut	Yes	0.0	(0.0)	Hall and others 1987
	Clearcut	No	30.1	(16.7) d	
Oregon	Clearcut Rx burn	No	13.0	(7.2) sm	Levno and Rothacher 1969
Oregon	Wildfire: 26% slope	No	18.0	(10.0) mx	Amaranthus and others 1989
	Wildfire: 54% slope	No	5.9	(3.3) mx	
Washington	Wildfire	No	18.0	(10.0) sm	Helvey 1980
M331 SOUTHERN ROCKY MOUNTAINS STEPPE-WOODLAND-CONIFER PROVINCE					
Wyoming	Wildfire*	No	18.0	(10.0) mx	Hungerford and others 1991

[1] sm = summer mean temperature, mn = mean temperature, d = daily temperature, mx = maximum.
* = Yellowstone Fires 1988.

pH of Water

The *pH* of water at a point in time is an indication of the balance of chemical equilibria in a water body. Its level affects the presence of some chemicals in the water. The pH of water can be affected by ash depositions immediately after a fire. In the first year after fire, increased pH values of the soil (Wells and others 1979, DeBano and others 1998, Landsberg and Tiedemann 2000) can also contribute to increased values of streamflow pH. A secondary drinking water quality standard for pH is 6.5 to 8.5 (U.S. Environmental Protection Agency 1999). In the investigations reviewed by Landsberg and others (1999), there was only one study that reported pH values outside the U.S. Environmental Protection Agency standards (table 6.4). During the first 8 months after the Entiat Fires in eastern Washington, Tiedemann (1973) detected transient pH values up to 9.5 in streamflow water and, 2 days after fertilizer application, a transient pH value of 9.2. These latter values are generally reflective of toxic limits in the water.

Chemical Characteristics of Water

Watershed-scale studies provide an integrated view of the effects of fire on chemical (ionic) concentrations and losses. Some investigators have reported little

effects of burning on the chemical concentrations in water following burning of a watershed and, therefore, have concluded that reported increases in streamflow resulting from fire-caused transpirational reductions had likely masked concentration effects (Helvey and others 1976, Tiedemann and others 1978, Gottfried and DeBano 1990, DeBano and others 1998). Other investigators observed higher concentrations of some chemicals in streamflow from burned watersheds (Snyder and others 1975, Campbell and others 1977). However, these elevated concentrations often return to prefire levels after the first flush of flow.

Dissolved Chemical Constituents

The main sources of dissolved chemical constituents (nutrients) in the water flowing from watersheds are geologic weathering, decompositions of photosynthetic products into inorganic substances, and large storm events. Vegetative communities accumulate and cycle large quantities of nutrients in their biological role of linking soil, water, and atmosphere into a biological continuum (Tiedemann and others 1979, DeBano and others 1998, Brooks and others 2003). Nutrients are cycled in a largely orderly (tight) and often predictable manner until a disturbance alters the form of their distribution. One such disturbance is fire.

The effects of fire on the nutrient capital (status) of a watershed ecosystem are largely manifested by a rapid mineralization and dispersion of plant nutrients

124

USDA Forest Service Gen. Tech. Rep. RMRS-GTR-42-vol. 4. 2005

Table 6.4—The pH of water after fire alone or in combination with other treatments (Adapted from Landsberg and Tiedemann 2000).

Treatment	Habitat	Location	Treatment Pre	Treatment Post	Reference
			- - - - - pH - - - - -		
Wildfire	Ponderosa pine and Douglas-fir	Washington	—	7.2-8.5	Tiedemann 1973
Wildfire + N			—	7.1-9.5	
Wildfire + N	Mixed conifer	California	6.2-7.0	6.7-7.0	Hoffman and Ferreira 1976
Pile burn	Juniper	Texas Slopes			Wright and others 1976
		3-4%	7.3	7.3	
		8-20%	7.6	7.7	
		37-61%	7.4	7.7	
Wildfire	Pine, spruce	Minnesota	6.2	6.3	Tarapchak and Wright 1977
Rx fire	Ponderosa pine	Arizona	6.2	6.4	Sims and others 1981
Pile burn	Juniper	Texas	7.1	7.3	Wright and others 1982
Slash burn	Hemlock/cedar	British Columbia	6.8	7.8	Feller and Kimmins 1984
Wildfire	Mixed conifer	Wyoming	7.4	7.5	Lathrop 1994

from an intrabiotic to an extrabiotic state (Grier 1975, Tiedemann and others 1979, DeBano and others 1998, Brooks and others 2003). Part of the plant- and litter-incorporated N, P, K, Ca, Mg, copper (Cu), Fe, manganese (Mn), and zinc (Zn) are volatilized and, through this process, evacuated from the system. Metallic nutrients such as Ca, Mg, and K are converted into oxides and deposited as ash layers on the soil surface. Oxides are low in solubility until they react with carbon dioxide (CO_2) and water in the atmosphere and, as a result, are converted into bicarbonate salts. In this form, they are more soluble and vulnerable to loss through leaching or overland flow than they are as oxides, or they are incorporated into plant tissues or litter.

In a postfire situation where, compared to prefire conditions, there are less vegetative cover and lower accumulations of litter and other organic materials, the result is an increase in susceptibility to nutrient loss from a watershed through erosion. With a reduced vegetative cover, soil-plant cycling mechanisms reduce nutrient uptake, further increasing the potential nutrient loss by leaching. Responses of ionic concentrations of the more important nutrients to burning are discussed in the following paragraphs.

Nitrogen

NO_3-N, NH_4-N, and organic-N are the nitrogen forms most commonly studied as indicators of fire disturbance. Most of the attention of hydrologists and watershed managers relative to water quality responses to fire focuses on NO_3-N because it is highly mobile. The potential for increased NO_3-N in streamflow after burning is attributed mainly to accelerated mineralization and nitrification (Vitousek and Melillo 1979, Covington and Sackett 1986, 1992, DeBano and others 1998) and reduced plant demand (Vitousek and Melillo 1979). This increase results from the conversion of organic N to available forms (Kovacic and others 1986), mineralization (Covington and Sackett 1992, Ojima and others 1994), or mobilization by microbial biomass through the fertilizing effect of ash nutrients and improved microclimate (Koelling and Kucera 1965, Hulbert 1969, Ojima and others 1995). These postfire effects are short lived, however, usually lasting only a year or so (Kovacic and others 1986, Monleon and others 1997).

The response of NO_3-N to burning is varied. Some investigators have found no significant change in the postfire levels of NO_3-N in streamflows, while others

USDA Forest Service Gen. Tech. Rep. RMRS-GTR-42-vol. 4. 2005

125

report increases of NO_3-N in either the soil solution or streamflow (Hibbert and others 1974, Tiedemann and others 1979). Examples of increases in NO_3-N (table 6.5) with fire and followup applications of herbicides causing the largest increases are shown in most studies of forest disturbances such as fire.

The most striking response of NO_3-N concentration in streamflow to fire (and the only case not related to herbicides where the primary water quality standard was exceeded) was observed in southern California, where N loadings from atmospheric deposition are relatively high, and the frequent wildfires in the chaparral shrublands are often high in their fire severities (Riggan and others 1994). Severe burning of a watershed in this Mediterranean-type resulted in a maximum NO_3-N level of 15.3 ppm (mg/L) in streamflow compared to 2.5 ppm (mg/L) in streamflow from an unburned control watershed. Maximum concentration for a moderately burned watershed was 9.5 ppm (mg/L). These results are likely to represent an "unusual response" because the watersheds studied were subject to a chronic atmospheric deposition of pollutants. Regardless of the treatment or treatment combinations on these watersheds, levels of NO_3-N in the streamflows were well below maximum allowable concentrations in most other studies. Beschta (1990) reached the same conclusion in his assessment of streamflow NO_3-N responses to fire and associated treatments.

Table 6.5—Effect of forest disturbances on maximum NO_3-N levels in streamflow (Adapted from Neary and Hornbeck 1994, Neary and Michael 1997, Landsberg and Tiedemann 2000).

Location	Forest type	Treatment	Maximum NO_3-N	Reference
			ppm (mg/L)	
1. Harvesting				
M212 ADIRONDACK-NEW ENGLAND MIXED FOREST PROVINCE				
New Hampshire	Hardwoods	Clearcut	6.1	Hornbeck and others 1987
M242 CASCADE MIXED-CONIFER-MEADOW FOREST PROVINCE				
Oregon	Douglas-fir	Clearcut	2.1	Brown and others 1973
2. Herbicides				
M212 ADIRONDACK-NEW ENGLAND MIXED FOREST PROVINCE				
New Hampshire	Hardwoods	Cut, herbicide	17.8	Aubertin and Patric 1974
313 COLORADO PLATEAU SEMIDESERT PROVINCE				
Arizona	Chaparral	Herbicide	15.3	Davis 1984
3. Fire				
231 SOUTHEASTERN MIXED FOREST PROVINCE				
South Carolina	Loblolly pine	Rx fire	<0.1	Richter and others 1982
South Carolina	Loblolly pine	Rx fire	<0.1	Douglass and Van Lear 1983
M242 CASCADE MIXED-CONIFER-MEADOW FOREST PROVINCE				
Oregon	Douglas-fir	Clearcut, burn	0.6	Fredriksen and others 1975
Washington	Ponderosa	Wildfire	<0.1	Tiedemann 1973
Washington	Ponderosa	Wildfire + N	0.3	
Washington	Douglas-fir			
M262 CALIFORNIA COASTAL RANGE WOODLAND-SHRUB-CONIFER PROVINCE				
California	Chaparral	Unburned	2.5	Riggan and others 1994
		Moderate burn	9.5	
		Severe burn	15.3	
	Chaparral	Wildfire	<0.1	Taylor and others 1993
313 COLORADO PLATEAU SEMIDESERT PROVINCE				
Arizona	Chaparral	Herbicide, burn	18.4	Davis 1987
		Rx fire alone	12.0	
M313 AZ-NM MOUNTAINS SEMIDESERT-WOODLANDS-CONIFER PROVINCE				
Arizona	Ponderosa	Wildfire	0.2	Campbell and others 1977
Arizona	Ponderosa	Rx fire	<0.1	Gottfried and DeBano 1990
M313 SOUTHERN ROCKY MOUNTAIN STEPPE-CONIFER-MEADOW PROVINCE				
Idaho	Douglas-fir	Slash burn	<0.1	Clayton and Kennedy 1985
M333 NORTHERN ROCKY MOUNTAIN STEPPE-CONIFER-MEADOW PROVINCE				
Montana	Lodgepole	Wildfire	0.3	Hauer and Spencer 1998

Tiedemann (1973) and Tiedemann and others (1978) show that fertilizer application following fire resulted in higher concentrations of NO_3-N in streamflows than fire alone. It is probably safe to conclude, however, that neither fire nor fertilization after fire will have adverse effects on NO_3-N in drinking water (Tiedemann and others 1978, Landsberg and others 1999, Scatena 2000).

NO_2-N was reported by itself rather than in combination with NO_3-N in two studies on the effects of fire on the chemical concentrations of streamflows in California and Washington. At the Lexington Reservoir in southern California, Taylor and others (1993) found NO_2-N levels of 0.03 ppm (mg/L) in streamflows after a wildfire occurred in a subwatershed that drains into the reservoir, while control levels were 0.01 ppm (mg/L). Tiedemann (1973) reported that NO_2-N concentrations were below that of analytical detection after a wildfire occurred in eastern Washington.

Other studies on the effects of wildfire or prescribed burning on both N and other dissolved chemical constituents are summarized in the following paragraphs. The observed elevated postfire ionic concentrations observed in these studies returned to prefire levels after the first flush of flows following the fire in most cases.

The maximum concentrations of NO_3-N in streamflows increased from less than 0.016 to 0.56 ppm (mg/L) in the 3 years after a wildfire destroyed the mixed conifer forest cover on a 1,410 acre (564 ha) watershed in eastern Washington (Tiedemann and others 1978). These increases appeared to be a result of increased nitrification. Concentrations of total P in the streamflow from the burned watershed were 1.5 to three times greater than those from an undisturbed watershed. Despite the lack of prefire information, elevated levels of these magnitudes indicate that the wildfire significantly affected the P levels in streamflow, at least in the short term. The combined concentrations of Ca, Mg, K, and Na in streamflow before the fire ranged from 12.0 to 14.9 ppm (mg/L). The concentrations declined to a range of 7.4 to 10.5 ppm (mg/L) in the second postfire year because of the dilution caused by increased streamflow. These losses were insignificant relative to the total capital of these nutrients, however.

The chemical quality of streamflows from two small watersheds that originally supported ponderosa pine forests in northern Arizona was not greatly affected by a wildfire (Campbell and others 1977). One watershed of 20.2 acres (8.1 ha) was severely burned while a smaller watershed of 10 acres (4 ha) was moderately burned. While concentrations of Ca, Mg, and K increased with the first postfire streamflow events from these watersheds, concentrations of these ions decreased rapidly in subsequent flow events. Sodium (Na) concentrations were largely unaffected by the fire

or the observed changes in streamflow discharges. Combined organic-inorganic N concentrations also increased in the initial postfire streamflow event and then quickly decreased to the level that was found in streamflow from the unburned watershed.

The magnitude of stream water quality changes for N after prescribed fire are normally less than those observed after wildfires. It is unlikely that prescribed burning would consume as much of the litter and other organic materials, understory herbaceous vegetation, or overstory trees as severe wildfires (McNabb and Cromack 1990, DeBano and others 1998). Stream chemistry responses to prescribed fire in an undisturbed 1,163 acre (470 ha) Southwestern ponderosa pine forest watershed (Gottfried and DeBano 1990) support the speculation of minimal changes in N following a prescribed fire. Surface fuels were burned on 43 percent of this watershed, and 5 percent of the trees were killed. The prescribed fire resulted in relatively small increases in NO_3-N in the streamflow, the levels of which did not approach the primary water quality standard. Measures taken to protect streams and riparian corridors with unburned buffers could also minimize effects of fire on stream chemistry.

Several studies show that increased streamflow discharges and increased concentrations of N in the streamflow from burned areas can cause an accelerated loss of N from watershed lands in the short term. While these losses in N have not been referenced to the total N capital on a watershed, these losses likely pose little threat to continued onsite productivity (DeBano and others 1998, Brooks and others 2003).

The primary drinking water standard for NO_3-N is 10 ppm (mg/L), and the standard for NO_2-N is 1 ppm (m/L). There is no standard for dissolved organic N, NH_4-N, or urea-N in drinking water. The combined concentrations of these N forms seldom exceed 1 ppm (mg/L), and dissolved organic N is usually the most abundant form.

Phosphorus

Phosphorus is present in several forms in soil solution and streamflow. These are reactive orthophosphate (inorganic phosphate), dissolved complex organic phosphate, particulate organic phosphate, and other inorganic forms (Ice 1996). Total phosphate (PO_4-P) is reported as total P in most of the studies of the P response to fire. PO_4-P is not as readily leached as NO_3-N because it complexes with organic compounds in the soil (Black 1968). Studies of soil leachates have reported increased levels of total phosphorous due to burning, indicating accelerated mobilization of phosphorous after burning (McColl and Grigal 1975, Knighton 1977). PO_4-P concentrations in streamflows prior to fire can range from 0.007 to 0.17 ppm (mg/L)

USDA Forest Service Gen. Tech. Rep. RMRS-GTR-42-vol. 4. 2005

127

(Longstreth and Patten 1975, Hoffman and Ferreira 1976, Wright and others 1976, Tiedemann and others 1978, Tiedemann and others 1988). After wildfire, prescribed burning, or the clearcutting of timber followed by broadcast burning of the slash, Longstreth and Patten (1975) reported that PO_4-P concentrations stayed the same or increased only as high as 0.2 ppm (mg/L).

Phosphorus concentrations in overland flow from the hillslopes of a watershed can increase as a result of burning, although these increases are not always sufficient to alter the quality of the watershed's streamflow regime (Longstreth and Patten 1975, Gifford and others 1976). There is no established standard for PO_4-P in drinking water. Phosphorus can be limiting in aquatic habitats, but once in the water, it is taken up quickly by aquatic organisms, especially algae.

Sulfur

Sulfate (SO_4-S) is relatively mobile in soil-water systems (Johnson and Cole 1977). Although not as well studied as those for N, the mineralization processes for S are essentially similar. Observed levels of SO_4-S in the streamflow from most wildland watersheds are inherently low. Prefire concentrations of the ion can range from as low as 1.17 ppm (mg/L) to as high as 66 ppm (mg/L), while postfire values range from 1.7 to 76 ppm (mg/L). All of the SO_4-S concentrations reported by Landsberg and Tiedemann (2000) were below the secondary water quality standard for drinking water of 250 ppm (mg/L) (U.S. Environmental Protection Agency 1999).

Chloride

Chloride ion (Cl) responses to fire have been documented in several studies (Landsberg and Tiedemann 2000). All responses were significantly lower (less than 5 ppm, or mg/L) than the water quality standard of 250 ppm (mg/L) established by the U.S. Environmental Protection Agency (1999). Concentrations of some natural sources of Cl (such as geothermal areas) are reportedly larger than those produced by wildfires.

Bicarbonate

Bicarbonate ions in soil solutions and streamflows are often increased as a consequence of burning. The bicarbonate ion represents the principal anion in soil solution, the end product of root respiration and a product of oxide conversion after fire (McColl and Cole 1968, Tiedemann and others 1979, DeBano and others 1998). Concomitant fluctuations of bicarbonate and cation concentrations indicate that bicarbonate is a main carrier of cations in the soil solution (Davis 1987).

Total Dissolved Solids

In their synthesis of the scientific literature on the quality of drinking water originating from natural ecosystems, Landsberg and Tiedemann (2000) found only two studies that reported concentrations of total dissolved solids (TDS), although the investigators in other studies have measured some of the constituents of TDS but not TDS per se. Hoffman and Ferreira (1976) detected TDS concentrations of about 11 ppm (mg/L) in the streamflow from an unburned area in Kings Canyon National Park of California and 13 ppm (mg/L) in the streamflow from an adjacent burned area, which had been a mixed conifer and shrub stand. Lathrop (1994) found that Yellowstone Lake and Lewis Lake in Yellowstone National Park had prefire TDS concentrations of 65.8 and 70 ppm (mg/L), respectively, and that these concentrations were similar to those observed after the fires. These values were significantly below the secondary standard of TDS for drinking water of 500 ppm (mg/L) recommended by the U.S. Environmental Protection Agency (1999).

Nutrients and Heavy Metals

Heavy metals are a growing concern in some forested areas of the United States. Particular concern exists about the release of heavy metals to the air and eventually to streams by increased prescribed fire programs and large wildfires (Lefevre, personal communication). Information on this aspect of water quality and fire use and management is relatively scarce.

One potentially important source of nutrient and heavy metal loss from a watershed that is often ignored is that transported by sediment particles (Gifford and Busby 1973, Fisher and Minckley 1978, Gosz and others 1980, Brooks and others 2003). Sediment has been reported to transport relatively high levels of nutrients and heavy metals (Angino and others 1974, Potter and others 1975), although most of these investigators have looked at large river basins consisting of numerous combinations of bedrock and vegetation.

Few studies have focused directly on the effects of fire on nutrient and heavy metal losses that are transported by sediment from smaller upland watersheds. One study reported that sediment losses of N, P, and ions in streamflow from burned watersheds in chaparral shrublands of southern California chaparral can substantially exceed those lost in solution (DeBano and Conrad 1978). Nitrogen and P losses of 13.5 and 3.0 lb/acre (5.1 and 3.4 kg/ha), respectively, were found in sediments, as compared to only trace levels in solution. Furthermore, losses of Ca, Mg, Na, and K in solution were only one-fourth of the losses in sediment. However, sediment and solution losses of N, P, K, Mg, Ca, and Na were only a small fraction of the total prefire nutrient capital of plants, litter accumulations, and the upper 4 inches (10 cm) of soil.

Biological Quality of Water

Accumulations of litter and of other decomposed organic matter on the soil surface often function as a filter that removes bacteria and other biological organisms from overland flow. Rainfall-induced runoff and snowmelt that percolates through a litter layer or strip of organic matter can contain fewer bacteria than water that had not passed through the strip (DeBano and others 1998, Brooks and others 2003). It follows, therefore, that the destruction of this layer or strip by burning might result in higher concentrations of bacterial and other biological organisms flowing overland to a stream channel.

Fire Retardants

Fire retardants are frequently used in the suppression of wildfires. Although their effects on the soil-water environment are not a direct effect of fire, their use in the control of wildfires can produce adverse environmental impacts. A brief discussion of fire retardant effects on water quality is provided here to acquaint the reader with the topic. More detailed reviews and studies have been completed by Labat and Anderson Inc. (1994), Adams and Simmons (1999), Kalabokidis (2000), and Gimenez and others (2004). The main environmental concerns with fire retardant use are: (1) effects on water quality and aquatic organisms, (2) toxicity to vegetation, and (3) human health effects.

Ammonium-based fire retardants (diammonium phosphate, monoammonium phosphate, ammonium sulfate, or ammonium polyphosphate) play an important role in protecting watershed resources from destructive wildfire (fig. 6.4; table 6.6). However, their use can affect water quality in some instances, and they can also be toxic to aquatic biota (table 6.7; see also chapter 7).

Nitrogen-containing fire retardants have the potential to affect the quality of drinking water, although the research on the applications of these retardants to streams has largely focused on their impacts on aquatic environments (Norris and Webb 1989). For example, in an in vitro study to determine the toxicity of some retardant formulations to stream organisms, McDonald and others (1996) evaluated the impacts of two nonfoam retardants containing SO_4-S, PO_4-P, and ammonium compounds (Fire-trol GST-R and Phos-Chek D75-F), a retardant containing ammonium and phosphate compounds (Fire-Trol LCG-R), and two foam suppressant compounds that contained neither S, P, nor ammonium compounds (Phos-Chek WD-881 and

Figure 6.4—Fire retardant drop from an contract P-3 Orion, San Bernardino National Forest, California. (Photo by USDA Forest Service).

Silv-Ex). These investigators found that concentrations of NO_3-N in water rose from (0.08) to 3.93 ppm (mg/L) after adding the nonfoam retardants. They also discovered that NO_2-N reached concentrations as high as 33.2 ppm (mg/L), well above the primary water quality standard of 1 ppm (mg/L). However, the solutions they tested were much less concentrated than that which is used in firefighting.

The main chemical of concern in streams 24 hours after a retardant drop is ammonia nitrogen (NH_3 + NH_4^+; table 6.6). Un-ionized ammonia (NH_3) is the principal toxic component to aquatic species. The distances downstream in which potentially toxic conditions persist depend on stream volume, the number of retardant drops, and the orientation of drops to the stream long-axis (table 6.8). While concentrations of

Table 6.6—Composition of forest fire retardants (Adapted from Johnson and Sanders 1977).

Trade name and chemical components	Amount of component	
	Concentrate	Field mix
	- - - - - - - ppm(mg/L) - - - - -	
FIRE-TROL 100		
$(NH_4)_2SO_4$	635,000	178,624
NH_4	173,353	48,764
FIRE-TROL 931		
Ammonium polyphosphate	930,000	268,122
NH_4	119,235	34,376
PHOS-CHEK 202		
$(NH_4)_2HPO_4$	833,500	114,085
NH_4	241,372	31,168
PHOS-CHEK 259		
$(NH_4)_2HPO_4$	919,500	155,358
NH_4	251,207	42,443

USDA Forest Service Gen. Tech. Rep. RMRS-GTR-42-vol. 4. 2005

129

Table 6.7—Fire retardant lethal levels for aquatic organisms (Adapted from Johnson and Sanders 1977).

Organism	24 hour LC50 concentration[1]			
	FIRE-TROL 100	FIRE-TROL 931	PHOS-CHEK 202	PHOS-CHEK 259
	----------------- *(ppm) mg/L* -----------------			
Coho salmon				
Yolk-sac fry	160	> 500	175	> 200
Swim-up fry	1,100	1,050	210	175
Fingerling	> 1,500	1,050	320	250
Rainbow trout				
Yolk-sac fry	158	> 500	140	> 200
Swim-up fry	900	780	210	175
Fingerling	> 1,000	> 1,000	230	175
Bluegill	> 1,500	> 1,500	840	600
Fathead minnow	> 1,500	> 1,500	820	470
Largemouth bass	> 1,500	> 1,500	840	720
Scud	> 100	> 100	100	> 100

[1]LC50 concentration is the concentration needed to kill 50 percent of the exposed individuals in a given time period (24 hr).

$NH_3 + NH_4^+$ can reach 200 to 300 ppm (mg/L) within 164 to 328 feet (50-100 m) below drop points, toxic levels may persist for over 3,280 feet (1,000 m) of stream channel.

Inadvertent applications of fire retardants into a stream could have water quality consequences for NO_3-N, SO_4-S, and possibly trace elements. However, information about these potential effects of retardants on water quality is limited.

Another source of concern is fire retardants that contain sodium ferrocyanide (YPS) (Little and Calfee 2002). Photo-enhanced YPS has a low lethal concentration in water for aquatic organisms. Fortunately, YPS, like the other retardant chemicals, is adsorbed onto organic and mineral cation exchange sites in soils. Thus, its potential for leaching out of soils is reduced.

Gimenez and others (2004) concluded that the most significant environmental impact of fire retardants is the toxic effect on aquatic organisms in streams. They noted that the amount of fire retardant used and its placement on the landscape are the two main factors determining the degree of environmental impact. Thus, placement planning and operational control of fire retardant aircraft are critical for minimizing impacts on streams and lakes and their biota.

Table 6.8—Fish mortality related to a hypothetical fire retardant drop orientation (Adapted from Norris and Webb 1989).

Angle to long axis of a stream		Distance for 100% mortality	
Degrees	Position	Standard drop	2 standard drops
		----------- *ft (m)* -----------	
90	Perpendicular	164 (50)	1,575 (480)
67		164 (50)	1,837 (560)
45		328 (100)	3,281 (1,000)
22		787 (240)	>3,281 (>1,000)
0	Over stream	3,281 (1,000)	>3,281 (>1,000)

Rodeo-Chediski Fire, 2002: A Water Quality Case History

The Rodeo-Chediski Fire of 2002 was the largest wildfire in Arizona's history (Neary and Gottfried 2002, Ffolliott and Neary 2003). This fire damaged, destroyed, or disrupted the hydrologic functioning and ecological structure of the ponderosa pine forest ecosystems at the headwaters of the Salt River, a major river supplying the city of Phoenix's main water supply reservoir, Lake Roosevelt. The Rodeo-Chediski Fire was actually two fires that ignited on lands of the White Mountain Apache Nation and merged into one. Arson was the cause of the Rodeo Fire, which began a few miles from Cibecue, a small streamside village, on June 18, 2002. The Chediski Fire was set on the Reservation as a signal fire by a seemingly lost person a few days later. This second fire spread out of control, moving toward and eventually merged with the ongoing and still out of control Rodeo Fire. Burning northeastwardly, the renamed Rodeo-Chediski Fire then burned onto the Apache-Sitgreaves National Forest, along the Mogollon Rim in central Arizona, and into many of the White Mountain tourist communities scattered along the Mogollon Rim from Heber to Show Low. Over 30,000 local people were eventually forced to flee the inferno.

The Rodeo-Chediski Fire had burned 276,507 acres (111,898 ha) of Apache land, and the remainder of the total of 467,066 acres (189,015 ha) were on the Apache-Sitgreaves National Forest. Nearly 500 buildings were destroyed; more than half of the burned structures were houses of local residents or second homes of summer visitors. Rehabilitation efforts began immediately after the fire was controlled and after it was declared safe to enter into the burned area. Two BAER teams operated out of the White Mountain Apache Tribe headquarters at White River, AZ, and the other for the Apache-Sitgreaves National Forest at Show Low, AZ. Watershed protection was of prime importance because of the municipal watershed values of these lands.

Culverts were removed from roads to help mitigate the anticipated flash flooding that is often caused by high-intensity, short-duration monsoonal rainfall events that commonly occur in Arizona from early July through August and, occasionally, into September. Temporary detention dams and other diversions were constructed to divert intermittent water flows initiated by these storms away from critical infrastructures. In a major rehabilitation activity, helicopters ferried bales of straw to burned sites susceptible to erosion, and the straw was spread onto the ground to alleviate the erosive impact of the monsoonal rain. Seeding of rapidly established grasses and other herbaceous plants accompanied this rehabilitative activity, which continued into the late summer and early autumn.

Water-quality constituents of primary interest to hydrologists, land managers, and watershed managers in Southwestern ecosystems are sediment concentrations and dissolved nutrients, specifically nitrogen and phosphorus (DeBano and others 1996). In July, monsoon thunderstorms initiated storm runoff from the wildfire-burned area. While there was not an opportunity to collect samples to determined the quality characteristics of the water flows from several watersheds within the Rodeo-Chediski Fire area (Stermer Ridge watersheds; Ffolliott and Neary 2003), samples were taken of the ash and sediment-laden streamflow farther downstream where the Salt River enters into Lake Roosevelt. The storm runoff streamflows contained large amounts of organic debris, dissolved nutrients (including nitrogen, phosphorus, and carbon), and other chemicals that were released by the fire. Some of the elevated concentrations of nitrogen and phosphorus may have originated from the fire retardants dropped to slow the advancement of the fire.

The sediment- and organic-rich water significantly increased the flow of water into the Salt River, the major tributary to the Theodore Roosevelt Reservoir, a primary source of water for Phoenix and its surrounding metropolitan communities. The Salt River Project provides drinking and irrigation water, as well as power, to 2 million residential, business, and industrial customers in a 2,900 mile^2 (7,511 km^2) service area in parts of Maricopa, Gila, and Pinal Counties, Arizona (Autobee 1993). Water is furnished primarily by the Salt and Verde Rivers, which drain a watershed area of 13,000 miles^2 (33,670 km^2). Four storage reservoirs on the Salt River form a continuous chain of lakes almost 60 miles (96 km) long. An important supplemental supply is obtained from well pumping units. Capacity assigned to flood control is 556,000 acre-feet (685.8 million m^3). Total storage capacity of Salt River reservoirs is more than 2.4 million acre-feet (3.0 trillion m^3). Total hydroelectric generating capacity is 232 megawatts, including power from pumped storage units.

What made this situation more serious than would be expected were the size of the Rodeo-Chediski Fire and the critically low level of the drought-impacted Roosevelt reservoir, which was less than 15 percent of its capacity at the time of the fire. The reservoir has a drainage area of 5,830 miles^2 (3.7 million acres or 1.5 million ha) of which about 12 percent was burned in the Rodeo-Chediski Fire. Even with a small percent burned, there was a concern that the flow of ash and debris might threaten the aquatic life inhabiting the reservoir, leaving it lifeless for months to come. However, this dire situation failed to materialize. Even

USDA Forest Service Gen. Tech. Rep. RMRS-GTR-42-vol. 4. 2005

131

though some fish died upstream of the reservoir and a few carcasses showed up at the diversion dam above Roosevelt Dam, the reservoir water body itself suffered little permanent environmental damage.

The largest pulses in water quality parameters were for suspended sediment, conductivity, and turbidity. Nutrient levels (particularly P and N) in the water shot off the chart in the first few days of monsoon-induced storm runoff (U.S. Geological Survey 2002, Tecle, personal communication; table 6.9) but fell quickly, and therefore, the large algae blooms that were predicted to form and consume much of the water's oxygen did not form. The postfire debris in the water was never a health risk to people as most of the pollutants were easily removed at water treatment plants. The outcome would be different in a situation where a much larger percentage of a municipal watershed is burned by a wildfire.

Management Implications _____

A number of management considerations relate to water quality and the use or management of prescribed fires and wildfires. These considerations are tied to both Federal and State regulations and laws such as the National Forest Management Act of 1976, the Clean Water Act of 1972 as amended, the Code of Federal Regulations (36 CFR 219.13), and State laws. The purpose of these laws and regulations is to conserve soil and water resources, minimize serious or long-lasting hazards from flood, wind, wildfire, erosion, or other natural forces, and to protect streams, lakes, wetlands, and other bodies of water. The approach of these regulations is to use Best Management Practices to achieve water quality goals and protection. Special attention is given to land and vegetation in recognizable areas dominated by riparian vegetation.

Several sources of conservation measures guidelines relate to the use of prescribed fire (table 6.10) and the management of wildfire (table 6.11) to maintain water quality. These include the U.S. Environmental Protection Agency (1993), the USDA Forest Service (1988, 1989a,b, 1990a), California Department of Forestry (1998), Georgia Forestry Association (1995), to mention just a few.

In some instances conservation measures may be prohibitive or at least impractical because of local terrain conditions, hydrology, weather, fuels, or fire behavior. The ground rule is to use common sense when applying these guidelines. If conservation measures or actual State-mandated Best Management Practices are to be used, it is important to clearly communicate objectives and goals at briefings and on operational period fire plans. Assistance from staff hydrologists and soil scientists is also crucial in successfully applying these water quality protection guidelines.

Table 6.9—Water quality at the Salt River entrance to Lake Roosevelt, Arizona, Salt River Basin, before and after the Rodeo-Chediski Fire, 2002, compared to existing water quality standards (U.S. Geological Survey 2002).

Water quality parameter	Unit	Drinking water standard	Prefire level	Postfire peak
Arsenic	mg/L	0.05	0.05-0.350	0.685
Bicarbonate	mg/L	380.0	80-250	312.0
Calcium	mg/L	50.0	30-85	144.0
Chloride	mg/L	250.0	100-1,100	2,110.0
Copper	mg/L	1.0	<1	0.375
Iron	mg/L	0.3	<0.3	90.6
Lead	mg/L	0.05	0.0-0.10	0.69
Magnesium	mg/L	20.0	8-40	45.0
Phosphorus	mg/L	0.1	0.0-0.1	39.0
Potassium	mg/L	5.0	2-15	26.0
Sulfate	mg/L	100.0	20-140	170.0
Total nitrogen	mg/L	10.0	<10	220.0
Dissolved oxygen	mg/L	5.0	8.2-8.6	7.4
Sediment	mg/L	500.0	10-500	25,800
Conductivity	usiemens/cm	1,650	800-4,000	6,970
Turbidity	NTUs	1	8-110	51,000

Table 6.10—Suggested conservation measures for prescribed fires.

Fire type	Item	Best Management Practice
Prescribed	1	Carefully plan burning to adhere to weather, time of year, and fuel conditions that will help achieve the desired results and minimize impacts on water quality.
	2	Evaluate ground conditions to control the pattern and timimg of the burn.
	3	Intense prescribed fire for site preparation should not be conducted in Streamside Management Zones (SMZ) except to achieve riparian vegetation management objectives.
	4	Piling and burning for slash removal should not be conducted in SMZs.
	5	Avoid construction of firelines in SMZs.
	6	Avoid conditions requiring extensive blading of firelines by heavy equipment.
	7	Use handlines, firebreaks, and hoselays to minimize blading of firelines.
	8	Use natural or in-place barriers to minimize the need for fireline construction.
	9	Construct firelines in a manner that minimizes erosion and prevents runoff from directly entering watercourses.
	10	Locate firelines on the contour whenever possible, and avoid straight up-downhill placement
	11	Install grades, ditches, and water bars while the line is being constructed.
	12	Install water bars on any fireline running up-down slope, and direct runoff onto a filter strip or sideslope, not into a drainage.
	13	Construct firelines at a grade of 10 percent or less where possible.
	14	Adequately cross-ditch all firelines at time of construction.
	15	Construct simple diversion ditches or turnouts at intervals as needed to direct surface runoff off a plowed line and onto undisturbed forest floor or vegetation for dispersion of water and soil particles.
	16	Construct firelines only as deep and wide as necessary to control the spread of the fire.
	17	Maintain the erosion control measures on firelines after the burn.
	18	Revegetate firelines with adapted herbaceous species. Native plants are preferable when there are adequate sources of seed.
	19	Execute burns with a well-trained crew and avoid high-severity burning.

Table 6.11—Suggested conservation measures for wildfires.

Fire type	Item	Best Management Practice
Wildfire	1	Review and use BMPs listed for prescribed fires when possible.
	2	Whenever possible, avoid using fire-retardant chemicals in SMZs and over watercourses, and use measures to prevent their runoff into watercourses.
	3	Do not clean retardant application equipment in watercourses or locations that drain into waterways.
	4	Close water wells excavated for wildfire suppression activities as soon as practical following fire control.
	5	Provide advance planning and training for firefighters that considers water quality impacts when fighting wildfires. This can include increasing awareness so direct application of fire retardants to waterbodies is avoided and firelines are placed in the least detrimental position.
	6	Avoid heavy equipment use on fragile soils and steep slopes.
	7	Implement Burned Area Emergency Rehabilitation (BAER) team recommendations for watershed stabilization as soon as possible.

USDA Forest Service Gen. Tech. Rep. RMRS-GTR-42-vol. 4. 2005

133

Summary _____

When a wildland fire occurs, the principal concerns for change in water quality are: (1) the introduction of sediment; (2) the potential increasing nitrates, especially if the foliage being burned is in an area chronic atmospheric deposition; (3) the possible introduction of heavy metals from soils and geologic sources within the burned area; and (4) the introduction of fire retardant chemicals into streams that can reach levels toxic to aquatic organisms.

The magnitude of the effects of fire on water quality is primarily driven by fire severity, and not necessarily by fire intensity. Fire severity is a qualitative term describing the amount of fuel consumed, while fire intensity is a quantitative measure of the rate of heat release (see chapter 1). In other words, the more severe the fire, the greater the amount of fuel consumed and nutrients released, and the more susceptible the site is to erosion of soil and nutrients into the stream where they could potentially affect water quality. Wildfires usually are more severe than prescribed fires. As a result, they are more likely to produce significant effects on water quality. On the other hand, prescribed fires are designed to be less severe and would be expected to produce less effect on water quality. Use of prescribed fire allows the manager the opportunity to control the severity of the fire and to avoid creating large areas burned at high severity. The degree of fire severity is also related to the vegetation type. For example, in grasslands the differences between prescribed fire and wildfire are probably small. In forested environments, the magnitude of the effects of fire on water quality will probably be much lower after a prescribed fire than after a wildfire because of the larger amount of fuel consumed in a wildfire. We expect canopy-consuming wildfires to be the greatest concern to managers because of the loss of canopy coupled with the destruction of soil aggregates. These losses present the worst-case scenario in terms of water quality. The differences between wild and prescribed fire in shrublands are probably intermediate between those seen in grass and forest environments.

Another important determinant of the magnitude of the effects of fire on water quality is slope. Steepness of the slope has a significant influence on movement of soil and nutrients into stream channels where it can affect water quality. Wright and others (1976) found that as slope increased in a prescribed fire, erosion from slopes accelerated. If at all possible, the vegetative canopy on steep, erodible slopes needs to be maintained, particularly if adequate streamside buffer strips do not exist to trap the large amounts of sediment and nutrients that could be transported quickly into the stream channel. It is important to maintain streamside buffer strips whenever possible, especially when developing prescribed fire plans. These buffer strips will capture much of the sediment and nutrients from burned upslope areas.

Nitrogen is of concern to water quality. If soils on a particular site are close to N saturation, it is possible to exceed maximum contamination levels of NO_3-N (10 ppm or 10 mg/L) after a severe fire. Such areas should not have N-containing fertilizer applied after the fire. Review chapter 3 for more discussion of N. Fire retardants typically contain large amounts of N, and they can cause water quality problems where drops are made close to streams.

The propensity for a site to develop water repellency after fire must be considered (see chapter 2). Water-repellent soils do not allow precipitation to penetrate down into the soil and therefore are conducive to erosion. Severe fires on such sites can put large amounts of sediment and nutrients into surface water.

Finally, heavy rain on recently burned land can seriously degrade water quality. Severe erosion and runoff are not limited to wildfire sites alone. But if postfire storms deliver large amounts of precipitation or short-duration high intensity rainfalls, accelerated erosion and runoff can occur even after a carefully planned prescribed fire. Conversely, if below-average precipitation occurs after a wildfire, there may not be a substantial increase in erosion and runoff and no effect on water quality.

Fire managers can influence the effects of fire on water quality by careful planning before prescribed burning. Limiting fire severity, avoiding burning on steep slopes, and limiting burning on potentially water-repellent soils will reduce the magnitude of the effects of fire on water quality.

John N. Rinne
Gerald R. Jacoby

Chapter 7: Aquatic Biota

Fire Effects on Fish _____

Prior to the 1990s, little information existed on the effects of wildfire on fishes and their habitats (fig. 7.1). Severson and Rinne (1988) reported that most of the focus of fire effects on riparian-stream ecosystems—that is, habitat for fishes—was on hydrological and erosional responses to vegetation removal and resultant effects on sedimentation and water quality. Most of these studies were conducted in the 1970s (Anderson 1976, Tiedemann and others 1978) and examined water quality and quantity effects, algae, and aquatic micro- and macroinvertebrates. Accordingly, they addressed the potential "indirect effects" of fire on fishes; none addressed the "direct effects" of fire on fishes.

The Yellowstone Fires in 1988 ushered in an extensive effort to examine both the direct and indirect effects of wildfire on aquatic ecosystems (Minshall and others 1989, Minshall and Brock 1991). Other studies on fishes were conducted following some of the historically worst wildfires in the Southwestern United States

Figure 7.1—Crowning Clear Creek Fire, Salmon, Idaho. (Photo by Karen Wattenmaker).

USDA Forest Service Gen. Tech. Rep. RMRS-GTR-42-vol. 4. 2005

135

(Rinne 1996, Rinne and Neary 1996). The result is that most of the information available on fire effects on fishes and their habitats has been generated in the 1990s and on a regional basis. By the late 1990s, summary and review papers on the topic of fire and aquatic ecosystems, including fishes, were drafted (Reiman and Clayton 1997, Gresswell 1999), which brought 20th century information together and suggested future research and management direction for both corroborating aquatics-fisheries and fire management and conservation of native, sensitive species (fig. 7.2). This state-of-our-knowledge documentation is what we use to base 21st century fisheries and aquatic management relative to both wild and prescription fires.

Because of the increased number, size, and intensity of fires commencing in 2000 in the Western and Southwestern United States, we now see a marked increase in data describing the impact of post-wildfire events on fishes (Rinne and Carter, in press, Rinne 2003a,b). While before the turn of the century we had information on fire effects on salmonid species only, now the data base includes a dozen native fish species and several nonnative fish species (table 7.1). Key messages from these data are discussed below.

Direct Fire Effects

Fire can result in immediate mortalities to fishes, but few studies have documented direct mortality following wildfire (McMahon and de Calesta 1990, Minshall and Brock 1991, Reiman and others 1997). High severity fire and heavy fuel and slash accumulations in the riparian zone are the common predisposing factors for direct fish mortality. Moring and Lantz (1975) found some fish mortality after prescribed fire conducted after harvesting in the Alsea Basin of

Figure 7.2—Wildfire encroaching on a riparian area, Montana, 2002. (Photo courtesy of the Bureau of Land Management, National Interagency Fire Center, Image Portal).

Oregon. The fish kill was confined to stream heads where slash accumulations were heavy and the prescribed fire reached levels of high severity. Rinne (1996) found no significant reduction in densities of fishes in three streams as a direct result of a large fire (Dude Fire, Arizona, 1990) that burned across the watersheds encompassing them. Although the Dude Fire had large areas of high severity fire, many of the riparian areas either suffered only low severity fire or did not burn at all. Reiman and others (1997) reported both dead fish and reaches of stream with no live fish after a high severity fire burned through two riparian corridors in Idaho. On the other hand, no mortalities of the endangered Gila trout were reported immediately after the Divide Fire in southwestern New Mexico (Propst and others 1992).

Key factors in immediate postfire fish mortality are the size of the riparian area, the riparian fuel load, fire severity, and stream size. Small streams with high fuel loads and high severity fire are the ones most likely to suffer immediate aquatic organism mortality from fire.

Fire retardants can also be a source of fish mortality (chapter 6 this volume, Van Meter and Hardy 1975). Dead fish have been reported following fire retardant application (Jones and others 1989); however, documentation is poor (Norris and Webb 1989). The number of retardant drops and orientation to the stream are key factors determining fish mortality.

Indirect Fire Effects

The indirect effects of large fires on fishes are better documented and can be significant (Reiman and Clayton 1997). Within 2 weeks after the Divide Fire in New Mexico, a single Gila trout was collected from Main Diamond Creek. Sampling 3 months later suggested extirpation of this endangered species from this headwater stream (Rinne and Neary 1996). Similarly, Rinne (1996) reported dead fishes on streambanks 2 weeks postfire and documented only a single brook trout at spring outflow in Dude Creek remaining 3 months after the Dude Fire. In both cases, "slurry ash flows" appeared responsible for fish mortality within onset of summer monsoons and basically local extirpation after several months of sustaining flooding of stream corridors resulting from heavy monsoon precipitation and vegetation removal (fig. 7.3).

Rinne and Carter (in press) documented the complete loss of four species and more than 2,000 individual fishes from the fire impacted reaches of Ponil Creek, New Mexico, following the Ponil Complex 2002 wildfire (table 7.2). Drought conditions and stream intermittency combined with ash and flood flows synergistically resulted in this total loss of fishes Rinne and Carter (in press). By comparison, postfire ash and flood flows

136

USDA Forest Service Gen. Tech. Rep. RMRS-GTR-42-vol. 4. 2005

Table 7.1—Species that have been affected by wildfire in the Southwestern United States from 1989 to 2003. Species are listed by respective-named fires and locations. Nonnative species are denoted by an asterisk.

Fire	State	Common name	Scientific name	Stream
Divide	AZ	Gila trout	*Oncorhynchus gilae*	Main Diamond
Dude	AZ	*Rainbow trout	*Oncorhynchus mykiss*	Dude and Ellison
		*Brook trout	*Salvelinus fontinalis*	Bonita Creek
Ponil	NM	*Rainbow trout	*Oncorhynchus mykiss*	Ponil
		Blacknose dace	*Rhinichthys atratulus*	Ponil
		Creek chub	*Semotilus atromaculatus*	Ponil
		White sucker	*Catostomus commersoni*	Ponil
Borrego	NM	*Brown trout	*Salmo trutta*	Rio Medio
Cub Mtn.	NM	Desert sucker	*Catostomus clarki*	West Fork Gila
		Sonora sucker	*Catostomus insignis*	West Fork Gila
		Longfin dace	*Agosia chrysogaster*	West Fork Gila
		Speckled dace	*Rhinichthys osculus*	West Fork Gila
		Spikedace	*Meda fulgida*	West Fork Gila
		Loach Minnow	*Rhinichthys cobitis*	West Fork Gila
		Roundtail chub	*Gila robusta*	West Fork Gila
Picture	AZ	Headwater chub	*Gila nigra*	Turkey (T), R, S
		Speckled dace	*Rhinichthys osculus*	T, Rock (R), S
		Desert Sucker	*Catostomus clarki*	T, R, Spring (S)
		*Green sunfish	*Lepomis cyanellus*	R, S
		*Yellow bullhead	*Ameiurus natalis*	R, S
		*Brown trout	*Salmo trutta*	R, S
Dry Lakes	NM	Gila chub	*Gila intermedia*	Turkey

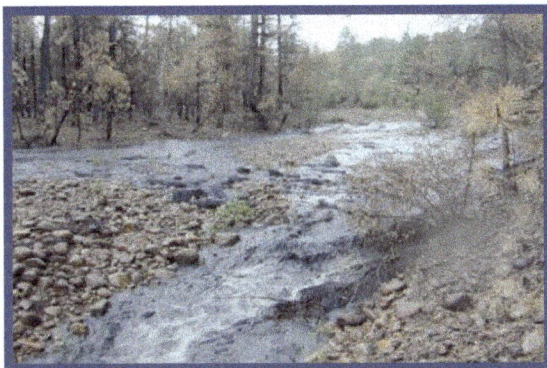

Figure 7.3—Slurry ash flow after the Rodeo-Chediski Fire, Apache-Sitgreaves National Forest, Arizona, 2002. (Photo by Daniel Neary).

Table 7.2—Comparison of fish species abundance and total fish numbers in Ponil Creek, site one, at initial (June) and final autumn (October) sampling, 2002.

Species	Month	
	June	October
	- - - - *Number of fish* - - - -	
Site 1: (Unburned)		
Rainbow trout	18	13
White sucker	8	5
Creek chub	6	2
Longnose dace	15	29
TOTAL: Site 1	47	49
TOTAL: Sites 2-8	1,910	0

USDA Forest Service Gen. Tech. Rep. RMRS-GTR-42-vol. 4. 2005

137

in Rio Medio, New Mexico, following the Borrego Fire, reduced brown trout *Salmo trutta* populations by 70 percent from June to October 2002 (table 7.3). Similarly, six native cypriniform (minnow and sucker) species were reduced 70 percent in the West Fork of the Gila River following the Cub Mountain Fire (table 7.4). Both of these streams had summer, base-flushing flows that apparently diluted the impacts of ash and flood flows from July to October 2002 and enabled fishes to survive in fire-impacted reaches of the river. In 2003, the Picture Fire in Arizona impacted three streams and six species of fishes (including three native species). It markedly altered stream habitat on the Tonto National Forest (Carter and Rinne in press). Overall, fish numbers were reduced by 90 percent (fig. 7.4). Finally, two fires, the Lake Complex in New Mexico and the Aspen Fire in Arizona, probably resulted in total loss of the endangered Gila chub (*Gila intermedia*) in two streams. In summary, short-term data suggest from 70 percent to total loss of fishes in wildfire-impacted reaches of streams in the Southwest (Rinne 2003a,b).

Bozek and Young (1994) noted mortalities of four species of salmonids 2 years postfire following heavy precipitation and flooding on a burned watershed in Wyoming. Death was attributed to the increase in suspended sediment loading. In Montana and Idaho, Novak and White (1989) reported that flooding and debris flows extirpated fish from stream reaches below forest fires

Other Anthropogenic Influencing Factors

Watershed disturbances have occurred on the landscapes of National Forests for the past century. Roads to support timber harvest, timber harvest itself, and grazing of watersheds have cumulatively disturbed and altered watersheds. Dams, diversions, and road culverts, in concert with introduction of nonnative species of fishes, have fragmented and isolated native fish populations (Rinne 2003a,b). Fire suppression itself has greatly altered the vegetative and litter component of forested landscapes (Covington and Moore 1992, Covington and others 1994), increasing tree densities and litter loads. These combined have facilitated intense crowning fires that can be more devastating to fishes and aquatic habitats (Rinne and Neary 1996).

Temporal-Spatial Scales

Most fires that burn on the landscape are less than several acres (1 ha) in size because they are normally suppressed through fire management. By comparison, only 1 percent of the wildfires are responsible for more than 90 percent of the landscaped burned (DeBano and others 1998). Most studies of the effects of fire on fishes and aquatic systems are of short term (less than 5 years) (Gresswell 1999). Although attempts have been made to extrapolate the effects of fire on aquatic systems and organisms to a watershed scale, this in reality has not been achieved to date (Gresswell 1999). Attempts to connect the effects of fire 25 to 50 years previous have been made (Rinne and Neary 1996, Rinne 2003a,b, Albin 1979) but are wanting at best. Sampling of two streams impacted by the Dude Fire (1990), a decade after the fire, suggests impacts of fire on stream fish populations may be

Table 7.3—Pre- and postfire comparison of brown trout densities per 50-m reaches of stream, Rio Medio, Santa Fe National Forest. Percent reductions between June and October are in parentheses. Sites 2 and 3 were not sampled in August.

Site	Month		
	June	August	October
	- - - - - - *Number of fish per 50 m reach* - - - - - -		
1	74	33	21 (72)
2	77	—	19 (75)
3	97	—	18 (86)

Table 7.4—Total numbers of fishes in 50-meter reaches of stream in the West Fork of the Gila River, July and October, 2002, after the Cub Mountain Fire.

Date	Stream condition	Site 1	Site 2
		- - *Number of Fish* - -	
Early July	Post-Wildfire	168	560
Late July	Post-Wildfire, After 1st Storm	278	481
October	Post-Wildfire, After 2nd Storm	50	118

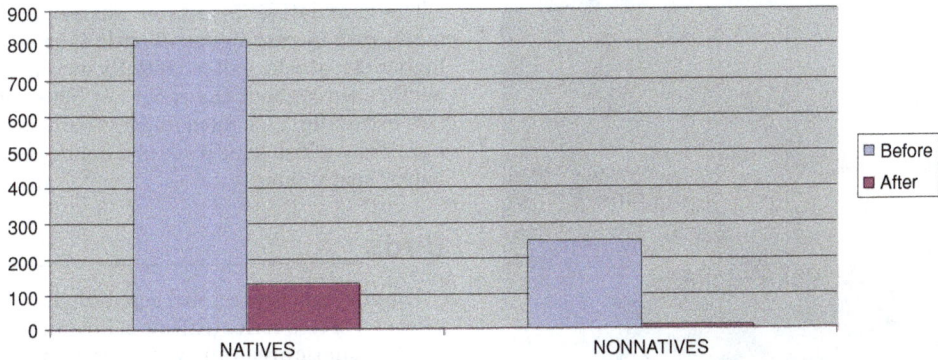

Figure 7.4—Impact of the Picture Fire on fishes in Spring Creek, Tonto National Forest, 2003.

chronic (fig. 7.5). Further, attempting to simplify these effects in time and space is not advisable.

Species Considerations

Recovery of fishes to a stream can occur quickly depending on connectivity of populations and refugia populations to replenish locally extirpated populations (Propst and others 1992, Reiman and Clayton 1997, Gresswell 1999, Rinne and Calamusso in press). In the Southwest, fragmentation of streams as a result of land uses, coupled with fisheries management such as the introduction of nonnative fishes, most often preclude access and replenishing of fishes to a fire-extirpated stream. A big consideration is whether the species is a threatened or endangered

one, such as the Gila trout in the Southwest or the Bull or redband trout in the Pacific Northwest. These cases may necessitate quick removal of surviving fishes prior to ash flows or intense flooding resulting from watershed denudation. In case of put and take, introduced sport fishes, these can always be replenished through stocking. Managers should be vigilant of opportunities to restore native stocks or races of fishes to streams where introduced species have been removed through the aftermath of wildfire. This has been successfully completed for the Gila trout in Dude Creek a decade after extirpation of a brook trout population by the Dude Fire of 1990.

Summary and General Management Implications for Fish

The effects of wildland fire on fish are mostly indirect, with most studies demonstrating the effects of ash flows, changes in hydrologic regimes, and increases in suspended sediment on fishes. These impacts are marked, ranging from 70 percent to total loss of fishes. There are some documented instances of fires killing fish directly (Reiman and others 1997).

The largest problems arise from the longer term impact on habitat that includes changes in stream temperature due to plant understory and overstory removal, ash-laden slurry flows, increases in flood peakflows, and sedimentation due to increased landscape erosion (fig. 7.6).

Most of information on the effects of wildfire on fishes has been generated since about 1990. The Yellowstone Fires of 1988 resulted in extensive study of fire effects on aquatic ecosystems including fishes. The effects of fire retardants on fishes are observational and not well documented. Further, all information is from forested biomes as opposed to grasslands.

Anthropogenic influences, largely land use activities over the past century, cumulatively influence fire

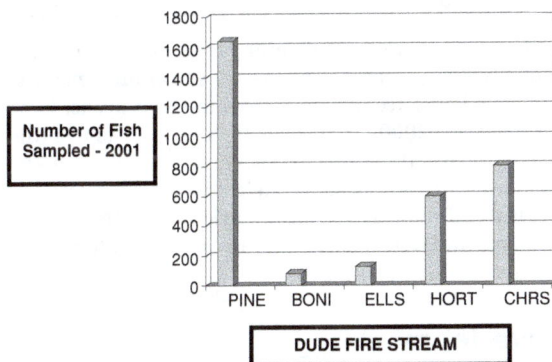

Figure 7.5—Response of fish numbers in two streams, Bonita (BONI) and Ellison (ELLS) Creeks, compared to three nonfire impacted streams, Pine (PINE), Horton (HORT), and Christopher (CHRS) Creeks, 11 years after the Dude Fire, 1990, Tonto National Forest, Arizona.

USDA Forest Service Gen. Tech. Rep. RMRS-GTR-42-vol. 4. 2005

139

Figure 7.6—Four-Corners Fire encroaching on a lake near Crane Prairie, Oregon. (Photo by Tom Iraci).

effects on fishes. Fire suppression alone has affected vegetation densities on the landscape, and affected the severity and extent of wildfire and, in turn, its effects on aquatic ecosystems and fishes. Most studies of fire effects on fishes are short term (less than 5 years) and local in nature. A landscape approach to analyses has not been made to date. Fish can recover rapidly from population reductions or loss but can be markedly limited or precluded by loss of stream connectivity because of human-induced barriers. Fisheries management postfire should be based on species and fisheries and their management status. Managers should be vigilant of opportunities to restore native fishes in event of removal of introduced, nonnative, translocated species.

Recent advances in atmospheric, marine, and terrestrial ecosystem science have resulted in the correlation of ocean temperature oscillations, tree ring data, and drought. These newer advances, along with current information on fire effects on fishes and data on climate change, drought, and recent insect infestations, all combine to show that the greatest impacts on fish may lie ahead. For example, new climate change analyses suggest that the Southeastern and parts of the Western United States have a high probability of continuing and future drought into the next 20 to 30 years, and that the potential impact of wildfire on fishes in the Southwest is marked. Also, the overarching global indicators are starting to unfold: tree ring data indicate 2002 was the driest year in the past 1,000 years (Service 2004); Atlantic and Pacific oceanic temperatures are in turn correlated with historic wet and dry cycles.

The cumulative impacts of warmer and drier climate, with increasing insect outbreaks at higher and higher elevations, will potentially increase the risk of wildfire throughout the ranges of Southwestern native fishes (fig. 7.7). Such impacts increase the future risk of loss of fish gene pools and Southwestern native fishes' sustainability.

Birds

The effects of fires on bird populations are covered in volume 1 of this series (Effects of Fire on Fauna, Smith 2000). Aquatic areas and wetlands often provide refugia during fires. However, wetlands such as cienegas, marshes, cypress swamps, spruce and larch swamps, and so forth do burn under the right conditions. The impacts of fires on individual birds and populations in wetlands would then depend upon the season, uniformity, and severity of burning (Smith 2000).

Reptiles and Amphibians

The effects of fires on reptile and amphibian populations are also covered in volume 1 of this series (Effects of Fire on Fauna, Smith 2000). Russell and others (1999) concluded that there are few reports of fire-caused injury to herpetofauna in general, much less aquatic and wetland species. They noted that aquatic and semiaquatic herpetofauna benefit from wetland fires due to vegetation structure improvement and increases in the surface area of open water. Excessive postfire sedimentation of streams or small standing bodies of water could potentially reduce habitat for reptile and amphibian populations. However, the effects are not well documented.

Mammals

Aquatic and wetland dwelling mammals are usually not adversely impacted by fires due to animal mobility and the lower frequency of fires in these areas. Lyon and others (2000a,b) discuss factors such as fire uniformity, size, duration, and severity that affect mammals. However, most of their discussion relates to terrestrial mammals, not aquatic and wetland ones. Aquatic and wetland habitats also provide safety zones for mammals during fires.

Invertebrates

The fauna volume in this series deals mainly with terrestrial invertebrates (Smith 2000). As with fishes, little information existed on the effects of wildfire on aquatic macroinvertebrates prior to 1980, although some studies were conducted in the 1970s (Anderson

140

USDA Forest Service Gen. Tech. Rep. RMRS-GTR-42-vol. 4. 2005

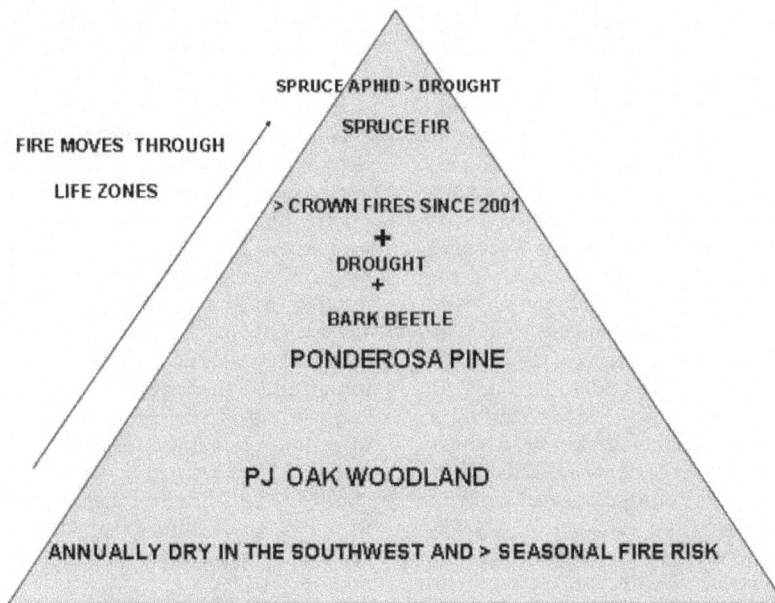

Figure 7.7—Multiple, cumulative impacts are occurring in the West and Southwest to greatly increase the risk of wildfire and further the marked loss of native fishes.

1976). Tiedemann and others (1978) examined water quality and quantity effects, algae, and aquatic micro- and macroinvertebrates. As with fishes, these studies addressed the potential "indirect effects" of fire on aquatic macroinvertebrates rather than any "direct effects."

Parallel with information generation on fishes, the Yellowstone Fires in 1988 ushered in an extensive effort to examine both the direct and indirect effects of wildfire on aquatic ecosystems (Minshall and others 1989, Minshall and Brock 1991). Isolated studies on aquatic macroinvertebrates have been conducted following some of the historically worst wildfires in the Southwestern United States (Rinne 1996). By the late 1990s, a summary and review paper on the topic of fire and aquatic ecosystems, including aquatic macroinvertebrates, was produced (Gresswell 1999). As for fishes, the Gresswell paper is the state-of-the-art reference for the effects of fire on aquatic macro- invertebrates, collates a comprehensive review of infor- mation from the 20th century, and suggests future research and management direction.

Response to Fire

Similar to fishes, the direct effects of fire on macroinvertebrates have not been observed or re- ported. Albin (1979) reported no change in aquatic macro invertebrates abundance during and after fire and no dead macroinvertebrates observable. Rinne (1996) could not document any changes in mean aquatic macroinvertebrate density in three head- water streams from prefire to immediately follow- ing the then-worst wildfire Arizona history, the Dude Fire (table 7.5). However, sampling after initial "ash slurry flows" (in the 2 weeks postfire) revealed an

Table 7.5—Aquatic macroinvertebrate densities following the Arizona Dude Fire, Tonto National Forest, 1990 (Adapted from Rinne 1996).

Stream	Date				
	Prefire	07/03/90	07/26/90	05/20/91	06/08/91
			Number/m² x 1,000		
Dude Creek	5.4	5.0	0.1	2.8	2.5
Bonita Creek	8.6	3.8	0.1	6.0	3.8
Ellison Creek	9.7	10.8	0.0	6.5	2.9

80 to 90 percent reduction in mean densities. Further sampling after several significant flood events determined that macroinvertebrate populations were near zero. Sampling over the next 2 years revealed dramatic fluctuations in density and diversity (number of species) of macroinvertebrates in all three streams.

In general, responses to fire by aquatic macroinvertebrates, as with fishes, are indirect and vary widely in response. Studies in the 1970s by Lotspeich and others (1970) and Stefan (1977) suggested the effects of fire were minimal or undetectable. La Point and others (1996) also reported no difference in macroinvertebrate distribution in streams encompassed by burned and unburned sites nor shifts in water chemistry. Changes in functional feeding groups (Cummins 1978) by aquatic macroinvertebrates have been noted and attributed to substrate stability differences between fire impacted and nonimpacted streams. By comparison, Richards and Minshall (1992) reported that in the first 5 years postfire, macroinvertebrate diversity in streams affected by fire exhibited greater annual variation in diversity than in unaffected streams. Annual fluctuations or variation in diversity did decline with time; however, greater species richness was sustained in streams within burned areas.

As indicated in the introduction, the Yellowstone Fires provided an opportunity to study the effects of fire on aquatic ecosystems and to generate most of the available information on aquatic macroinvertebrate response to fire. In the first year postfire, minor declines in macroinvertebrate abundances, species richness, and diversity were recorded (Robinson and others 1994, Lawrence and Minshall 1994, Minshall and others 1995, Mihuc and others 1996). Two years postfire, these indices had increased but less so in smaller order streams (Minshall and others 1997). Food supply, in the form of unburned coarse particulate material, was suggested to be the factor most important to macroinvertebrates.

Jones and others (1993) reported macroinvertebrate abundance fluctuations in four larger (fifth and sixth order) streams in Yellowstone National Park, yet species diversity and richness did not decline. A change in functional feeding groups from shredder-collector species to scraper-filter feeding species occurred (Jones and others 1993).

Temporally, the effects of fire on macroinvertebrates can be sustained for up to a decade (Roby 1989) or possibly longer (greater than three decades; Albin 1979). Roby and Azuma (1995) studied changes in benthic macroinvertebrate diversity in two streams after a wildfire in northern California (fig. 7.8).

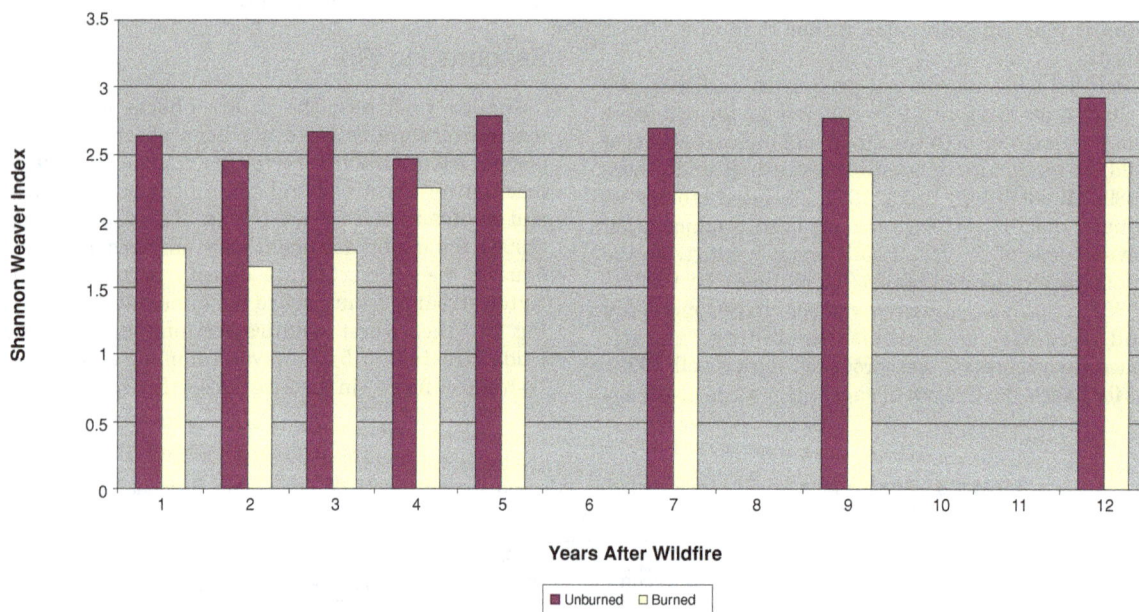

Figure 7.8—Invertebrate diversity changes after a wildfire in northern California. (Adapted from Roby, K.B.; Azuma, D.L. 1995. Changes in a reach of a northern California stream. Figure 2. Environmental Management. Copyright © 1995. With kind permission of Springer Science and Business Media).

Invertebrate diversity, density, and taxa richness in the stream of the burned watershed were low immediately after the fire compared to the unburned watershed. Within 3 years, mean density was significantly higher in the burned watershed (fig. 7.9). A decade after the wildfire, taxa richness and species diversity were still lower in the stream of the burned watershed. Albin (1979) reported lower diversity of macroinvertebrates in reference compared to streams in burned areas. Macroinvertebrate density also was greater in burned streams; however, Chironomidae (immature midges adapted to disturbed areas characterized by fine sediments) were the most abundant group represented in samples. Mihuc and others (1996) suggested changes in physical habitat and availability of food supply are the primary factors affecting postfire response of individual macroinvertebrate populations.

Invertebrate Summary

Similar to fishes, the recorded effects of fire on aquatic macroinvertebrates are indirect as opposed to direct. Most data have been produced during and since the 1990s. Regarding abundance, diversity, or richness, response varies from minimal to no changes, and from considerable to significant changes. Abundance of macroinvertebrates may actually increase in fire-affected streams, but diversity generally is reduced. These differences are undoubtedly related to landscape variability, burn size and severity, stream size, nature and timing of postfire flooding events, and postfire time. Temporally, changes in macroinvertebrate indices in the first 5 years postfire can be different from ensuing years, and long-term (10 to 30 years) effects have been suggested.

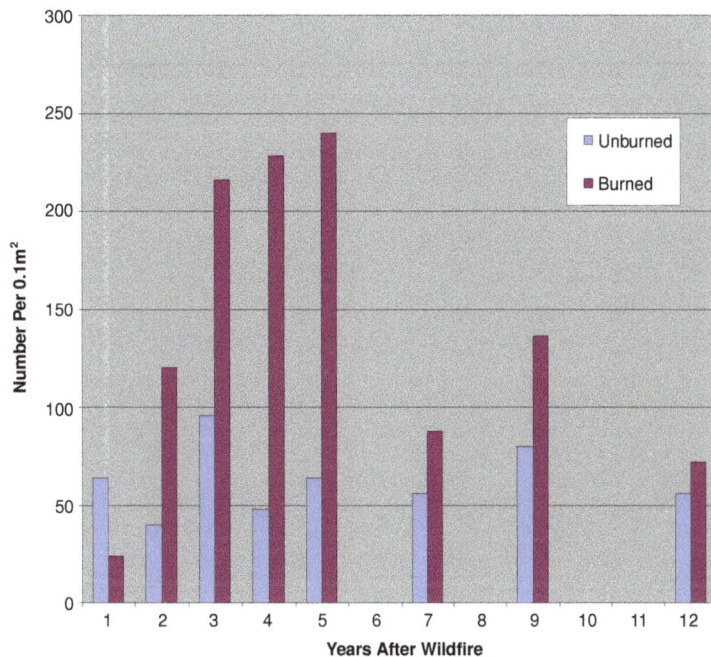

Figure 7.9—Invertebrate density changes in streams after a wildfire in northern California. (Adapted from Roby, K.B.; Azuma, D.L. 1995. Changes in a reach of a northern California stream. Figure 2. Environmental Management. Copyright © 1995. With kind permission of Springer Science and Business Media).

Notes

Part C

Other Topics

Daniel G. Neary

Part C—Other Topics

Chapters 8 through 11 in this part of the volume deal with special topics that were not considered in the original "Rainbow Series" on the effects of fire on soils (Wells and others 1979) and water (Tiedemann and others 1979). These topics include wetlands and riparian ecosystems, fire effects models and soil erosion models, the Burned Area Emergency Rehabilitation (BAER) program with its various treatments and results, and information sources such as databases, Web sites, journals, books, and so forth. Part C concludes with the References section and an appendix of glossary terms.

Wetlands science has advanced considerably since the original "Rainbow Series" was published in 1979, and the role of fire in wetlands ecosystems is now understood more completely. Tiedemann and others (1979) briefly discussed everglades wetlands of southern Florida and cypress wetlands of northern Florida. Chapter 8 of this volume recognizes the extent of wetlands in North America beyond the swamps of Florida. The authors go into considerable detail on the effects of fire in wetlands and their associated riparian areas.

Fire effects models and soil erosion models were in their infancy in the mid 1970s, and researchers and land managers then only dreamed of the compact, high speed desktop and laptop computers we now have that are able to do the millions of calculations needed to run these models. Chapter 9 briefly examines how these models increase our understanding of the effects of fire on soils and water. While this discussion is not intended as inclusive of the complete scope of fire effects models, we do intend to illustrate some of the more commonly available ones.

Rehabilitation of burned landscapes has long been recognized as a necessary step for healing severely burned watersheds to reduce erosion and mitigate adverse changes in hydrology. However, the BAER program did not start until the mid 1970s and was not initially well organized nor well understood. Chapter 10 provides a much needed synopsis of the BAER program based on a Robichaud and others (2000) literature and research review publication.

The numbers of textbooks, research reports, journal articles, users guides, and databases with specific focus on wildland fire have rapidly increased in the past quarter century. And now in the 21st century, we have readily accessible sources of fire effects information such as databases and Web sites that did not exist in the 1970s. The Web sites facilitate rapid dissemination of fire effects information to managers, researchers, and the general public. Their use and importance will continue to grow. Chapter 11 details these information sources.

Chapter 12 is an overall summary of this volume and contains suggestions for research needs and priorities.

USDA Forest Service Gen. Tech. Rep. RMRS-GTR-42-vol. 4. 2005

147

Notes

_____

_____

_____

_____

_____

_____

_____

James R. Reardon
Kevin C. Ryan
Leonard F. DeBano
Daniel G. Neary

Chapter 8:

Wetlands and Riparian Systems

Introduction

Wetlands and riparian ecosystems contain biotic communities that develop because of the regular presence of water (Brooks and others 2003). Undisturbed or well-managed wetlands and riparian ecosystems provide benefits that are proportionally much more important compared to the relatively small portion of the land area that they occupy because these ecosystems support a diversity of plants and animals that are not found elsewhere. Wetlands and riparian ecosystems also play important roles relating to enhanced water quality, flood peak attenuation, and reduced erosion and sediment transport. At the same time, these ecosystems are often fragile and often easily disturbed, and both wildfires and prescribed burns can affect the soil, water, litter, and vegetation in wetlands and riparian ecosystems.

Although the terms "wetland" and "riparian" are sometimes used interchangeably, it is important to distinguish between the two systems both ecologically and in terms of their individual responses to fire. While the presence of water is an essential feature of both ecosystems, wetlands are more typically represented by large accumulations of surface organic matter that are waterlogged for long periods during the year, producing persistent anaerobic soil conditions; riparian areas, on the other hand, are characterized more as communities of water obligate plants that are typically found along river systems where moisture is readily available but the soils are saturated for only short periods. These differences in morphology and hydric environment strongly affect the type of fire that burns in each ecosystem. When the water table is low in wetlands, surface and crown fires often ignite smoldering grond fires in think layers of humus, peat, and muck resulting in deep depth of burn. Conversely, when the water table is high, fires exhibit low depth of burn regardless of the fireline intensity. In riparian areas fires often back down hill at low intensity leaving irregular mixed severity burns. Periodically, however, riparian areas experience high intensity crown fires due to the high fuel continuity and the chimney effect of the terrain. Thus mixed severity fire regimes are common in wetland and riparian communities.

Wetlands

Wetlands are widely distributed from tropical to arctic regions and found in both arid and humid environments. They cover approximately 4 to 6 percent

USDA Forest Service Gen. Tech. Rep. RMRS-GTR-42-vol. 4. 2005

149

of Earth's land area and display a broad ecological range with characteristics of both aquatic and terrestrial ecosystems (Mitsch and Gosselink 1993). The wide distribution and diversity of wetlands make forming generalizations and widely adaptable management solutions difficult (Lugo 1995) (fig. 8.1).

Wetland functions are crucial to environmental quality. They supply the habitat requirements for a diverse range of plants, animals, and other organisms, some of which are threatened and endangered. Wetlands regulate water quality and quantity and are an integral element of global climate control. Improved understanding of wetlands and the relationship between fire and wetlands has become increasingly important for a number of reasons, including concerns about wetland habitat loss, global warming, toxic metal accumulation, and the treatment of hazardous fuels.

While the connection between wetlands and fire appears tenuous, fire plays an integral role in the creation and persistence of wetland species and ecosystems. Changes in climate and vegetation are recorded in the deposition of macrofossils, pollen, and peat, and disturbances such as fire are also recorded in this depositional record (Jasieniuk and Johnson 1982, Cohen and others 1984, Kuhry and Vitt 1996). Paleoecological evidence of fire is derived from a number of sources including pollen analysis and charcoal distribution. Charcoal, which is produced when plant materials are pyrolized, is a record of the influence of fire on the structure of ecosystems (Cope and Chaloner 1985). The distribution of pollen and charcoal within wetland soils reflect both fire history and environmental conditions (Kuhry 1994).

Paleoecological studies show consistent evidence of frequent fire in a wide distribution of wetlands from the Southeastern United States to the boreal regions of Canada and Alaska. For example, peat cores from the Okefenokee Swamp in Georgia exhibited distinct charcoal bands indicating a fire history (Cohen and others 1984, Hermann and others 1991). Patterns in pollen and charcoal distribution in Wisconsin and Minnesota wetlands show changes that reflect variation in environmental conditions and fire frequency (Davis 1979, Clark 1998). Trees from mires of northern Minnesota show fire scars as evidence of fire history (Glaser and others 1981). In boreal regions, peat sediments from sphagnum-dominated wetlands show a fire history as distinct charcoal layers and pollen changes (Kuhry 1994).

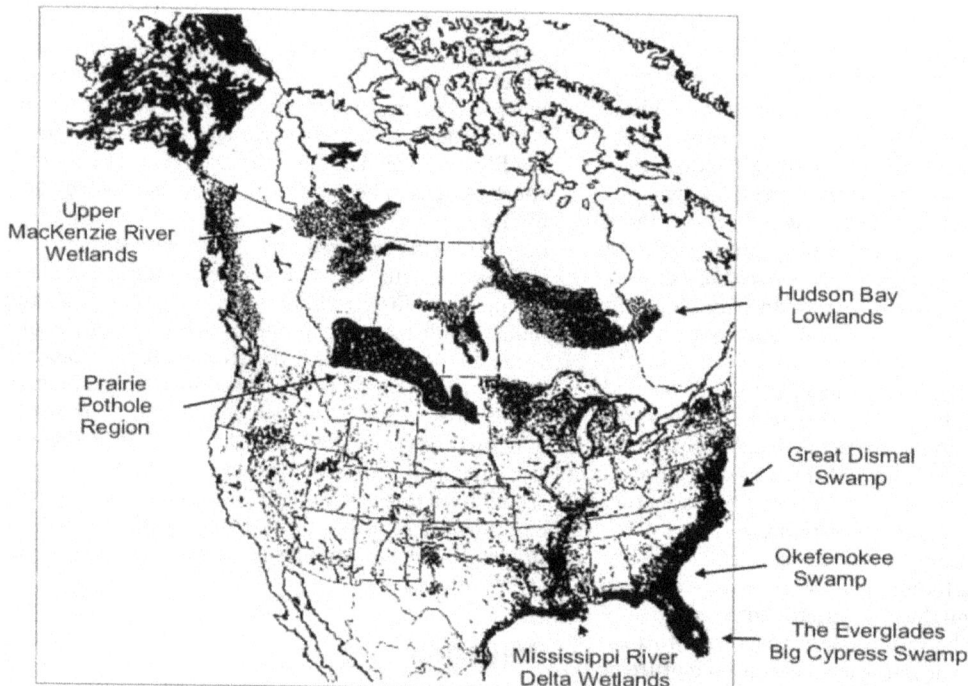

Figure 8.1—Distribution of wetland-dominated regions of North American. Shaded areas have a significant percentage of wetlands.

150

USDA Forest Service Gen. Tech. Rep. RMRS-GTR-42-vol. 4. 2005

Wetland and Hydric Soil Classification

Not only are there a confusing number of terms and definitions to describe organic soils and organic matter while classifying wetlands (table 8.1), but during the past three decades several new wetland classification systems have been developed and used by various government agencies. This array of classification systems reflects both our understanding of wetland processes and the missions of the individual agencies.

The system developed and currently in use by the U.S. Fish and Wildlife Service (Cowardin and others 1979; table 8.2) for wetland inventory is well suited for ecological applications (Mitsch and Gosselink 1993).

In this classification system, the divisions between the five wetland systems are based on hydrologic similarities in the water source and flow. Wetland systems of interest to fire managers in this classification include:

- Lacustrine wetlands, which are associated with lakes and reservoirs.
- Riverine wetlands, which are associated with flowing water of rivers.
- Estuarine wetlands, which include salt marshes.
- Palustrine wetlands, which include swamps, bogs, fens and other common freshwater wetlands.

Table 8.1—Terms and definitions used to describe organic soils and organic matter (Adapted from Soil Science Society of America 1997, and Mitsch and Gosselink 1993).

Term	Definitions
Bog	A peat-accumulating wetland that has no significant inflows and supports acidophilic mosses, particularly sphagnum.
Fen	A peat-accumulating wetland that receives some drainage from surrounding mineral soils and usually supports marsh-like vegetation. These areas are richer in nutrients and less acidic than bogs. The soils under fens are peat (Histosol) if the fen has been present for a while.
Fibric	A type of organic soil where less than one-third of the material is decomposed, and more than two-thirds of plant fibers are identifiable.
Marsh	A frequently or continually inundated wetland characterized by emergent herbaceous vegetation adapted to saturated soil conditions.
Muck	Organic soil material in which the original plant remains is not recognizable. Contains more mineral matter and is usually darker in color than peat.
Muskeg	Large expanses of peatlands or bogs; particularly used in Canada and Alaska.
Oligotrophic	Describes a body of water (for example, a lake) with a poor supply of nutrients and a low rate of formation of organic matter by phototsynthesis.
Ombrotrophic	True raised bogs that have developed peat layers higher than their surroundings and which receive nutrients and other minerals exclusively by precipitation.
Peat	Organic soil materials in which the original plant remains are recognizable (fibric material).
Peatlands	A generic term for any wetland that accumulated partially decayed plant matter.
Pocosin	Temperate zone evergreen shrub bogs dominated by pond pine, ericaceous shrubs, and sphagnum.
Sapric	Type of organic soil where two-thirds or more of the material is decomposed, and less than one-third of plant fibers are identifiable.
Wet meadow	Grassland with waterlogged soil near the surface but without standing water most of the year.
Wet prairie	Similar to marsh but with water levels usually intermediate between a marsh and a wet meadow.

Table 8.2—U.S. Fish and Wildlife Service classification hierarchy of wetland and deepwater habitats (Cowardin and others 1979). These are wetland classes of particular interest to fire management. Wetlands are shown in bold print.

System	Subsystem	Class
Marine	Subtidal	Rock Bottom, Unconsolidated Bottom, Aquatic Bed, Reef
Marine	Intertidal	Aquatic Bed, Unconsolidated Bottom, Aquatic Bed, Reef
Estuarine	Subtidal	Rock Bottom, Unconsolidated Bottom, Aquatic Bed, Reef
Estuarine	Intertidal	Aquatic Bed, Reef, Streambed, Rocky Shore, Unconsolidated Shore **Emergent Wetland** **Scrub-Shrub Wetland** **Forested Wetland**
Riverine	Tidal	Rock Bottom, Unconsolidated Streambed, Aquatic Bed Streambed, Rocky Shore, Unconsolidated Shore **Emergent Wetland**
Riverine	Low Perennial	Rock Bottom, Unconsolidated Bottom, Aquatic Bed Rocky Shore, Unconsolidated Shore **Emergent Wetland**
Riverine	Upper Perennial	Rock Bottom, Unconsolidated Bottom, Aquatic Bed Rocky Shore, Unconsolidated Shore
Riverine	Intermittent	Streambed
Lacustrine	Limnetic	Rock Bottom, Unconsolidated Bottom, Aquatic Bed
Lacustrine	Littoral	Rock Bottom, Unconsolidated Bottom, Aquatic Bed Rocky Shore, Unconsolidated Shore **Emergent Wetland**
Palustrine		Rock Bottom, Unconsolidated Bottom, Aquatic Bed Unconsolidated Shore **Moss-Lichen Wetland** **Emergent Wetland** **Scrub-Shrub Wetland** **Forested Wetland**

At a finer level, the division of wetland systems and subsystems into wetland classes is based upon the dominant vegetation type or substrate. The major wetland classes of interest in this system include:

- Forested
- Scrub-shrub
- Emergent
- Moss-lichen

Wetland classes are further modified by factors including water regime, water chemistry, and soil factors.

In addition to unique vegetation and hydrologic characteristics, wetlands are also distinguished by the presence of hydric soils. The definition of hydric soils has evolved during the last few decades, and the changes to the definition reflect our increased understanding of wetland soils genesis (Richardson and Vepraskas 2001). Although organic soils (Histisols) are commonly associated with wetlands, poorly drained mineral soils with wet moisture regimes are also common on wetland sites. An additional soil order has recently been added to the U.S soil classification system (Gelisols). This new soil order reflects the influence of low temperatures and soil moisture on soil processes. Thus, wetland soils of the boreal peatlands are now classified as Gelisols (Bridgham and others 2001).

The current definition of hydric soils includes both histisols and mineral soils (table 8.3). The primary factor contributing to the differences between hydric and nonwetland soils is the presence of a water table

152

USDA Forest Service Gen. Tech. Rep. RMRS-GTR-42-vol. 4. 2005

Table 8.3—Definitions of hydric soils (Adapted from Mausbach and Parker 2001).

1. All Histosols

2. Mineral Soils
   a. Somewhat poorly drained with a water table equal to 0.0 foot (0.0 m) from the surface during the growing season, or
   b. Poorly drained or very poorly drained soils that have either:

   (1) water table equal to 0.0 foot (0.0 m) from the surface during the growing season if textures are coarse sand, sand, or fine sand in all layers within 20 inches (51 cm), or for other soils, or
   (2) water table at less than or equal to 0.5 foot (0.15 m) from the surface during the growing season if permeability is equal to or greater than 6.0 in/hr (15 mm/hr), or
   (3) water table at less than or equal to 1.0 foot (0.3 m) from the surface during the growing season, or

3. Soils that are frequently ponded for long duration or very long duration during the growing season, or

4. Soils that are frequently flooded for long duration or very long duration during the growing season.

close to the soil surface or frequent long-duration flooding or ponding. The oxygen-limiting conditions found in these waterlogged environments affect micro- and macrofaunal organisms and nutrient cycling processes and result in decreased decomposition rates and different metabolic end products (Stevenson 1986).

Wetland Hydrology and Fire

Wetland hydrology is a complex cycle of water inflows and outflows that are balanced by the interrelationships between biotic and abiotic factors. The presence and movement of water are dominant factors controlling wetland dynamics, nutrient and energy flow, soil chemistry, organic matter decomposition, and plant and animal community composition.

Wetland systems are characterized by annual water budgets comprising storage, precipitation, evapotranspiration, interception, and surface and ground water components. In addition to water budgets, the seasonality and movement of water is characterized by hydroperiod and hydrodynamics. *Hydroperiod* is the time a soil is saturated or flooded, and results from water table movements caused by distinct seasonal changes in the balance between inflows and outflows of a particular wetland. *Hydrodynamics* is the movement of water in wetlands and is an important process affecting soil nutrients and productivity.

Alteration of ground water levels and flow induced by anthropogenic activities such as ditching and fire line construction is a serious concern in wetland fire management (Bacchus 1995). Long-term changes in hydroperiod or hydrodynamics affect productivity, decomposition, fuel accumulation, and the subsidence of organic soils. Changes in these wetland characteristics lead to changes in fire behavior and ground fire potential.

Water budgets, hydroperiods, and hydrodynamics are unique for each wetland type and, as a result, have substantial effect on site productivity and decomposition rates (Mitsch and Gosselink 1993). Significant ground and surface water inputs are characteristic of the water budgets of marshes and fens, while precipitation and evapotranspiration dominate the water budgets of bogs and pocosins. The timing and pattern of these inputs has a distinct influence on fire occurrence. For example, the water budget of pocosin ecosystems in the Southeastern United States is dominated by rainfall and evapotranspiration with minor surface and ground water components. Typically, the highest wildfire danger occurs between March and May because the evapotranspiration rate is higher than precipitation, thereby causing a water deficit in these soils (fig. 8.2).

In contrast to the pocosin wetlands, in the boreal black spruce/feather moss wetlands, frozen soil restricts drainage during spring and early summer, and the organic soil retains a higher percentage of water. Typically, soil drying in these communities is delayed until in June or July (Foster 1983). Early season fires in these wetlands consume little organic soil material, while fires later in the season or during extended droughts can consume significant amounts of organic soil material (Heinselman 1981, Johnson 1992, Duchesne and Hawkes 2000, Kasischke and Stocks 2000).

Although studies have identified interrelationships among hydrology, fire, and soil in wetland ecosystems (Kologiski 1977, Sharitz and Gibbons 1982, Frost 1995), the codominant influence of fire and hydrology is not clear. Hydrologic characteristics influence fire severity, and in turn the fire severity influences postburn hydrology and future fire severity. Both

USDA Forest Service Gen. Tech. Rep. RMRS-GTR-42-vol. 4. 2005

153

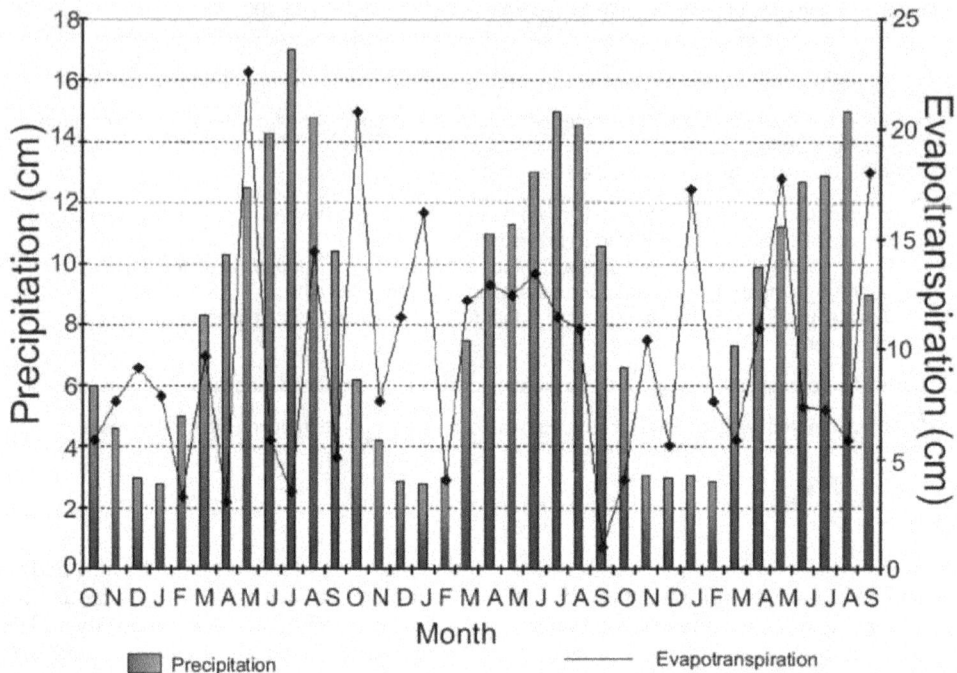

Figure 8.2—An example of precipitation and evapotranspiration dynamics in a pocosin habitat measured for a 3-year period. (Adapted from Daniel 1981, Pocosin Wetlands, Copyright © 1981, Elsevier B.V. All rights reserved).

these factors also affect numerous ecological variables including soil moisture, soil aeration, and soil temperature regime. The separation of disturbance effects from hydrological effects is difficult because wetland response to disturbance is a function of changes in numerous spatial and temporal gradients (Trettin and others 1996).

Wetland community responses to fire frequency, hydroperiod, and organic soil depth were investigated in a North Carolina pocosin (scrub-shrub wetland). The results showed that emergent sedge bogs on deep organic soils with long hydroperiods were associated with frequent fires. Deciduous bay forest communities with shallower organic soils and shorter hydroperiods were associated with decreases in fire frequency, while pine savanna communities with shallow organic soils and short hydroperiods were associated with a decrease in fire frequency (Kologiski 1977) (fig. 8.3). A similar relationship between fire and hydroperiod was found in a study of Florida cypress domes communities. Emergent/persistent wetlands with long hydroperiods were associated with frequent fires, while forested alluvial wetlands with shorter hydroperiods were associated with infrequent fires (Ewel 1990) (fig. 8.4).

Fire regimes are representative of long-term patterns in the severity and occurrence of fires (Brown 2000). They reflect the interdependence of several factors including climate, fuel accumulation, and ignition sources. Several fire regime classification systems have been developed using fire characteristics or effects produced by the fire (Agee 1993, Brown 2000).

Wetland fire regimes vary in frequency and severity. Light, frequent fires maintain the herbaceous vegetation in emergent meadows such as tidal marshes and sedge meadows (Frost 1995) (fig. 8.5). Light surface fires in *Carex stricta* dominated sedge meadows remove surface litter and resulted in increased forb germination and species diversity (Warners 1997). Infrequent stand-replacing fires are needed to regenerate Atlantic white cedar and black spruce/feather moss forested wetlands (Kasischke and Stocks 2000).

Surface fires associated with high water table conditions reduce shrub and grass cover and produce conditions favorable for Atlantic white cedar germination and survival (Laderman 1989). Experimental results have shown that when the organic soil horizons in black spruce/feather moss wetlands are removed by surface fires, the resulting elevated nutrient levels

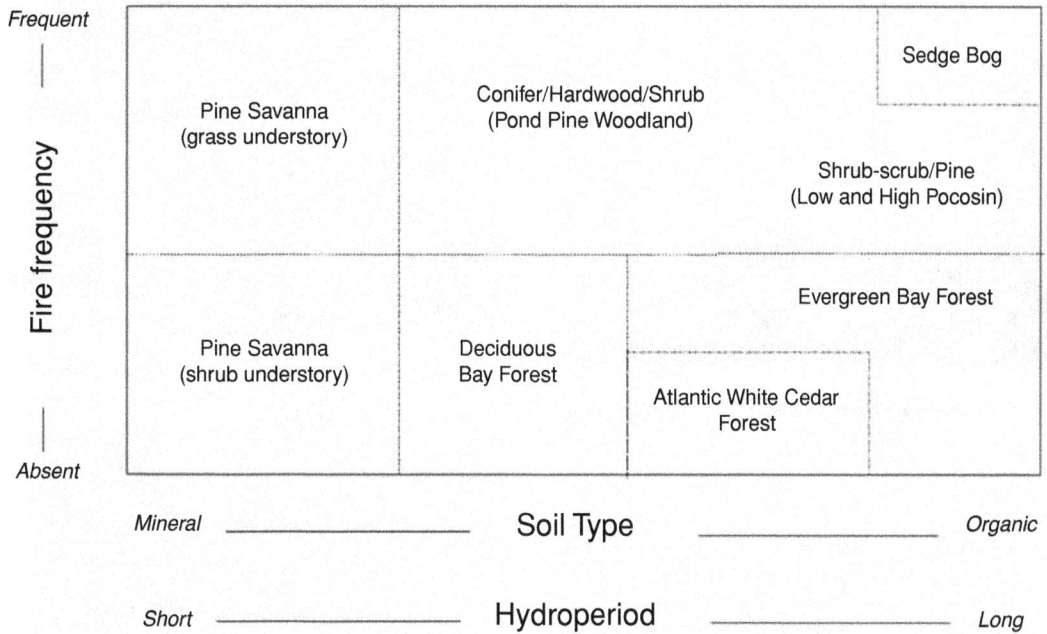

Figure 8.3—Idealized interrelationships between fire frequency, hydroperiod, and organic soil depth in a pocosin wetland in North Carolina. (Adapted from Kologski 1977).

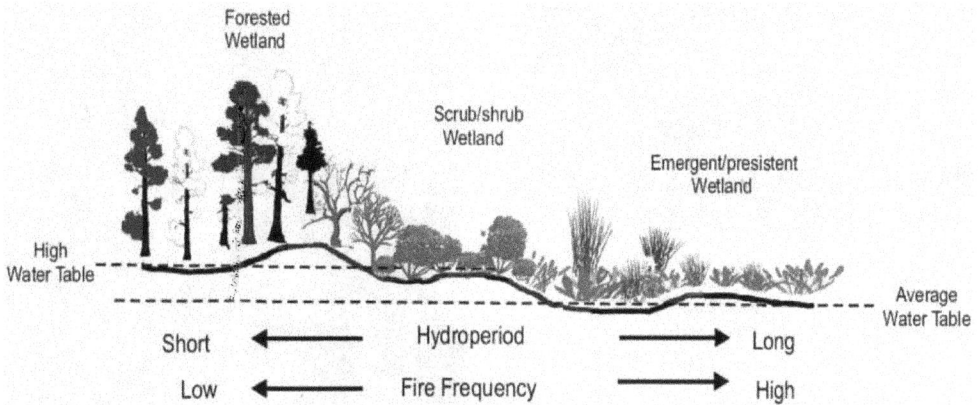

Wetland Type	Forested				Scrub/shrub		Emergent, persistent		
Hydroperiod	Temporarily and seasonally flooded				Temporarily and seasonally flooded; saturated		Temporarily, seasonally and semipermanently flooded; intermittently exposed		
Vegetation Type	Black Spruce	Atlantic White Cedar	Pond Pine Woodland	Pitch Pine	Low Pocosin	High Pocosin	Freshwater Marsh/ Sawgrass	Prairie Pothole	Wet Prairie
Fire Return Interval (yrs)	50-100	25-300	2-10	20-50	13-50	25-50	3-5	5-10	2-6
Citation	Heinselman, 1983; Viereck, 1983	Wade and others 2000	Little, 1946, 1973	Christensen and others, 1988	Sharitz and Gresham, 1998	Sharitz and Gresham, 1998	Wade and others, 1980	Wright and Bailey, 1982	Connelly and Kauffman, 1991

Figure 8.4—Fire frequency, hydrological, and ecological characteristics of palustrine wetlands.

USDA Forest Service Gen. Tech. Rep. RMRS-GTR-42-vol. 4. 2005

155

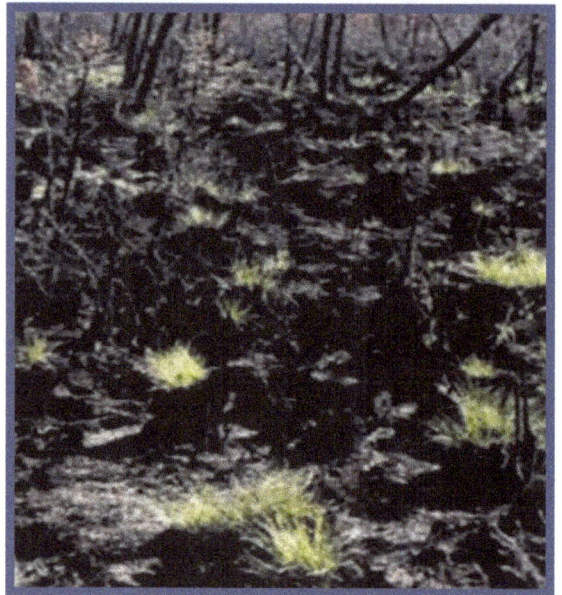

Figure 8.5—Emergent wetland in Agassiz Wildlife Refuge, Minnesota: (top) fall prescribed burning, (bottom) postburn ground surface. (Photos by USDA Forest Service).

and soil temperatures increase site productivity (Viereck 1983) (fig. 8.6).

Fuel accumulation in wetlands is a function of productivity and decomposition. Productivity is a function of hydroperiod and hydrodynamics, which regulate the inflows and outflows of water, nutrients, and oxygen. The seasonal or temporary flooding that is common in tidal marshes and alluvial forest wetlands

Figure 8.6—Black spruce/feather moss forested wetland, Tetlin Wildlife Refuge, Alaska: (top) prescribed burning (7/1993); (middle and bottom) postburn ground surfaces. (Photos by Roger Hungerford and James Reardon).

provides external inputs of water and nutrients and results in high productivity and high decomposition rates. In contrast, the long, stable hydroperiod and dependence on precipitation for water and nutrients common in bogs, pocosins, and other nutrient-limited wetlands result in lower productivity and slower decomposition rates.

On the nutrient-limited wetlands with long, stable hydroperiods, greater amounts of fuel accumulate because of slower decomposition rates in the relatively nutrient-poor litter and fine fuels. This slower decomposition rate leads to more frequent fires on the nutrient-limited sites compared to the productive sites that have inherently less litter and fuels production (Christensen 1985).

Surface Fire in Wetlands—In wetland soils, the effects produced by surface and ground fires are related to the intensity and duration of the energy at either the mineral soil surface or the interface between burning and nonburning organic soil materials. In general, surface fires can be characterized as short duration, variable intensity sources of energy, while ground fires can be characterized as longer duration and lower intensity sources of energy (see chapter 1). Fire severity, which is a measure of the immediate fire effects on plants and soils, is the result of both the intensity and duration of the flaming combustion of surface fires and the smoldering combustion of the organic materials (Ryan and Noste 1985, Ryan 2002).

Following ignition, surface fire behavior depends on weather, topography, and fuels (Albini 1976, Alexander 1982, Finney 1998). Behavior is characterized by parameters that include intensity, rate of spread, and flame geometry. One factor of significance in the prediction of surface fire behavior is fuel compactness (packing ratio). Compactness is the relationship between the amount of fuel in each size class (fuel loading) and the physical volume it occupies (fuel depth), and it is correlated with rate of spread and intensity (Burgan and Rothermel 1984). The compactness ratio of fuel loading to fuel depth reflects the structure and composition of the vegetation (Brown 1981).

Wetland vegetation types range from short sedge meadows to forested wetlands. The surface fuels in graminoid-dominated or scrub-shrub wetlands are primarily vertical, and these fuel types rapidly increase in depth as loading increases. In contrast, the fuels in forested wetlands, which are dominated more by overstory species, are primarily horizontal, and these fuel types slowly increase in depth as fuel loading increases (Anderson 1982). The fuel loading in many wetlands is a product of both the horizontal fuel component of the understory and the vertical component of the overstory.

Physiography and vegetation structure are important factors influencing localized surface fire behavior (Foster 1983). A wide range of fire behavior occurs in wetlands due to the diverse wetland plant communities and their associated fuel types. Fires in hardwood swamps will typically burn with low intensities and low flame lengths, while fires in scrub-shrub wetlands can exhibit extreme behavior (Wade and Ward 1973). Measurements of pocosin scrub-shrub fuel loading and depths show characteristics similar to the chaparral vegetation type with higher loading of fine live and dead materials (Scott 2001). In pocosins, high fuel loads in the small size classes and the presence of volatile oils and resins in the vegetation contribute to high rates of spread and high intensities, making suppression difficult (Anderson 1982).

Ground Fire in Wetlands—Due to the presence of large amounts of organic material, ground fires are a special concern in Histisols, Gelisols, and other soil types with thick organic horizons. Surface fires that result from periodic fluctuations in the water table are correlated with weather cycles. These fires generally consume the available surface fuels but little organic soil material (Curtis 1959). However, fires that occur during longer or more severe water table declines result in significant consumption of both surface fuels and organic soil material (Johnson 1992, Kasischke and Stocks 2000).

Sustained combustion and depth of burn are correlated with soil water content, which is a function of soil water holding capacity, hydroperiod, and evapotranspiration. While soil moisture and aeration in wetland landscapes are controlled by hydrology, the capacity of soil to store water is influenced by the organic matter content and its degree of decomposition. Soil water storage capacity is correlated with physical properties that include bulk density, organic matter content, and hydraulic conductivity. Slightly decomposed fibric materials in soils dominated by sphagnum moss material have low bulk density, high hydraulic conductivity, high porosity, and hold a high percentage of water at saturation. At saturation this fibric material can hold greater than 850 percent of its dry weight as water (USDA Natural Resources Conservation Service 1998). In contrast, the highly decomposed sapric materials found in soils formed in the pocosin wetlands have greater bulk density, finer pore structure, and lower hydraulic conductivity. At saturation, sapric material can hold up to 450 percent of its dry weight in water (USDA Natural Resources Conservation Service 1998).

Following the initial ignition of organic soil material, the probability of sustaining smoldering combustion (that is, ground fire) is a function of moisture and mineral content. Laboratory studies of the relationship

USDA Forest Service Gen. Tech. Rep. RMRS-GTR-42-vol. 4. 2005

157

between these two factors were conducted using a standardized ignition source and commercial peat moss as an organic soil/duff surrogate. Peat moss was used because of its physical and chemical uniformity.

The results show an inverse relationship between the smoldering moisture limit and moisture and inorganic content. Consequently, these results were used to derive a sustained smoldering probability distribution based upon moisture and mineral content factors. For comparison purposes, the results are reported at the moisture content for a 50 percent probability level of sustained combustion (Frandsen 1987, Hartford 1989) (table 8.4).

Organic soil materials from various wetlands show a similar relationship between smoldering moisture limit and moisture and mineral content. The smoldering limit of low mineral content soils from a North Carolina pocosin site is greater than the smoldering limit of high mineral content soils from a Michigan sedge meadow site. The results also show soil depth differences in smoldering limits. Smoldering limits decline with increasing depth and increasing mineral content in black spruce/sphagnum, sedge meadow, and other sites (Frandsen 1997) (fig. 8.7).

Although the smoldering limits for both peat moss and organic soil materials are the function of mineral and moisture contents, differences exist between the predicted smoldering limits derived from the peat

moss surrogate and the limits derived from organic soil material. With the exception of samples from black spruce/feather moss sites in Alaska, at the 50 percent probability level, organic soil materials from other sites smoldered at consistently higher moisture contents than the limits predicted for peat moss. Frandsen (1997) speculated that ignition methods were partially responsible for some of the difference.

Fire line intensity alone does not give the total heat input to the soil surface because it does not take into account the residence time of the flaming combustion or smoldering and glowing combustion (Peter 1992, Albini and Reinhardt 1995). The long duration of smoldering combustion is an important characteristic that differentiates the effects of ground fires and surface fires (Wein 1983).

In laboratory studies, the rate of smoldering combustion depended on both moisture and mineral content (Frandsen 1991a,b), with the laboratory estimates of smoldering rates in agreement with the 1.2 to 4.7 inches/hour (3 to 12 cm/hour) reported by Wein (1983). The results of laboratory studies of the relationship between consumption and moisture content have shown that sustained smoldering of sphagnum moss was possible at up to 130 percent moisture content, and that the percent consumption of organic material was linked to the variability in moisture distribution (Campbell and others 1995).

Table 8.4—Moisture limits of sustained combustion and physical properties of wetland soil materials.

Vegetation type	Wetland class	Average inorganic content	Average organic bulk density	Depth		Moisture content for 50 percent probability of sustained combustion
		Percent	*kg/m³*	*cm*	*in*	*Percent*
Sphagnum (upper)	Forested Wetland	12.4	22	0-5	0-2	118
Sphagnum (lower)	Forested Wetland	56.7	119	10-25	25-64	81
Spruce/Moss	Forested Wetland	18.1	43	0-25	0-64	39
Spruce/Moss/Lichen	Forested Wetland	26.1	56	0-5	0-2	117
Sedge Meadow (upper)	Emergent Wetland	23.3	69	5-15	2-6	117
Sedge Meadow (lower)	Emergent Wetland	44.9	92	15-25	6-64	72
White Spruce Duff	Upland	35.9	122	0-5	0-2	84
Peat	Peat Moss	9.4	222	17-25	7-64	88
Peat Muck	Peat Moss	34.9	203	12-20	5-8	43
Sedge Meadow (Seney)	Emergent Wetland	35.4	183	17-25	7-64	70
Pine Duff (Seney)	Upland	36.5	190	0-5	0-2	77
Spruce/Pine Duff	Upland	30.7	116	0-5	0-2	101
Grass/Sedge Marsh	Wetland	35.2	120	0-5	0-2	106
Southern Pine Duff	Upland	68.0	112	0-5	0-2	39
Pocosin	Shrub-Scrub	18.2	210	10-30	25-12	150

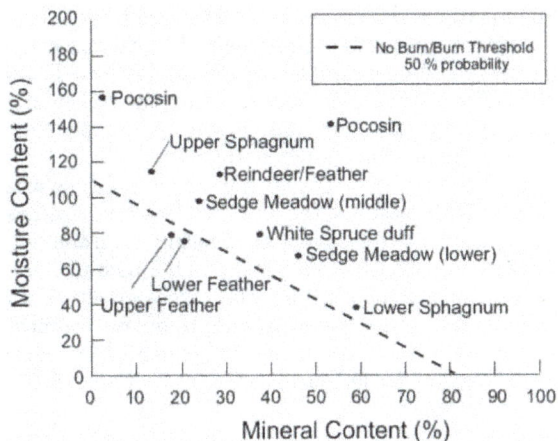

Figure 8.7—Comparison of the predicted moisture smoldering limits of organic soil materials in different wetlands with the smoldering limit of an organic soil surrogate. The dotted line is the 50 percent probability of burn limit. (Adapted from Hungerford and others 1995b).

Heat transfer in organic and mineral soils is a function of the energy produced by combustion and soil thermal properties. The energy produced by flaming and smoldering combustion depends on fuel properties and environmental conditions. Soil thermal properties are affected by soil characteristics and moisture content. Changes in thermal properties are primarily the result of soil drying in advance of the smoldering front (Peter 1992, Campbell and others 1994, 1995).

In addition to fire behavior characteristics, the depth of burn is an important factor in predicting and interpreting soil heating and the effects of fire on wetland soils. In mineral soil, the interface between the burnable organic materials and unburnable mineral soil is significant because of differences in nutrient pools and heat transfer properties. In wetland soils, the depth of burn, which is the interface between burning and nonburning organic soil materials, is important because it reflects the amount of organic material consumed. A number of factors are correlated with depth of burn including nutrient volatilization, ash deposition, and emissions.

Fall prescribed burns were conducted by the Nature Conservancy in pocosin scrub-shrub wetlands in 1997 and 1998 at Green Swamp, North Carolina. The soils were Histosols that were greater than 3 feet (0.9 m) deep. During average years the study site is temporally flooded a number of times, and the soils remain saturated most of the year. The vegetation and surface fuels of the burn units were dominated by shrubs

species (*Lyonia lucida* and *Ilex glabra*), scattered pond pine (*Pinus serotina*), and evergreen bays (*Persia borbonia, Gordonia lasianthus*). Fuel and soil moisture variability was influenced by the hummock and depression microtopography of the ground surface. The surface fires in both burns were active to running surface fires with passive crowning that was associated with single or small clusters of pond pine and evergreen bay species (Hungerford and Reardon, unpublished data).

The 1997 fall prescribed burn was conducted during a period of extended drought and dry soil conditions (fig. 8.8, top). At the time of burning, the water table depth was greater than 28 inches (71 cm) from the

Figure 8.8—Prescribed burn of a pocosin scrub-shrub wetland conducted with dry soil conditions, Green Swamp, North Carolina (September 1997): smoldering combustion (Top, photo by Gary Curcio); postburn surface, (Middle, photo by Roger Hungerford); postburn hydrologic and species changes (Bottom, photo by James Reardon).

USDA Forest Service Gen. Tech. Rep. RMRS-GTR-42-vol. 4. 2005

159

ground surface of the depression areas. Soil moisture in the upper profile ranged from 85 percent at the surface to 144 percent at 6 inches (15 cm). These moisture contents in the upper soil profile were less than the moisture limits predicted to sustain smoldering in pocosin soils. The moisture contents in the lower soil profile (greater than 18 inches or 46 cm) were greater than 200 percent and exceeded the moisture limits predicted to sustained smoldering (table 8.4).

The surface fire and extended smoldering resulted in significant consumption of soil organic material and surface fuels (fig. 8.8, middle). After the surface fire, the smoldering front advanced both laterally within the upper soil profile and downward. Smoldering in the downward direction was limited by soil moisture of the lower soil profile. The depth of burn varied between 18 and 24 inches (46 to 61 cm) and reflected both the soil moisture distribution within the soil profile and the hummock and depression microtopogaphy common in pocosin wetlands (fig. 8.9).

Soil heating below the burn/no-burn interface was primarily the result of the long duration temperature pulse associated with the smoldering environment. Maximum measured temperatures 0.2 inch (0.5 cm) below the burn/no-burn interface were greater than 680 °F (360 °C), while maximum temperature measured 2.6 inches (6.5 cm) below the burning interface in unburned soil was less than 158 °F (70 °C) (fig. 8.10).

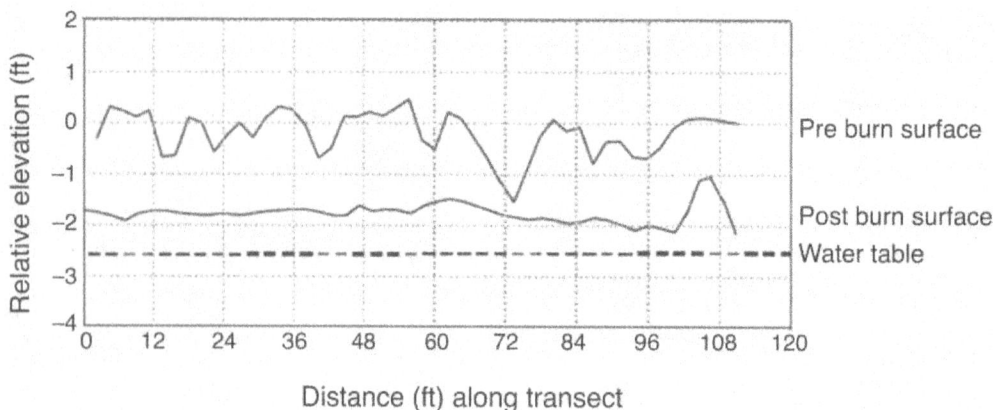

Figure 8.9—Pre- and postburn microelevation transect of a pocosin dry burn study site (Hungerford and Reardon, unpublished data).

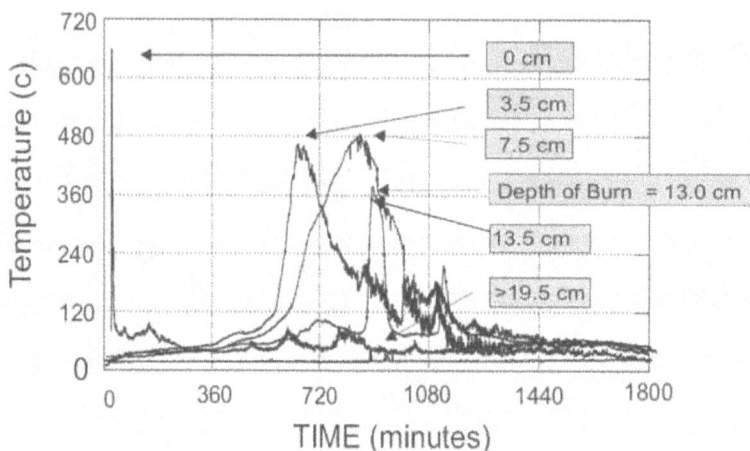

Figure 8.10—Soil profile temperature measurements from a pocosin prescribed burn conducted with dry soil conditions (Hungerford and Reardon, unpublished data).

In areas of high burn severity, the hydrologic changes associated with the consumption of organic soil have led to species composition changes (fig. 8.8, bottom).

The 1998 fall prescribed burn was conducted during a period of saturated soil conditions (fig. 8.11, top). This burn resulted in limited consumption of surface fuels and organic soil materials and limited soil heating (fig. 8.11, middle). At the time of burning, soil moisture in the upper soil profile exceeded 250 percent, and the water table was within 2 to 5 inches (5.1 to 12.7 cm) of the depression ground surface. The depth of burn was limited, and the consumption of organic soil material was restricted to the top of hummocks and microsites with dryer soil conditions. Litter and surface fuels were scorched but not consumed in much of the unit (fig. 8.11, bottom). Soil

heating below the burn/no-burn interface was limited and resulted from the flaming combustion primarily associated with the surface fire. The maximum measured temperatures of unburned soil at 2.8 and 4.3 inch (7.0 and 11.0 cm) depths were 171 °F (77 °C) and 189 °F (87 °C), respectively (fig. 8.12). The removal of dead and live fine fuel resulted in no significant species composition changes and regrowth, and fuel accumulation was sufficient to reburn the unit in autumn 2001.

The fire severity in wetlands is correlated with depth of burn. In addition to the direct effects caused by organic soil combustion, the physical removal of the soil material creates a number of interrelated physical, biological, and hydrological consequences. Removal of this material can result in soil moisture, temperature, and aeration changes that affect microorganism activity changes and ultimately decomposition rate changes (Armentano and Menges 1986).

In boreal wetland systems, the active and permafrost layers depend on the presence of an insulating moss layer (Viereck 1973, Van Cleve and Viereck 1983). A dynamic relationship exists between the depth to permafrost, organic layer thickness, and soil temperature. Removal of this insulating organic layer by burning leads to an increase in soil temperatures, changes in soil moisture regime, and an increase in depth to permafrost. Over time, successional processes and the reestablishment of the organic layer will cause a decrease in soil temperatures and a decrease in the depth of the permafrost layer (Ping and others 1992) (fig. 8.13).

Responses to litter consumption and soil exposure after spring burning of sedge meadow wetlands in Michigan also result in changes in soil temperature

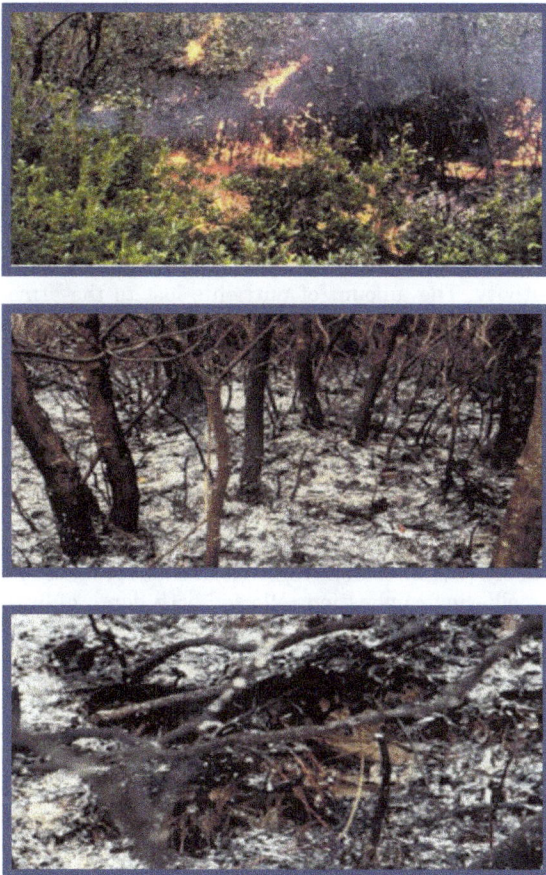

Figure 8.11—Prescribed burn of pocosin scrub-shrub wetland conducted with wet soil conditions, Green Swamp, North Carolina (September 1998): (top) surface fire behavior, (middle and bottom) fuel consumption and postburn ground surface. (Photos by Roger Hungerford and James Reardon)

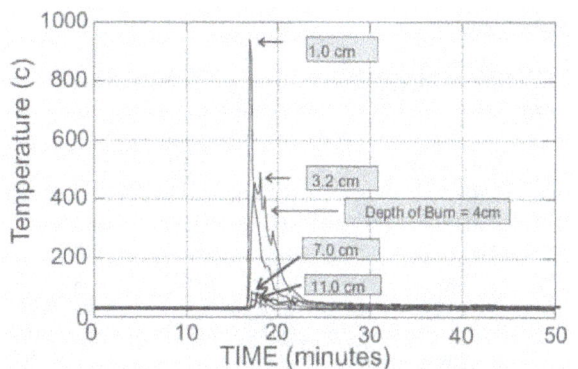

Figure 8.12—Soil profile temperature measurements from a pocosin prescribed burn conducted with wet soil conditions (Hungerford and Reardon, unpublished data).

USDA Forest Service Gen. Tech. Rep. RMRS-GTR-42-vol. 4. 2005

161

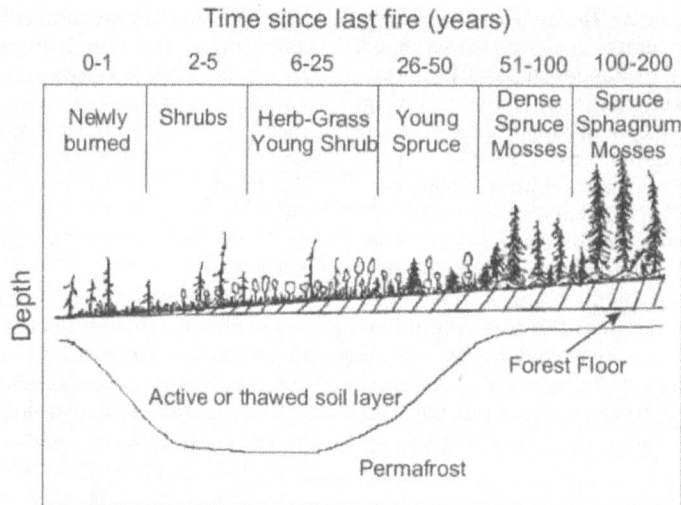

Figure 8.13—Fire cycle effects on soil microclimate, permafrost, and vegetation succession in Alaska. (adapted from Van Cleve and Viereck 1983, Ping and others 1992).

regime. Prescribed burns were conducted over a range of fire intensities and resulted in the removal of the litter and a blackened soil surface. The postburn soil surface had higher soil temperatures and produced increased germination rates of *Carex strica* (Warners 1997).

Wein (1983) suggested that low intensity fires favor bog-forming processes because the thick organic layers that remain after low intensity fires decompose slowly and cause temperature and moisture changes by insulating the soil and retaining moisture. In contrast, deep hot fires resulted in increased soil temperatures and nutrient cycling rates. These conditions favor plant communities dominated by vascular plants.

Wetland Fire Effects and Soil Nutrient Responses

Soil nutrient responses are the result of a number of processes including nutrient volatilization, condensation of combustion products on cool soil surfaces, ash deposition, and soil heating. Soil nutrients exhibit a wide range of sensitivity to temperature changes. Nitrogen-containing compounds are the most heat-sensitive and show changes at temperatures as low as 392 °F (200 °C) while cations such as magnesium (Mg) and calcium (Ca) are less sensitive and show changes at temperatures greater than 1,832 °F (1,000 °C) (see chapter 3; DeBano and others 1998).

Nutrient volatilization is the result of flaming and smoldering combustion and nutrient temperature sensitivity. The amount of nutrients lost to the atmosphere or remaining in the ash layer are dependent on combustion temperatures and preburn nutrient content of the fuels (Raison 1979, Soto and Diaz-Fierros 1993). Laboratory studies of smoldering combustion were conducted with wetland organic soil cores from black spruce/feather moss (Alaska) and sedge meadow (Michigan) sites. Different depth of burn treatments were simulated by controlling soil core moisture. The results demonstrated that ash differences in total carbon (C) and total nitrogen (N) content produced by smoldering combustion reflected both depth of burn treatment and preburn nutrient levels. The ash and char residue derived from the deep burn treatments showed the largest reductions in total N and C (Hungerford and Reardon unpublished data) (fig. 8.14).

In comparison with the flaming combustion of aboveground biomass, the smoldering of high N content organic soils from pocosin scrub-shrub wetlands in North Carolina produced larger ammonia (NH_3) and hydrocarbon emissions. Smoldering also produced a "tar-like" substance that condensed on cool surfaces. This material had a total N content that was six to seven times greater than the parent material (Yokelson and others 1997).

In addition to nutrient volatilization, combustion temperatures, and preburn fuel nutrient content, the

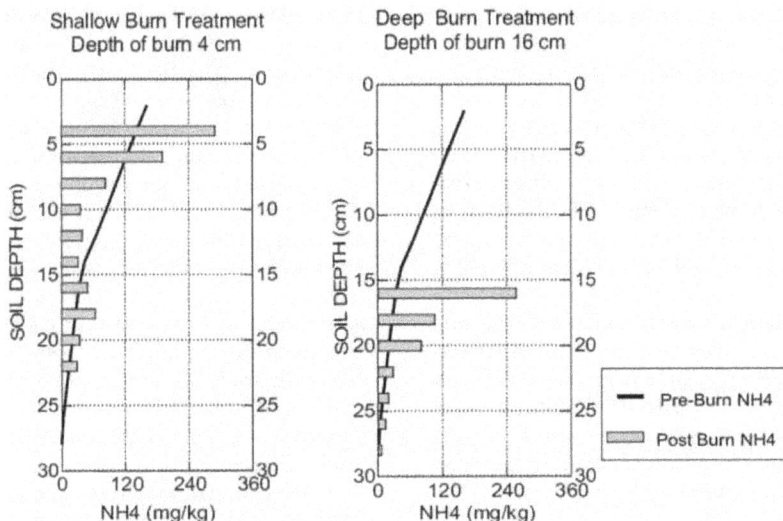

Figure 8.14—Comparison of preburn and postburn NH$_4$–N distribution from the laboratory burning of spruce/feather moss soil cores (Hungerford and Reardon, unpublished data).

significance of ash deposition is also dependent on the interaction between fire and hydrology. Postburn spatial and temporal water movement dynamics are the result of surface microtopography changes and reflect the extent and depth of burn. In wetlands, the duration and magnitude of the postburn ash-derived nutrient pulse is a function of preburn nutrient levels, burning conditions, and seasonal waterflow.

The significance of the relationship between ash deposition and increased nutrient availability was tested in Michigan *Carex stricta* dominated sedge meadow sites. No significant increase in productivity due to ash additions was found (Warners 1997). The loss of ash material due to water movement was not a factor in this study, and the results suggest that either productivity was not limited by the nutrients present in the ash residue or the ash produced from burnt sedge meadow litter was nutrient poor (Warners 1997).

Nitrogen and phosphorus (P) additions were made to a brackish marsh site in coastal North Carolina at rates intended to simulate ash deposition from burning. The nutrient additions were made along a salinity and hydroperiod gradient. A significant productivity increase to low levels of N and P addition was observed at intermediate salinity levels (Bryant and others 1991). In contrast to the Michigan sedge study (Warners 1997) these findings suggest that nutrient release from burning may stimulate growth.

Soil and tissue nutrient concentrations from *Carex* spp. dominated meadows showed complex interrelationships between nutrient concentrations, water depth, and fire incidence. Fire incidence and water depth were negatively correlated with plant tissue N, P, potassium (K), copper (Cu), manganese (Mn), and zinc (Zn) concentrations. The results suggest that volatilization and loss of these nutrients by snowmelt and runoff after spring fires may result in losses of these elements (Auclair 1977).

The site-specific studies of wetland systems presented in this chapter have shown mixed results in the relationship between burning and nutrient loss and retention. The interrelationships of nutrient loss, season of burning, and wetland system are not well understood, and generalizations across wetland systems and classes are limited at this time.

Nutrient Transformations and Cycling—In nutrient deficient wetlands with low decomposition rates, the increased availability of nitrate nitrogen (NO$_3$-N), ammonium nitrogen (NH$_4$-N), and phosphate (PO$_4$) is linked with burning because a high percentage of soil nutrients are stored in organic sediments and plant material (Wilbur and Christensen 1983). Nutrient transformations that result from soil heating are primarily dependent on soil temperatures (DeBano and others 1998). The destructive distillation of organic materials occurs at relatively low temperatures 392 to 572 °F (200 to 300 °C) (Hungerford and others 1991), and soil temperatures above 482 °F (250 °C) result in a decreases in available nutrients (Kutiel and Shaviv 1992).

Wetland soil nutrient cycling processes are similar to those of terrestrial environments, but oxygen (O_2) limitations in the wetland soil environment lead to important differences. Anaerobic reactions require both a supply of organic C and a soil saturated with slow moving or stagnant water. In this environment, aerobic organisms deplete the available O_2, and anaerobic organisms utilize other compounds as electron acceptors to respire and decompose organic tissues (Craft 1999). Soil organisms dominant in aerobic soils do not function as efficiently in low O_2 environments (Stevenson 1986), and soil chemical transformations are dominated by reduction reactions. Dominant reactions taking place in the soil include the denitrification of NO_3-N, the reduction of Mn, iron (Fe), sulfur (S), and methane production. These reactions depend on the presence of electron accepting compounds (such as NO^{-3}, Fe^{+3} Mn^{+4}, SO^{-4}), temperature, pH, and other factors (Craft 1999).

Aerobic conditions may dominate a shallow surface layer of waterlogged soils. The thickness of this layer depends on hydrology and depth of burn. The chemistry of the remaining soil profile is dominated by reduction reactions that result from the anaerobic or O_2 limiting conditions.

Burning has numerous direct influences on nutrient cycling processes. The removal of vegetation and litter material changes the surface soil moisture and temperature dynamics of bare soils, leading to changes in microbial composition and activity. Altered microbial activity can result in nutrient cycling changes in nitrogen mineralization, nitrification rates, and phosphorous mineralization rates.

Wetland Soil Nutrients—Carbon is stored in both living and dead plant materials. In wetland soils the percentage of the total C stored in dead materials is greater than in other terrestrial systems. Because the C sequestered in wetlands contains approximately 10 percent of the global C pool, wetlands play an important role in the global C cycle (Schlesinger 1991). The C balance in wetlands is sensitive to land management and environmental factors (Trettin and others 2001) and is a function of numerous interrelated factors including productivity, decomposition, and fire frequency.

Factors influencing C accumulation rates were studied in sphagnum-dominated boreal peatlands in Western Canada (Kuhry 1994). The results showed that increased frequency of peat surface fires led to decreases in C accumulation rates and organic layer thickness. Kuhry (1994) concludes that in these systems, the increased short-term productivity from postburn nutrient release may not compensate for the peat lost from frequent burning.

In contrast to sphagnum-dominated boreal systems, the burning of forested wetlands in the Southeastern United States can lead to increased C accumulation

rates. The loss of overstory vegetation that results from burning produces changes in the balance between transpiration from vegetation and surface evaporation from exposed soil. Evapotranspiration changes leading to higher water levels and reduced soil O_2 limitation may cause decreased decomposition rates and increased C accumulation rates (Craft 1999).

In freshwater wetlands, N is present in both organic and inorganic forms. Organic N is associated with plants, microbes, and sediments, while inorganic N is associated primarily with water and sediments (Craft 1999). Important transformations of available N involve the diffusion of NH_4-N between aerobic layers where mineralization and nitrification dominate N cycling and anaerobic layers where denitrification dominates N cycling (Patrick 1982, Schmalzer and Hinkle 1992).

Laboratory burning of wetland soil cores showed the effects of smoldering combustion on the distribution of available N. Different depth of burn treatments were simulated by manipulating soil core moisture. Comparisons of pre- and postburn available N distributions showed postburn NH_4-N enrichment below the ash deposition layer and the burn/no-burn interface (Hungerford and Reardon unpublished data). The findings support the conclusions of DeBano and others (1976) and suggest that the increase in available N was caused by soil heating and/or the condensation of N-rich combustion products on unburned soil surfaces (fig. 8.15).

Phosphorous limitation is common in wetlands because unlike N and C, it has no significant biologically induced or atmospheric inputs (Paul and Clark 1989). In wetlands the vegetation and organic sediments are the major storage sites for P, and cycling in wetlands is dominated by vegetation and microbes (Craft 1999). Phosphorus is present in wetland soils in a number of forms: organic forms in live and partially decomposed plant materials, in mineral form bound to aluminum (Al), Fe, Ca, Mg, and in orthophosphate compounds (Craft 1999).

The spring burning of pocosin scrub-shrub wetland sites produced immediate increases in NO_3-N, NH_4-N, and PO_4. Surface temperatures were moderate, and only limited smoldering combustion was observed. Postburn NO_3-N levels increased and remained high throughout the 18-month study. Postburn PO_4 concentrations were initially elevated but returned to preburn levels by the end of the first growing season. The NH_4-N levels were high throughout the first growing season following burning but returned to prefire levels by the end of that growing season. The limited soil heating from the surface fire and additional results from laboratory soil incubation studies suggested that postburn N and P nutrient increases were primarily the result of ash deposition (Wilbur 1985).

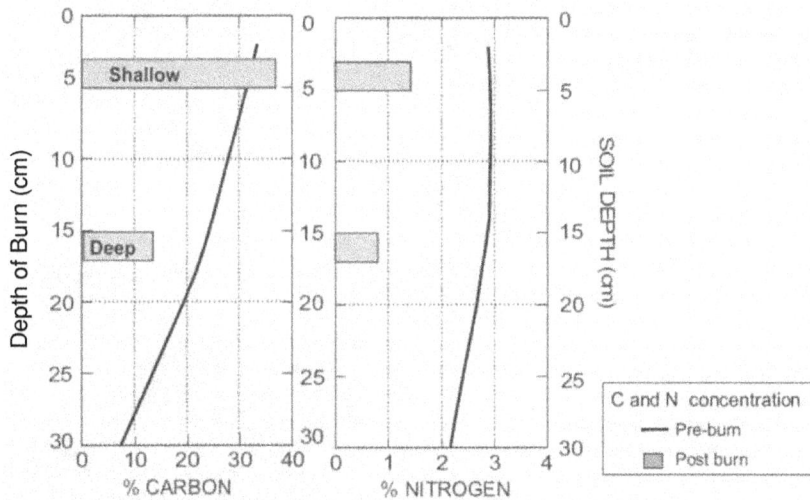

Figure 8.15—Comparison of the preburn and postburn percent C and N in ash and char material from the laboratory burning of black spruce/feather moss soil cores (Hungerford and Reardon, unpublished data).

The spring burning of *Juncus* spp. and *Spartina* spp. marshes in central Florida resulted in significant soil nutrient changes. Increases in PO_4 and soil cations were attributed to ash deposition (Schmalzer and Hinkle 1992). In contrast, the results of winter burning of *Juncus* and *Spartina* marsh communities on the Mississippi Gulf Coast suggested that increases in extractable P were the result of mild heating of organic sediments, and the retention of ash-derived nutrients may be a function of meteorological factors, tidal regimes, and topography (Faulkner and de la Cruz 1982).

Ammonium nitrogen levels of the *Juncus* and *Spartina* Florida marsh study sites were elevated for a period after the burn when the water table declined and soil temperatures increased. The results suggest that the NH_4-N changes were the result of increased soil O_2, soil temperatures, and higher N mineralization rates. Postburn soil NO_3-N levels were high and eventually declined because of a return of anaerobic conditions 6 to 9 months after burning. The authors suggest that NO_3-N declines were the caused by either decreases in nitrification or increases in denitrification rates. Soil heating was limited by standing water during the burning, and the results suggest that available nitrogen nutrient responses were linked with seasonal hydrology.

The role of fire in the biogeochemical cycles of trace metals is not well understood, and much research is currently being conducted in this area. Organic sediments accumulate metals such as mercury (Hg), arsenic, and selenium (Se) as a result of environmental processes including atmospheric deposition and water movement from upstream sources. The origins of these trace metals range from natural processes such as geochemical weathering to anthropogenic processes such as fossil fuel burning and agricultural runoff. These metals are transformed by microorganisms and enter the food web in wetland environments in methylated forms (Stevenson 1986). The accumulation and release of these metals is of concern because they are toxic to many organisms.

Selenium accumulation is of concern in wetlands in a number of Western States due to widespread geological sources. Wetland plants such as cattails and bullrush can accumulate Se, and the decomposition of the dead plant material produces organic sediments with large amounts of the element. The effect of burning on Se volatilization from wetland plants was studied at Benton Lakes National Wildlife Refuge in Montana (Zang 1997). The result showed that burning of wetland plants can volatilize up to 80 percent of the Se in leaves and stems, and the author suggests that fire may be an effective method of reducing the selenium concentrations of wetland sediments.

Mercury accumulation is a growing concern in wetland management. It is the result of atmospheric deposition and fossil fuel combustion. Inorganic Hg in sediments is transformed to methylated forms by microbial activity. The methylated forms of Hg are

USDA Forest Service Gen. Tech. Rep. RMRS-GTR-42-vol. 4. 2005

165

highly bioaccumlative and are readily incorporated into the food web. The effects of fire and prolonged drying on methylation rates were studied in the Florida Everglades (Krabbenhoft and others 2001). The results showed that burning and prolonged drying changed soil and ground water properties and resulted in Hg methylation rate increases. These rate increases were linked with postburn sulfate availability during reflooding. Increases in the average hydroperiod may result in a decrease in the occurrence and magnitude of conditions linked with high methylation rates.

Wetland Management Considerations

The wide distribution and critical ecological functions that are provided by wetlands make their continued health crucial to environmental quality. Contrary to the general perceptions of wetlands, ecological and paleoecological evidence suggests that creation and maintenance of these important ecosystems depends on fire and other disturbances. Land managers dealing with wetland ecosystems need to understand the ecological role of fire. Experience gained by the renewed use of prescribed fire and prescribed natural fire in wetlands will improve our understanding of important wetland processes and vegetation development. Increased knowledge will enable us to deal more effectively with wetland management challenges such as habitat restoration, threatened and endangered species management, and hazardous fuel reduction.

Riparian Ecosystems _____

Riparian areas are an integral and important component of watersheds throughout most of the United States (Baker and others 1998, Brooks and others 2003). They occur in both arid and humid regions and include the green-plant communities along the banks of rivers and streams (National Academy of Science 2002) (fig. 8.16). Riparian areas are found not only along most major waterways in arid and humid areas throughout the United States, but they are also important management areas on small perennial, ephemeral, and intermittent streams.

Riparian areas are a particularly unique and important part of the landscape in the Western United States where they represent the interface between aquatic and adjacent terrestrial ecosystems and are made up of unique vegetative and animal communities that require the regular presence of free or unbound water. Riparian vegetation in the United States ranges widely and includes high mountain meadows, deciduous and evergreen forests, pinyon juniper and encinal woodlands, shrublands, deserts, and desert grasslands. In most watersheds, riparian areas make up only a small percentage of the total land area (for example, only about 1 percent or less in the Western United States), yet their hydrologic and biological functions must be considered in the determination of water and other resource management goals for the entire watershed.

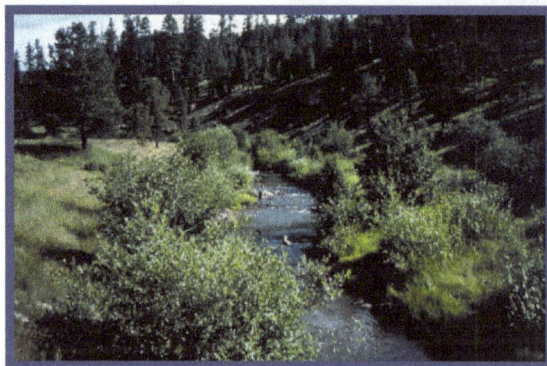

Figure 8.16—Examples of riparian areas: Gila River, New Mexico (Top, photo by Daniel Neary); Black River, Arizona (Middle, photo by Alvin Medina); Black River, Arizona (Bottom, photo by Alvin Medina).

Riparian areas have high value in terms of diversity because they share characteristics with both adjacent upland and aquatic ecosystems (Crow and others 2000). The kinds of biological and physical diversity vary widely, spatially and temporally, and thereby contribute greatly to the overall diversity of numbers, kinds, and patterns in the landscape and waterscape ecosystems along with a multitude of ecological processes associated with these patterns (Lapin and Barnes 1995). Also, riparian areas support a variety of plants and animals that are not found elsewhere, including threatened and endangered species. They play other important, but less obvious, roles relating to enhanced water quality, flood peak attenuation, and reduced erosion and sediment transport.

Riparian Definition and Classification

Riparian areas can be defined in many ways depending upon individual viewpoints or purposes. These approaches can reflect agency concepts, the disciplines involved, or the particular functional role that a riparian area plays in the total ecosystem (Ilhardt and others 2000). For example, a technical definition developed by the Society of Range Management and the Bureau of Land Management (Anderson 1987, p. 70) is:

> A *riparian area* is a distinct ecological site, or combination of sites, in which soil moisture is sufficiently in excess of that otherwise available locally, due to run-on and/or subsurface seepage, so as to result in an existing or potential soil-vegetation complex that depicts the influence of that extra soil moisture. Riparian areas may be associated with lakes; reservoirs; estuaries; potholes; springs; bogs; wet meadows; muskegs; and intermittent and perennial streams. The characteristics of the soil-vegetation complex are the differentiating criteria.

Disciplinary definitions are oriented to the particular disciplines involved, whereas functional definitions focus on using the flow of energy and materials as their basis rather than being based on static state variables.

Classifying riparian systems has involved schemes of hierarchical approaches based on geomorphology, soils, moisture regimes, plant succession, vegetation, and stream channels, or combinations of these parameters. Resource managers are increasingly interested in classifying riparian systems because of their important and unique values. A basic understanding of the ecology of riparian systems is complicated by extreme variation in geology, climate, terrain, hydrology, and disturbances by humans. Geomorphology is especially useful on riparian sites where the natural vegetation composition, soils, or water regimes, or a combination of these, have been altered by past disturbance, either natural or human induced. Other classification schemes are based on stream classification systems such as that developed by Rosgen (1994).

Hydrology of Riparian Systems

Riparian ecosystems are controlled by water, vegetation, soils, and a host of biotic organisms extending in size from microorganisms to large mammals. Water and its hydrologic processes, however, are a central functional force affecting the riparian-watershed system and are the main component that links a particular stream and associated riparian vegetation to the surrounding watershed. The magnitude and direction of the riparian-watershed relationship is further tempered by geology, geomorphology, topography, and climate—and their interactions. This interrelationship between riparian area and surrounding watershed is the most sensitive to natural and human-related disturbances, including fire (DeBano and Neary 1996, DeBano and others 1998).

The important hydrologic processes include runoff and erosion, ground water movement, and streamflow. As the boundary between land and stream habitats, streambanks occupy a unique position in the riparian system (Bohn 1986). Bank and channel profiles affect stream temperature, water velocity, sediment input, and hiding cover and suitable living space for fish. Streamside vegetation on stable banks provides food and shade for both fish and wildlife. As a result, streambank condition and the quality of the fish habitat are closely linked.

Riparian systems in arid environments are hydrologically different from those found in more humid areas. In the arid West, intermittent streamflow and variable annual precipitation are more common than in the more humid areas of the Eastern United States. As a result, riparian systems in humid regions are less dependent on annual recharge by episodic events to sustain flow and are more dependent on a regular and dependable ground water flow accompanied by interflow (subsurface flow) and, to a lesser extent, on overland flow. Watersheds in humid regions also commonly have lakes, ponds, and other systems that sustain flows over much of the year. As a result of variable precipitation, sediment transport is more episodic in the arid West, where pulses of aggradation and degradation are punctuated by periods of inactivity. Unlike on humid watersheds, side-slope erosion in drylands is discontinuous, and there are often long lag periods between watershed events and sediment delivery (Heede and others 1988). Channels in humid regions tend to be more stable over the long term, and the macrotopography of these systems, though still responsive to flooding events, tends to evolve much more slowly.

Riparian Fire Effects

Fire is a common disturbance in both riparian ecosystems and the surrounding hillslopes, and both

USDA Forest Service Gen. Tech. Rep. RMRS-GTR-42-vol. 4. 2005

167

wildfires and prescribed burns occur in many of these riparian-watershed systems. Fire affects riparian areas both directly and indirectly. The direct effects consist mainly of damage to the vegetation (trees, shrubs, and grasses) that intercepts precipitation, and the partial consumption of the underlying litter layer. The severity of the damage to the riparian vegetation depends upon the severity of the fire, which could consume part or all of the vegetation. Severe wildfires can cause profound damage to plant cover and can increase streamflow velocity, sedimentation rates, and stream water temperatures, as contrasted to low severity, cool-burning prescribed fires, which have less severe consequences. When fire burns the surrounding watershed, the indirect effect on the riparian area is that it decreases basin stability, and in steep erodible topography, debris flows along with dry ravel and small landslides off hillslopes are common. Therefore, the recovery of vegetation following fire reflects the combined disturbance of both the fire and flooding, and together they can impact the time required for revegetation and postfire rehabilitation efforts.

Role of Large Woody Debris

Large organic debris is a major component of watersheds and river systems because of its important role in hydraulics, sediment routing, and channel morphology of streams flowing through riparian systems (Smith and others 1993, DeBano and Neary 1996). An additional benefit of large woody debris is in nutrient cycling and the productivity of forest vegetation. Through time, entire live and dead trees and shrubs, or parts of them, are likely to fall into stream channels within riparian ecosystems. This woody debris increases the complexity of stream habitats by physically obstructing water flow. For example, trees extending partially across the channel deflect the current laterally, causing it to widen the streambed. Sediment stored by debris also adds to hydraulic complexity, especially in organically rich channels that are often wide and shallow and possess a high diversity of riffles and pools in low gradient streams of alluvial valley floors. Stream stabilization after major floods, debris torrents, or massive landslides is accelerated by large woody debris along and within the channel. After wildfire, while the postfire forest is developing, the aquatic habitat may be maintained by large woody debris supplied to the stream by the prefire forests.

Coarse woody debris accumulations are usually larger in the first few years following a wildfire than in prefire years, but they generally decline to below prefire conditions thereafter. The recovery time to prefire conditions ranges from 25 to 300 years depending on the stability of the ecosystem burned. While in stream channels, this added debris can have a beneficial effect on aquatic habitats in the short term by providing structure to the streams. Long-term impacts can be disruptive to stream morphology and the consequent streamflow and sediment transport regimes. Carefully prescribed fire should not affect the accumulations of coarse woody debris either on the watershed or in the channel. Recommended amounts of woody debris to be left on the watershed in Arizona, Idaho, and Montana are given earlier in chapter 4 (see also Graham and others 1994).

Coarse woody debris in the stream channels of burned riparian ecosystems is made up of components of larger individual size, and forms larger accumulations, than that of unburned systems. Postfire debris is also likely to be moved more frequently over longer distances in subsequent streamflow events (Young 1994). Managers frequently debate the merits of leaving burned dead debris in stream channels following fire. Some managers feel that this material could jam culverts and bridges, causing these structures to wash out and cause flooding, and, as a consequence, these managers feel the debris should be removed (Barro and others 1989). However, removal of this debris can also result in changes in channel morphology, a scouring of the channel bed, increases in streamflow velocities and sediment loads, an export of nutrients out of the ecosystem, and a deterioration of biotic habitats. Removal of postfire accumulations of coarse woody debris, therefore, is likely to best be decided on a site-specific basis.

Riparian Management Considerations

Fire is an important factor to consider in the management of riparian ecosystems. Most fire in riparian areas can be intense and cause extensive damage to the vegetation. However, even after severe fires, recovery can be rapid within a couple years to prefire conditions in some environments, but not all. The recovery of vegetation following fire reflects the combined disturbance of both the fire and flooding.

Riparian areas are particularly important because they provide buffer strips that trap sediment and nutrients that are released when surrounding watersheds are burned. The width of these buffer strips is critical for minimizing sediment and nutrient movement into the streams. The best available guidelines for buffer width associated with prescribed fire are that for low intensity fires, less than 2 feet (0.6 m) high, that do not kill stream-shading shrubs and trees. Such fire can be used throughout the riparian area without creating substantial damage. Where fire damages woody vegetation, the width should be proportional to the size of the contributing area, slope, cultural practices in the upslope area, and the nature of the drainage below. A general rule of thumb is that a width of 30 feet (9 m) plus (0.46 x percent slope) be left along the length of the stream to protect the riparian resource (DeBano and others 1998).

Managers need to consider the mixed concerns about leaving downed large woody debris in, or near, channel following fire. On one hand, large woody debris plays an important role in hydraulics, sediment routing, and channel morphology of streams flowing through riparian systems, thereby enhancing these systems. On the other hand, to protect life and property, channel clearing treatments following fire is usually desirable. The USDA Forest Service suggests that the best management practice balances downstream value protection with the environmental implications of the treatment.

Summary

While the connection between wetlands and riparian ecosystems and fire appears incongruous, fire plays an integral role in the creation and persistence of hydric species and ecosystems. Wetland and riparian systems have been classified with various classification systems that reflect both the current understanding of biogeochemical processes and the missions of the individual land management agencies.

This chapter examines some of the complex water inflows and outflows that are balanced by the interrelationships between biotic and abiotic factors. The presence and movement of water are dominant factors controlling the interactions of fire with wetland dynamics, nutrient and energy flow, soil chemistry, organic matter decomposition, and plant and animal community composition.

In wetland and riparian soils, the effects produced by surface and ground fires are related to the intensity of fire at either the soil surface or the interface between burning and nonburning organic soil materials. In general, surface fires can be characterized as short duration, variable intensity ones, while ground fires can be characterized as having longer duration and lower intensity. However, the latter type of fires can also produce high severity fires with profound physical and chemical changes.

USDA Forest Service Gen. Tech. Rep. RMRS-GTR-42-vol. 4. 2005

169

Notes

Kevin Ryan
William J. Elliot

Chapter 9:
Fire Effects and Soil Erosion Models

Introduction

In many cases, decisions about fire have to be made in short timeframes with limited information. Fire effects models have been developed or adapted to help land and fire managers make decisions on the potential and actual effects of both prescribed fires and wildfires on ecosystem resources (fig. 9.1). Fire effects models and associated erosion and runoff models apply the best fire science to crucial management decisions. These models are undergoing constant revision and update to make the latest information available to fire managers using the state-of-the-art computer hardware and software. Use of these models requires a commitment to understand their assumptions, benefits, and shortcomings, and a commitment to constant professional development.

First Order Fire Effects Model (FOFEM)

FOFEM (First Order Fire Effects Model) is a computer program that was developed to meet the needs of resource managers, planners, and analysts in

Figure 9.1—Wildland fires such as the Rodeo-Chediski Fire of 2002 affect the complete range of physical, chemical, and biological components of ecosystems. (Photo by USDA Forest Service).

USDA Forest Service Gen. Tech. Rep. RMRS-GTR-42-vol. 4. 2005

171

predicting and planning for fire effects. FOFEM provides quantitative predictions of fire effects for planning prescribed fires that best accomplish resource needs, for impact assessment, and for long-range planning and policy development. FOFEM was developed from long-term fire effects data collected by USDA Forest Service and other scientists across the United States and Canada (fig. 9.2).

Description, Overview, and Features

First order fire effects are those that concern the direct or indirect immediate consequences of fire. First order fire effects form an important basis for predicting secondary effects such as tree regeneration plant succession, soil erosion, and changes in site productivity, but these long-term effects generally involve interaction with many variables (for example, weather, animal use, insects, and disease) and are not predicted by this program. FOFEM predicts fuel consumption, smoke production, and tree mortality. The area of applicability is nationwide on forest and nonforest vegetation types. FOFEM also contains a planning mode for prescription development.

Applications, Potential Uses, Capabilities, and Goals

FOFEM makes fire effects research results readily available to managers. Potential uses include wildfire impact assessment, development of salvage specifications, design of fire prescriptions, environmental

Figure 9.2—Development of FOFEM and other fire effects models stemmed from long-term fire effects data collected by USDA Forest Service and other scientists across North America. (Photo by USDA Forest Service).

assessment, and fire management planning. FOFEM can also be used in real time, quickly estimating tree mortality, smoke generation, and fuel consumption of ongoing fires.

Scope and Primary Geographic Applications

FOFEM—national in scope—uses four geographic regions: Pacific West, Interior West, Northeast, and Southeast. Forest cover types provide an additional level of resolution within each region, and SAF and FRES vegetation types to stratify data and methods. Geographic regions and cover types are used both as part of the algorithm selection key, and also as a key to default input values. FOFEM contains data and prediction equations that apply throughout the United States for most forest and rangeland vegetation types that experience fire.

Input Variables and Data Requirements

FOFEM was designed so that data requirements are minimal and flexible. Default values are provided for almost all inputs, but users can modify any or all defaults to provide custom inputs.

Output, Products, and Performance

FOFEM computes the direct effects of prescribed fire or wildfire. It estimates fuel consumption by fuel component for duff, litter, small and large woody fuels, herbs, shrubs and tree regeneration, and crown foliage and branchwood. It also estimates mineral soil exposure, smoke production of CO, PM10, and PM2.5, and percent tree mortality by species and size class. Alternatively, if the user enters desired levels of these fire effects, FOFEM computes fuel moistures and fire intensities that should result in desired effects.

Advantages, Benefits, and Disadvantages

FOFEM is easy to use, applies to most vegetation types and geographic areas, synthesizes and makes available a broad range of available research results, incorporates planning and prediction modes, and provides a wide range of data in the form of default inputs for different vegetation and fuel types. The main disadvantage is that FOFEM is not currently linked to any other models (fire behavior, smoke dispersion, postfire succession).

System and Computer Requirements

FOFEM version 5.0 is available for IBM-compatible PCs with Windows 98 and Windows 2000 operating systems. FOFEM is supported by the Fire Effects Research Work Unit, Intermountain Fire Sciences

172

USDA Forest Service Gen. Tech. Rep. RMRS-GTR-42-vol. 4. 2005

Lab, Missoula, MT 59807. Additional information can be obtained from:

http://www.fire.org/
http://www.firelab.org/

FOFEM includes embedded help and user's information. The current version (5.21) can be downloaded for use with WINDOWS® 98, 2000, and XP at:

http://www.fire.org/index.php?option=com_content&task=view&id=58&Itemid=25

POWERPOINT® tutorials provide a FOFEM overview and information for basic and advanced users. FOFEM 4.0 should be replaced with FOFEM 5.21. For more detailed information contact Elizabeth Reinhardt at: ereinhardt@fs.fed.us

Models for Heat and Moisture Transport in Soils

Transfer of heat into the soil beneath a fire produces a large number of onsite fire effects to the physical, chemical, and biological properties of soils (Hungerford 1990) that include:

- plant mortality and injury
- soil organism mortality and injury
- thermal decomposition of organic matter
- oxidation or volatilization of chemical components of the upper soil profile
- other physiochemical changes

To predict the nature and extent of these effects, we need to understand temperature profiles within the soil beneath burned areas (Albini and others 1996). Temperature profiles are rarely measured in actual fires, so some type of model is needed to predict soil temperatures and the response of soils to the thermal input. Albini and others (1996) reviewed a number of existing models to determine their applicability and recommend future development goals.

The Albini and others (1996) review of heat transfer models from the soil science, engineering, and geophysics fields concluded that the only useful models for describing heat transfer phenomena for wildland fires come from the soil science arena. The models of Campbell and others (1992, 1995) seem to function well in predicting temperature histories and profiles of soils heated at rates and temperatures consistent with wildland fires. Their model did not perform as well with soil moisture contents as with temperatures. Because many of the heating effects are a function of soil moisture, this is an important ability for heat transfer prediction models.

Albini and others (1996) identified the omission of a number of important features in the soil science models. These include diffusive transport of water as a vapor or liquid, momentum equations, predictions of the transient movement of phase-change boundaries, lateral nonhomogeneity of soils, and the rapid decline of wetting attraction of liquid water to quartz near 149 °F (65 °C).

Finally, Albini and others (1996) made recommendations for further model development and simplifications of the existing models. They believed that some simplification would improve the use of the existing models without much sacrifice in the fidelity of their predictions.

WEPP, WATSED, and RUSLE Soil Erosion Models

Following a fire, it is often necessary to use some standard prediction technology to evaluate the risk of soil erosion. For forests that tend to regenerate rapidly, the risk of erosion decreases quickly after the first year, at a rate of almost 90 percent each year. For example, the year following a fire may experience 0.4 to 0.9 tons/acre (1 to 2 Mg/ha) erosion, the second year less than 0.04 to 0.10 tons/acre (0.1 to 0.3 Mg/ha), and the third year may be negligible (Robichaud and Brown 1999). The erosion rate depends on the climate, topography, soil properties (including hydrophobicity), and amount of surface cover. Surface cover may include unburned duff, rock, and needle cast following fire.

Three models are commonly used after soil erosion. In USDA Forest Service Regions 1 and 4, the WATSED and similar models have frequently been used (USDA Forest Service 1990b). The Universal Soil Loss Equation (USLE) has been used widely for many years, and more recently, the Revised USLE, or RUSLE, has become common (Renard and others 1997). The Water Erosion Prediction Project (WEPP) model has recently been parameterized for predicting erosion after fire, and an interface has been developed to aid in that prediction. Improvements in the usability of both the RUSLE and WEPP prediction technologies are ongoing. The WATSED model is intended to be a cumulative affects model, to be applied at watershed scale. RUSLE and WEPP are hillslope models. WEPP has a watershed version under development, but it has received little use outside of research evaluation.

WATSED is intended as a watershed model to combine the cumulative effects of forest operations, fires, and roads on runoff and sediment yield for a given watershed. Factors that account for burned area within the watershed, soil properties, topography, and delivery ratios are identified, and an average sediment delivery is calculated. This sediment delivery is reduced over a 15-year period following a fire before the impact is assumed to be zero. Within the Western geographical territory of the Forest Service Regions, some of the factors have been adjusted to calibrate the

USDA Forest Service Gen. Tech. Rep. RMRS-GTR-42-vol. 4. 2005

173

model for local conditions, leading to the development of models such as NEZSED and BOISED. The erosion predictions are based on observations in the mountains in Regions 1 and 4, and are not intended for use elsewhere. Table 9.1 provides the erosion rates predicted in WATSED, corrected for a USLE *LS* factor of 11.2 (Wischmeier and Smith 1978). These rates are adjusted for topography, landscape, and soil properties before arriving at a final prediction. A variation of the technology in WATSED has been adopted by the State of Washington for its Watershed Analysis procedure (Washington Forest Practices Board 1997).

The Revised USLE was developed not only for agriculture, but also included rangeland conditions. The RUSLE base equation is:

$$A = R\,K\,LS\,C\,P \qquad (1)$$

Where *A* is the average annual erosion rate, *R* is the rainfall erosivity factor, *K* is the soil erodibility factor, *LS* is the slope length and steepness factor, *C* is the cover management factor, and *P* is the conservation management factor. Although it has not been widely tested, the RUSLE values appear to give reasonable erosion values for rangelands (Renard and Simanton 1990, Elliot and others 2000) and will likely do the same for forests. There are no forest climates available in the RUSLE database. Table 9.1 provides some assumptions about rate of vegetation regeneration and typical erosion rates estimated for burned and recovering forest conditions based on those assumed cover values. The RUSLE *LS* factor was about 6.54, almost half of the USLE C-factor used for WATSED. The RUSLE *LS* factor is based on more recent research and the analysis of a greater number of plots (McCool and others 1987, 1989), so it should probably be used with the WATSED technology to adjust for slope length and steepness. The RUSLE *R* factor was estimated as 20 from the documentation (Renard and others 1997). This is a relatively low value because much of the precipitation

in the Northern Rockies comes as snowfall, and snowmelt events cause much less erosion than rainfall events.

The most recent erosion prediction technology is the Water Erosion Prediction Project (WEPP) model (Flanagan and Livingston 1995). WEPP is a complex process-based computer model that predicts soil erosion by modeling the processes that cause erosion. These processes include daily plant growth, residue accumulation and decomposition, and daily soil water balance. Each day that has a precipitation or snow melt event, WEPP calculates the infiltration, runoff, and sediment detachment, transport, deposition, and yield.

WEPP was released for general use in 1995, with an MS DOS text-based interface. Currently a Windows interface is under development and is available for general use (USDA 2000). Elliot and Hall (1997) developed a set of input templates to describe forest conditions for the WEPP model, for the MS DOS interface. The WEPP model allows the user to describe the site conditions with hundreds of variables, making the model extremely flexible, but also making it difficult for the casual user to apply to a given set of conditions. To make the WEPP model run more easily for forest conditions, Elliot and others (2000) developed a suite of interfaces to run WEPP over the Internet using Web browsers. The forest version of WEPP can be found at:

http://forest.moscowfsl.wsu.edu/fswepp

One of the interfaces is Disturbed WEPP, which allows the user to select from a set of vegetation conditions that describe the fire severity and recovering conditions. The Disturbed WEPP alters both the soil and the vegetation properties when a given vegetation treatment is selected. Table 9.2 shows the vegetation treatment selected for each of the years of recovery. In all cases, the cover input was calibrated to ensure that WEPP generated the desired cover given

Table 9.1—Erosion rates observed and predicted by WATSED, RUSLE, and WEPP for the cover shown, for a 30 percent steepness, 60-m long slope.

Year after fire	Estimated cover	Observed erosion rate	Predicted erosion rate		
			WATSED	RUSLE	WEPP
	Percent	*Mg/ha*	- - - - - - - - - *Mg/ha* - - - - - - - - -		
1	50	2.2	1.92	3.35	1.74
2	65	0.02	1.64	1.30	0.37
3	80	0.01	0.96	0.54	0.02
4	95	0.00	0.48	0.20	0.00
5	97	0.00	0.29	0.16	0.00
6	99	0.00	0.15	0.09	0.00
7	100	0.00	0.06	0.07	0.00

[1] From Robichaud and Brown (1999)

Table 9.2—Vegetation treatment selected for each year of recovery with the Disturbed WEPP interface.

Years since fire	Disturbed WEPP vegetation treatment
0	High severity fire
1	Low severity fire
3	Short grass
4	Tall grass
5	Shrubs
6	5-year-old forest (99 percent cover)
7	5-year-old forest (100 percent cover)

in table 9.1. The Disturbed WEPP interface has access to a database of more than 2,600 weather stations to allow the user to select the nearest station to the disturbed site. The values in table 9.3 were predicted for Warren, ID, climate. Warren climate is similar to the climate for Robichaud and Brown's study, and also near the site where the WATSED base erosion rates were developed in central Idaho.

An important aspect of soil erosion following a fire is that the degree of erosion depends on the weather the year immediately following the fire. Table 9.1 shows the rapid recovery of a forest in the years after fire. If the year after the fire has a number of erosive storms, then the erosion rate will be high. If the year after the fire is relatively dry, then the erosion rate will be low. The values presented in table 9.1 are all average values. There is a 50 percent chance that the erosion in this most susceptible year will be less than the average value. To allow managers to better evaluate the risk of a given level of erosion following a fire, the Disturbed WEPP interface includes some probability analyses with the output, giving the user an indication of the probability associated with a given level of erosion. Table 9.3 shows that there is a one in 50, or 2 percent, chance that the erosion rate from the specified hill will exceed 3.18 tons/acre (7.12 Mg/ha), and the sediment delivery will exceed 2.88 tons/acre

(6.45 Mg/ha). There is a one in 10, or 10 percent, chance that the erosion and sediment delivery rate will exceed 2.11 tons/acre (4.72 Mg/ha), and so forth. This feature will allow users to evaluate risks of upland erosion and sediment delivery to better determine the degree of mitigation that may be justified following a given fire. In California, for example, erosion is often estimated for a 5-year condition, which in this case is 1.47 tons/acre (3.3 Mg/ha). Disturbed WEPP also predicted that there was an 80 percent chance that there would be erosion on this hillslope the year following the fire.

The variability of erosion following a fire due to the climate makes any measurements difficult to evaluate. Note in table 9.1 the large drop from year 1 to year 2 in erosion rate. This decline was likely due not only to regeneration but also to the lower precipitation in 1996. In the nearby Warren, ID, climate, the average precipitation is 696 mm; the year following the fire it was 722 mm, and the second year after the fire only 537 mm. These variations from the mean also help explain why the Disturbed WEPP predicted erosion rates in table 9.1 for "average" conditions were below the observed value the first year but above the observed value the second year.

The variability in erosion observations and predictions is influenced not only by climate but also by spatial variability of soil and topographic properties. In soil erosion research to determine soil properties, it is not uncommon to have a standard deviation in observations from identical plots greater than the mean. A rule of thumb in interpreting erosion observations or predictions is that the true "average" value is likely to be within plus or minus 50 percent of the observed value. In other words, if a value of 0.9 tons/acre (2 Mg/ha) is observed in the field from a single observation, the true "average" erosion from that hillside is likely to be between 0.4 and 1.3 tons/acre (1 and 3 Mg/ha). Following this rule leads to the conclusions that WATSED, RUSLE, and WEPP predictions in table 9.1 are not different from the observed erosion rates.

Table 9.3—Exceedance probabilities associated with different levels of precipitation, runoff, and soil erosion for the year following a severe wild fire in central Idaho.

Return period	Precipitation		Runoff		Erosion		Sediment	
Years	mm	in	mm	in	Mg/ha	tons/ac	Mg/ha	tons/ac
50.0	973.60	38.33	31.82	1.25	7.12	3.18	6.45	2.88
25.0	892.80	35.15	31.79	1.25	6.45	2.88	6.32	2.82
10.0	811.50	31.95	27.65	1.09	4.72	2.11	4.72	2.11
5.0	756.10	29.77	20.56	0.81	3.30	1.47	3.30	1.47
2.5	671.80	26.45	14.74	0.58	1.80	0.80	1.80	0.80
Average	670.92	26.41	12.47	0.49	1.74	0.78	1.74	0.78

In the years of regeneration, it appears that both WATSED and RUSLE are overpredicting observed erosion rates, whereas the Disturbed WEPP predictions are nearer to the observed values. WATSED, as a cumulative effects model, is considering the impact of the disturbance on a watershed scale. Frequently eroded sediments following a disturbance may take several years to be routed through the watershed, whereas WEPP is only considering the hillslope in its predictions. RUSLE is also a hillslope model but considers only the upland eroding part of the hillside and does not consider any downslope deposition. This means that RUSLE values will frequently be overpredicted unless methods to estimate delivery ratio are considered. A RUSLE2 model currently under development addresses downslope deposition and sediment delivery.

Model Selection

Managers must determine which model most suits the problem at hand. The WATSED technology is geographic specific, as is the Washington Forest Practices model. These models should not be used outside of the areas for which they were developed. The WATSED technology is intended to assist in watershed analysis and not necessarily intended for estimating soil erosion after fires. RUSLE is intended to predict upland erosion and is best suited for estimating potential impacts of erosion on onsite productivity. It is less well suited for predicting offsite sediment delivery. The WEPP technology provides estimates of both upland erosion for soil productivity considerations and sediment delivery for offsite water quality concerns. The WEPP DOS and Windows technology requires skill to apply and should be considered only by trained specialists. The Disturbed WEPP interface requires little training, and documentation with examples is included on the Web site, making it available to a wider range of users.

DELTA-Q and FOREST Models

Two other models warrant brief mentioning. They can assist fire managers in dealing with watershed scale changes in water flow and erosion. These are DELTA-Q and FOREST. Both programs require an ESRI Arc 8.x license. Further documentation can be found at:

http://www.cnr.colostate.edu/frws/people/faculty macdonald/model.htm.

One of the difficult tasks facing land managers, fire managers, and hydrologists is quantifying the changes in streamflow after forest disturbances such as fire.

The changes of interest are alteration of peak, median, and low flows as well as the degradation of water quality due to increased sediment delivery to channels or channel degradation.

DELTA-Q is a model designed to calculate the cumulative changes in streamflow on a watershed scale from areas subjected to the combination of harvesting and road construction. Flow changes due to forest cover removal by wildfire can also be calculated. A current data limitation in the model is that it evaluates only changes due to vegetation combustion, not the possible effects of alterations to runoff and streamflow generation processes. The objective of DELTA-Q is to provide fire and watershed managers with a GIS-based tool that can quickly approximate the sizes of changes in different flow percentiles. The model does not estimate the increases in streamflow from extreme events (see chapters 2 and 5). The model was designed to be used for planning at watershed scales of 5 to 50 mi^2 (3,200 to 32,000 acres, or 1,300 to 13,000 ha).

The FOREST (FORest Erosion Simulation Tools) model functions with DELTA-Q. It calculates changes in the sediment regime due to forest disturbances. It consists of a hillslope model that uses a polygon GIS layer of land disturbances to calculate sediment production. Road-related sediment is treated separately because roads are linear features in the landscape. Input values for the road segment can be generated by several means including WEPP.Road. FOREST does not deal with changes in channel stability.

Models Summary

This chapter is not meant as a comprehensive look at simulation models. Several older modeling technologies commonly used estimate fire effects during and after fire (FOFEM, WATSED, WEPP, RUSLE, and others). New ones such as DELTA-Q, FOREST have been recently developed, and others are under construction. These process-based models provide managers with additional tools to estimate the magnitude of fire effects on soil and water produced by land disturbance. FOFEM was developed to meet needs of resource managers, planners, and analysts in predicting and planning for fire effects. Quantitative predictions of fire effects are needed for planning prescribed fires that best accomplish resource needs, for impact assessment, and for long-range planning and policy development. FOFEM was developed to meet this information need. The WATSED technology was developed for watershed analysis. The RUSLE model was developed for agriculture and rangeland hillslopes and has been extended to forest lands. The WEPP model was designed as an improvement over RUSLE that can either be run as a stand-alone computer

176

USDA Forest Service Gen. Tech. Rep. RMRS-GTR-42-vol. 4. 2005

model by specialists, or accessed through a special Internet interface designed for forest applications, including wild fires.

All of these models have limitations that must be understood by fire managers or watershed specialists before they are applied. The models are only as good as the data used to create and validate them. Some processes such as extreme flow and erosion events are not simulated very well because of the lack of good data or the complexity of the processes. However, they do provide useful tools to estimate landscape changes to disturbances such as fire. Potential users should make use of the extensive documentation of these models and consult with the developers to ensure the most appropriate application of the models.

USDA Forest Service Gen. Tech. Rep. RMRS-GTR-42-vol. 4. 2005

177

Notes

178

USDA Forest Service Gen. Tech. Rep. RMRS-GTR-42-vol. 4. 2005

Peter R. Robichaud
Jan L. Beyers
Daniel G. Neary

Chapter 10:
Watershed Rehabilitation

Recent large, high severity fires in the United States, coupled with subsequent major hydrological events, have generated renewed interest in the linkage between fire and onsite and downstream effects (fig. 10.1). Fire is a natural and important disturbance mechanism in many ecosystems. However, the intentional human suppression of fires in the Western United States, beginning in the early 1900s, altered natural fire regimes in many areas (Agee 1993). Fire suppression can allow fuel loading and forest floor material to increase, resulting in fires of greater intensity and extent than might have occurred otherwise (Norris 1990).

High severity fires are of particular concern because the potential affects on soil productivity, watershed response, and downstream sedimentation often pose threats to human life and property. During severe fire seasons, the USDA Forest Service and other Federal and State land management agencies spend millions of dollars on postfire emergency watershed rehabilitation measures intended to minimize flood runoff, onsite erosion, and offsite sedimentation and hydrologic damage. Increased erosion and flooding are certainly the most visible and dramatic impacts of fire apart from the consumption of vegetation.

Burned Area Emergency Rehabilitation (BAER)

The first formal reports on emergency watershed rehabilitation after wildfires were prepared in the 1960s and early 1970s, although postfire seeding with

Figure 10.1—Flood flow on the Apache-Sitgreaves National Forest, Arizona, after the Rodeo-Chediski Fire of 2002. (Photo by Dave Maurer).

USDA Forest Service Gen. Tech. Rep. RMRS-GTR-42-vol. 4. 2005

179

grasses and other herbaceous species was conducted in many areas in the 1930s, 1940s, and 1950s (Christ 1934, Gleason 1947). Contour furrowing and trenching were used when flood control was a major concern (Noble 1965, DeByle 1970b). The Forest Service and other agencies had no formal emergency rehabilitation program. Funds for fire suppression disturbance were covered by fire suppression authorization. Watershed rehabilitation funding was obtained from emergency flood control programs or, more commonly, restoration accounts. Prior to 1974, the fiscal year had ended June 30 of each year, allowing year-end project funds to be shifted to early season fires. After July 1, fires were covered by shifts in the new fiscal year funding. The shift to an October 1 to September 30 fiscal year made it difficult to provide timely postfire emergency treatments or create appropriated watershed restoration accounts.

In response to a Congressional inquiry on fiscal accountability, in 1974 a formal authority for $2 million in postfire rehabilitation activities was provided in the Interior and Related Agencies appropriation. Called Burned Area Emergency Rehabilitation (BAER), this authorization was similar to the fire fighting funds in that it allowed the Forest Service to use any available funds to cover the costs of watershed treatments when an emergency need was determined and authorized. Typically, Congress reimbursed accounts used in subsequent annual appropriations. Later, annual appropriations provided similar authorities for the Bureau of Land Management and then other Interior agencies. The occurrence of many large fires in California and southern Oregon in 1987 caused expenditures for BAER treatments to exceed the annual BAER authorization of $2 million. Congressional committees were consulted and the funding cap was removed. The BAER program evolved, and policies were refined based on determining what constituted a legitimate emergency warranting rehabilitation treatments.

The BAER-related policies were initially incorporated into the Forest Service Manual (FSM 2523) and the Burned Area Emergency Rehabilitation (BAER) Handbook (FSH 2509.13) in 1976. These policies required an immediate assessment of site conditions following wildfire and, where necessary, implementation of emergency rehabilitation measures. These directives delineated the objectives of the BAER program as:

1. Minimizing the threat to life and property onsite and offsite.
2. Reducing the loss of soil and onsite productivity.
3. Reducing the loss of control of water.
4. Reducing deterioration of water quality.

As postfire rehabilitation treatment increased, debates arose over the effectiveness of grass seeding and its negative impacts on natural regeneration. Seeding was the most widely used individual treatment, and it was often applied in conjunction with other hillslope treatments, such as contour-felled logs and channel treatments.

In the mid 1990s, a major effort was undertaken to revise and update the BAER handbook. A steering committee, consisting of regional BAER coordinators and other specialists, organized and developed the handbook used today. The issue of using native species for emergency revegetation emerged as a major topic, and the increased use of contour-felled logs (fig. 10.2) and mulches caused rehabilitation expenditures to escalate. During the busy 1996 fire season, for example, the Forest Service spent $11 million on BAER projects. In 2000, 2001, and 2002 the average annual BAER spending rose to more than $50 million.

Improvements in the BAER program in the late 1990s included increased BAER training and funding review. Increased training needs were identified for BAER team leaders, project implementation, and on-the-ground treatment installation. Courses were developed for the first two training needs but not the last. Current funding requests are scrutinized by regional and national BAER coordinators to verify that funded projects are minimal, necessary, reasonable, practicable, cost effective, and a significant improvement over natural recovery.

In the late 1990s, a program was initiated to integrate national BAER policies across different Federal agencies (Forest Service, Bureau of Land Management, National Park Service, Fish and Wildlife Service, and Bureau of Indian Affairs) as each agency had different authorities provided in the Annual Appropriations Acts.

Figure 10.2—Installing contour-felled logs for erosion control after a wildfire. (Photo by Peter Robichaud).

The U.S. Department of Agriculture and Department of the Interior approved a joint policy for a consistent approach to postfire rehabilitation in 1998. The new policy broadened the scope and application of BAER analysis and treatment. Major changes included:

1. Monitoring to determine if additional treatment is needed and evaluating to improve treatment effectiveness.
2. Repairing facilities for safety reasons.
3. Stabilizing biotic communities.
4. Preventing unacceptable degradation of critical known cultural sites and natural resources.

BAER Program Analysis

Early BAER efforts were principally aimed at controlling runoff and consequently erosion. Research by Bailey and Copeland (1961), Christ (1934), Copeland (1961, 1968), Ferrell (1959), Heede (1960, 1970), and Noble (1965) demonstrated that various watershed management techniques could be used on forest, woodland, shrub, and grassland watersheds to control both storm runoff and erosion (fig. 10.3). Many of these techniques were developed from other disciplines (such as agriculture and construction) and refined or augmented to form the set of BAER treatments in use today (table 10.1).

In spite of the improvements in the BAER process and the wealth of practical experience obtained over the past several decades, the effectiveness of many emergency rehabilitation methods have not been systematically tested or validated. Measuring erosion and runoff is expensive, complex, and labor intensive (fig. 10.4). Few researchers or management specialists have the resources or the energy to do it. BAER team

Figure 10.3—Straw bale check dams placed in channel by the Denver Water Board after the Hayman Fire, 2002, near Deckers, CO. (Photo by Peter Robichaud).

leaders and decisionmakers often do not have information available to evaluate the short- and long-term benefits (and costs) of various treatment options.

In 1998, a joint study by the USDA Forest Service Rocky Mountain Research Station and the Pacific Southwest Research Station evaluated the use and effectiveness of postfire emergency rehabilitation methods (Robichaud and others 2000).

The objectives of the study were to:

1. Evaluate the effectiveness of rehabilitation treatments at reducing postwildfire erosion, runoff, or other effects.
2. Assess the effectiveness of rehabilitation treatments in mitigating the downstream effects of increased sedimentation and peakflows.
3. Investigate the impacts of rehabilitation treatments on natural processes of ecosystem recovery, both in the short and long term.
4. Compare hillslope and channel treatments to one another and to a no-treatment option.
5. Collect available information on economic, social, and environmental costs and benefits of various rehabilitation treatment options, including no treatment.
6. Determine how knowledge of treatments gained in one location can be transferred to another location.
7. Identify information gaps needing further research and evaluation.

Robichaud and others (2000) collected and analyzed information on past use of BAER treatments in order to determine attributes and conditions that led to treatment success or failure in achieving BAER goals. Robichaud and others (2000) restricted this study to USDA Forest Service BAER projects in the continental Western United States and began by requesting Burned Area Report (FS-2500-8) forms and monitoring reports from the Regional headquarters and Forest Supervisors' offices. The initial efforts revealed that information collected on the Burned Area Report forms and in the relatively few existing postfire monitoring reports was not sufficient to assess treatment effectiveness, nor did the information capture the knowledge of BAER specialists. Therefore, interview questions were designed to enable ranking of expert opinions on treatment effectiveness, to determine aspects of the treatments that lead to success or failure, and to allow for comments on various BAER-related topics.

Interview forms were developed after consultation with several BAER specialists. The forms were used to record information when BAER team members and regional and national leaders were interviewed. Onsite interviews were conducted because much of the supporting data were located in the Forest Supervisors' and District Rangers' offices and could be

USDA Forest Service Gen. Tech. Rep. RMRS-GTR-42-vol. 4. 2005

181

Table 10.1—Burned Area Emergency Rehabilitation (BAER) treatments (From Robichaud and others 2000).

Hillslope	Channel	Road and trail
Broadcast seeding	Straw bale check dams	Rolling dips
Seeding plus fertilizer	Log grade stabilizers	Water bars
Mulching	Rock grade stabilizers	Cross drains
Contour-felled logs	Channel debris clearing	Culvert overflows
Contour trenching	Bank/channel armoring	Culvert upgrades
Scarification and ripping	In-channel tree felling	Culvert armoring
Temporary fencing	Log dams	Culvert removal
Erosion fabric	Debris basins	Trash racks
Straw wattles	Straw wattle dams	Storm patrols
Slash scattering	Rock gabion dams	Ditch improvements
Silt fences		Armored fords
Geotextiles		Outsloping
Sand or soil bags		Signing

retrieved during the interviews. Because much of the information was qualitative, attempts were made to ask questions that would allow for grouping and ranking results.

BAER program specialists were asked to identify treatments used on specific fires and what environmental factors affected success and failure. For each treatment, specific questions were asked regarding

Figure 10.4—During a short duration high intensity rain event, this research sediment trap was filled. Using pre- and postsurveys, Hydrologist Bob Brown and Engineer Joe Wagenbrenner measure the sediment collected with the help of a skid-steer loader. Research plots are in high severity burned areas of the 2002 Hayman Fire, Pike-San Isabel National Forest near Deckers, CO. (Photo by J.Yost).

the factors that caused the treatment to succeed or fail, such as slope classes, soil type, and storm events (rainfall intensity and duration) affecting the treated areas. They were also asked questions regarding implementation of treatments and whether any effectiveness monitoring was completed. For cases where monitoring was conducted (either formal or informal), interviewees were asked to describe the type and quality of the data collected (if applicable) and to give an overall effectiveness rating of "excellent," "good," "fair," or "poor" for each treatment.

This evaluation covered 470 fires and 321 BAER projects, from 1973 through 1998 in USDA Forest Service Regions 1 through 6. A literature review, interviews with key Regional and Forest BAER specialists, analysis of burned area reports, and review of Forest and District monitoring reports were used in the evaluation. The resulting report, Evaluating the Effectiveness of Postfire Rehabilitation Treatments (Robichaud and others 2000), includes these major sections:

1. Information acquisition and analysis methods.
2. Description of results, which include hydrologic, erosion and risk assessments, monitoring reports, and treatment evaluations.
3. Discussion of BAER assessments and treatment effectiveness.
4. Conclusions drawn from the analysis.
5. Recommendations.

This chapter provides a synopsis of the findings in that report, as well as new information that has been determined since the report was published.

Postfire Rehabilitation Treatment Decisions

The BAER Team and BAER Report

As soon as possible (even before a fire is fully contained), a team of specialists is brought together to evaluate the potential effects of the fire and to recommend what postfire rehabilitation, if any, should be used in and around the burned area. Hydrology and soil science are the predominant disciplines represented on nearly all BAER teams. Depending on the location, severity, and size of the fire, wildlife biologists, timber, range, and fire managers, engineers, archeologists, fishery biologists, and contracted specialists may be included on the team.

The Burned Area Report filed by the BAER team describes the hydrologic and soil conditions in the fire area as well as the predicted increase in runoff, erosion, and sedimentation. The basic information includes the watershed location, size, suppression cost, vegetation, soils, geology, and lengths of stream channels, roads, and trails affected by the fire. The watershed descriptions include areas in low, moderate, and high severity burn categories as well as areas with water repellent soils. The runoff, erosion, and sedimentation predictions are then evaluated in combination with both the onsite and downstream values at risk to determine the selection and placement of emergency rehabilitation treatments. The BAER team uses data from previous fires, climate modeling, erosion prediction tools, and professional judgment to make the BAER recommendations.

Erosion Estimates from BAER Reports

Robichaud and others (2000) found a wide range of potential erosion and watershed sediment yield estimates in the Burned Area Report forms. Some of the high values could be considered unrealistic (fig. 10.5). Erosion potential varied from 1 to 6,913 tons/acre (2 to 15,500 Mg/ha), and sediment yield varied over six orders of magnitude. Erosion potential and sediment yield potential did not correlate well (r = 0.18, n = 117). Different methods were used to calculate these estimates on different fires, making comparisons difficult. Methods included empirical base models such as Universal Soil Loss Equation (USLE), values based on past estimates of known erosion events, and professional judgment. In recent years, considerable effort has been made to improve erosion prediction after wildfire through the development and refinement of new models (Elliot and others 1999, 2000). These models are built on the Water Erosion Prediction Project (WEPP) technology (Flanagan and others 1994), which has been adapted for application after wildfire. The model adaptation includes the addition

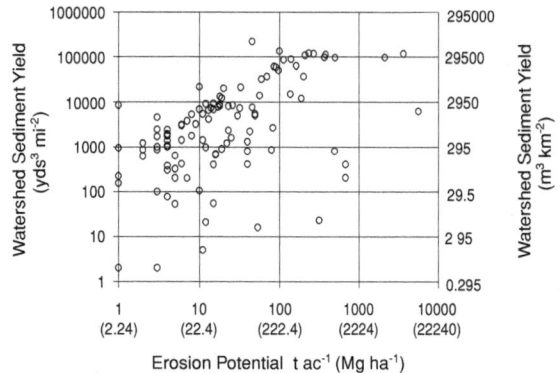

Figure 10.5—Estimated hillslope erosion potential and watershed sediment yield potential (log scale) for all fires requesting BAER funding. (From Robichaud and others 2000).

of standard windows interfaces to simplify use and Web-based dissemination for general accessibility at:

http://forest.moscowfsl.wsu.edu/fswepp/

and

http://fsweb.moscow.rmrs.fs.fed.us/fswepp.

Hydrologic Response Estimates

Evaluating the potential effects of wildfire on hydrologic responses is an important first step in the BAER process. This involves determining storm magnitude, duration, and return interval for which treatments are to be designed. Robichaud and others (2000) found that the most common design storms were 10-year return events (fig. 10.6, 10.7). Storm durations were

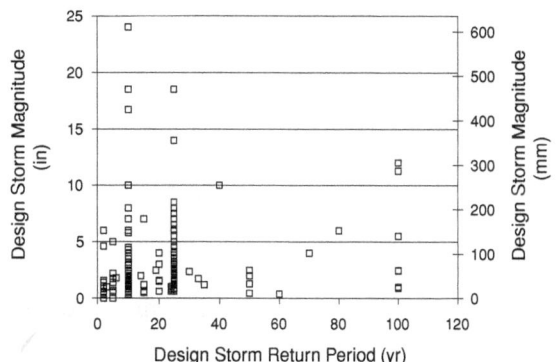

Figure 10.6—Design storm duration by return period for all fires requesting BAER funding. (From Robichaud and others 2000).

USDA Forest Service Gen. Tech. Rep. RMRS-GTR-42-vol. 4. 2005

183

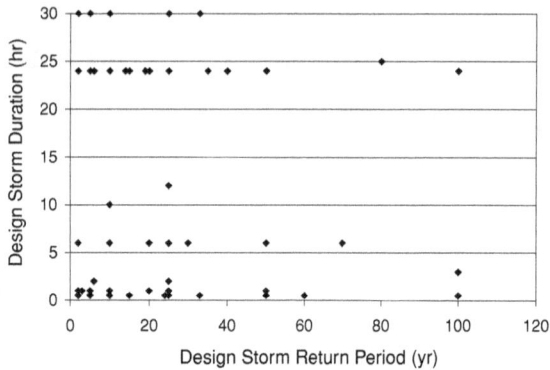

Figure 10.7—Design storm magnitude and return period for all fires in the Western United States requesting BAER funding. (From Robichaud and others 2000).

are often reported after wildfires and are expected to occur more commonly on coarse-grained soils, such as those derived from granite (fig. 10.8). However, no statistical difference was found in the geologic parent material and the percent of burned area that was water repellent. Robichaud and Hungerford (2000) also found no differences in the water repellant conditions with various soil types. BAER teams estimate a percentage reduction in infiltration capacity as part of the Burned Area Report. Comparison of reduction in infiltration rate to percentage of area that was water repellent showed no statistically significant relationship (fig. 10.9). However, Robichaud (2000) and Pierson and others (2001a) showed a 10 to 35 percent reduction in infiltration after the first year. Factors other than water repellent soil conditions, such as loss of the protective forest floor layers, obviously affect infiltration capacity.

Estimation methods for expected changes in channel flow due to wildfire were variable but primarily based on predicted change in infiltration rates. Thus, a 20 percent reduction in infiltration resulted in an estimated 20 percent increase in channel flows. Various methods were used to determine channel flow including empirical-based models, past U.S. Geological Survey records from nearby watersheds that had a flood response, and professional judgment. Some reports show a large percent increase in design flows (fig. 10.10).

usually less than 24 hours, with the common design storm magnitudes from 1 to 6 inches (25 to 150 mm). Five design storms were greater than 12 inches (305 mm), with design return intervals of 25 years or less. The variation in estimates reflects some climatic differences throughout the Western United States.

The Burned Area Report also contains an estimate of the percentage of burned watersheds that have water repellent soil conditions. Soils in this condition

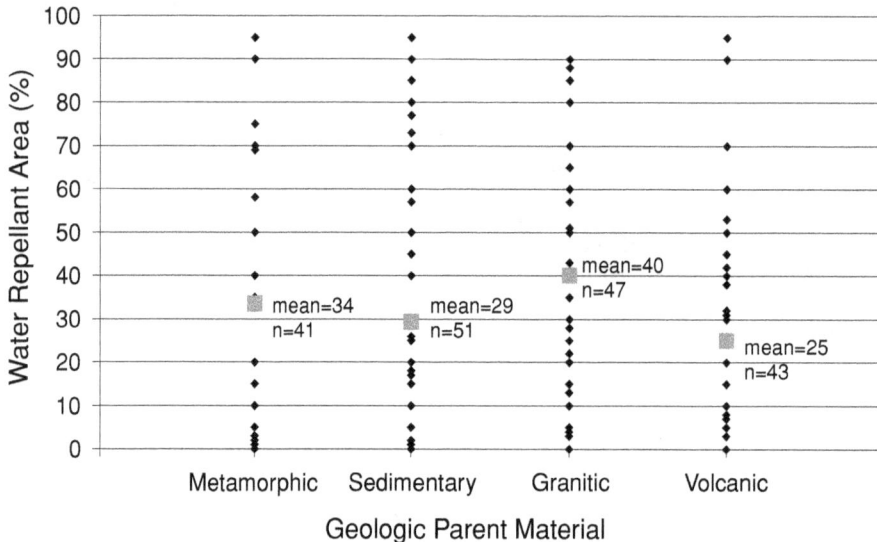

Figure 10.8—Fire-induced water repellent soil areas and their geologic parent material for all fires requesting BAER funding. Fire-induced water repellency was not significantly different by parent material (t-test, alpha = 0.05). (From Robichaud and others 2000).

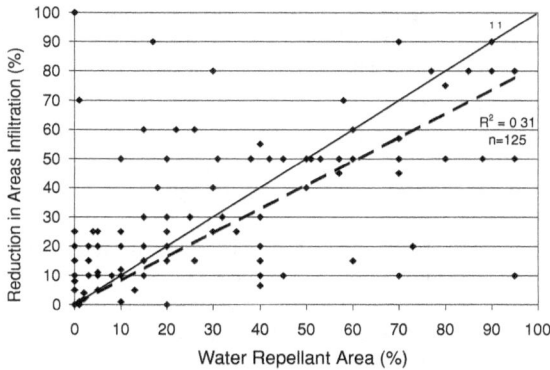

Figure 10.9—Fire-induced water repellent soil areas compared to the estimated reduction in infiltration for all fires requesting BAER funding. (From Robichaud and others 2000).

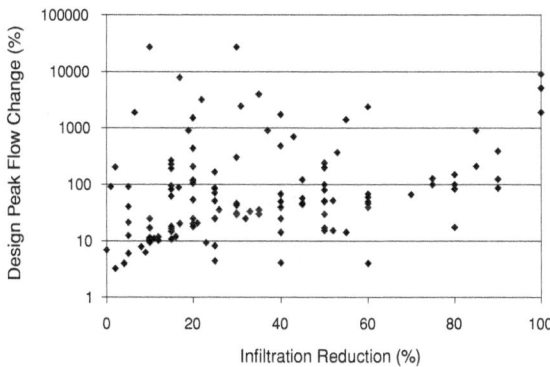

Figure 10.10—Estimated design peakflow change (log scale) due to wildfire burned areas relative to the estimated reduction in infiltration for all fires requesting BAER funding. (From Robichaud and others 2000).

Hillslope Treatments and Results

Hillslope Treatments

Hillslope treatments are intended to reduce surface runoff and keep postwildfire soil in place on the hillslope and thereby prevent sediment deposition in unwanted areas. These treatments are regarded as a first line of defense against postfire erosion and sediment movement. Hillslope treatments comprise the greatest portion of time, effort, and expense in most BAER projects. Consequently, more information is available on hillslope treatments than on channel or road treatments.

Broadcast Seeding— The most common BAER practice is broadcast seeding. Grass seeding after fire for range improvement has been practiced for decades, with the intent to gain useful products from land that will not return to timber production for many years (Christ 1934, McClure 1956). As an emergency treatment, rapid vegetation establishment has been regarded as the most cost-effective method to promote rapid infiltration of water and keep soil on hillslopes (Noble 1965, Rice and others 1965, Miles and others 1989).

Grasses are particularly desirable for this purpose because their extensive, fibrous root systems increase water infiltration and hold soil in place. Fast-growing nonnative species have typically been used. They are inexpensive and readily available in large quantities when an emergency arises (Barro and Conard 1987, Miles and others 1989, Agee 1993). Legumes are often added to seeding mixes for their ability to increase available nitrogen in the soil after the postfire nutrient flush has been exhausted, aiding the growth of seeded grasses and native vegetation (Ratliff and McDonald 1987). Seed mixes were refined for particular areas as germination and establishment success were evaluated. Most mixes contained annual grasses to provide quick cover and perennials to establish longer term protection (Klock and others 1975, Ratliff and McDonald, 1987). However, nonnative species that persist can delay recovery of native flora and alter local plant diversity. Native grass seed can be expensive and hard to acquire in large quantities or in a timely manner compared to cereal grains or pasture grasses. When native seed is used, it should come from a nearby source area to preserve local genetic integrity. When native seed is not available, BAER specialists have recommended using nonreproducing annuals, such as cereal grains or sterile hybrids that provide quick cover and then die out to let native vegetation reoccupy the site.

Application of seed can be done from the air or on the ground. In steep areas and in areas where access is limited, aerial seeding is often the only option. Effective application of seed by fixed-wing aircraft or helicopter requires global positioning system (GPS) navigation, significant pilot skill, and low winds for even cover. Ground seeding, applied from all-terrain vehicles or by hand, assures more even seed application than aerial seeding. Seeding is often combined with other treatments, such as mulching and scarifying, as these additional treatments help anchor the seeds and improve seed germination.

Effectiveness of seeding depends on timeliness of seed application, choice of seed, protection from grazing, and luck in having gentle rains to stimulate seed

USDA Forest Service Gen. Tech. Rep. RMRS-GTR-42-vol. 4. 2005

185

germination before wind or heavy rains blow or wash soil and seed away. Proper timing of seed application depends on location. In some areas, it is best to seed directly into dry ash, before any rain falls, to take advantage of the fluffy seedbed condition, while in other areas, seed is best applied after the first snow so that it will germinate in the spring. Both conditions also reduce loss to rodents. The potential advantage of seeded grass to inhibit the growth and spread of noxious weeds also depends on timely application and germination.

Mulch—Mulch is any organic material spread over the soil surface that functions like the organic forest floor that is often destroyed in high and moderate severity burn areas. Both wet mulch (hydromulch) and dry mulch (wheat straw, jute excelsior, rice straw, and so forth) are available; however, mulches have only recently been used as a postfire rehabilitation treatment. Mulch is applied alone or in combination to reduce raindrop impact and overland flow and, thereby, to enhance infiltration and reduce soil erosion. It is often used in conjunction with grass seeding to provide ground cover in critical areas. It also intercepts precipitation for subsequent infiltration. Mulch protects the soil and improves moisture retention underneath it, benefiting seeded plants in hot areas but not always in cool ones. Use of straw from pasture may introduce exotic grass seed or weeds, so BAER projects are now likely to seek "weed-free" mulch such as rice straw.

Mulches can be applied from the air or from the ground. Aerial dry mulching uses helicopters with attached cargo net slings carrying the straw mulch, which is released over the treatment area (San Dimas Technology Development Center 2003). Hydromulch can be applied from the air using helicopters fitted with hydromulch slurry tanks or buckets, which are released in controlled drops over the treatment areas. Both of these aerial applications are expensive. Ground application of dry mulch is done by hand using all terrain vehicles to carry the straw from a staging area into the treatment area. Ground application of hydromulch is done from spray trucks and is limited to an area 200 feet (61 m) of either side of a road. Given its expense, mulch is usually used in high value areas, such as above or below roads, above streams, or below ridge tops.

Mulching is most effective on gentle slopes and in areas where high winds are not likely to occur. Wind either blows the mulch off site or piles it so deeply that seed germination is inhibited. On steep slopes, rain can wash some of the mulch material downslope. Use of a tackifier or felling small trees across the mulch may increase onsite retention. Hydromulches often have tackifiers that help bind the mulch in the soil. Both hydromulch and dry mulch were used to stabilize soils on the Cerro Grande Fire of 2000 and Rodeo-Chediski and Hayman Fires of 2002. However, use of these treatments escalated the BAER treatment costs to $10 to $20 million per fire.

Contour Log Structures (Contour Log Basins, Log Erosion Barriers, Log Terraces, Terracettes)— This treatment involves felling logs on burned-over hillsides and laying them on the ground along the slope contour to provide mechanical barriers to water flow, promote infiltration, and reduce sediment movement. Contour-felled logs reduce water velocity, break up concentrated flows, induce hydraulic roughness to burned watersheds, and store sediment. The potential volume of sediment stored is highly dependent on slope, the layout design, the size and length of the felled trees, and the degree to which the felled trees are adequately staked and placed into ground contact. In some instances contour-felled log barriers have filled with sediment following the first several storm events after installation, while others have taken 1 to 2 years to fill (Robichaud 2000).

This treatment was originally designed to provide the same function as contour trenches and furrows. The primary function of the Contour Log Basins or Contour Log Terraces was to detain and infiltrate runoff from a design storm. To accomplish this, logs ranging generally from 6 to 12 inches (15 to 30 cm) in diameter were felled on the contour and staked in place. The treatment was begun at the top of the slope because each course of contour logs depends on the design spacing and capacity of the upslope courses to be effective. The spacing depends on the capacity of the structure to contain runoff according to the formula:

$$S = RO/12 \times C$$

Where: S = spacing of log courses down slope measured horizontally in feet.

RO = Storm runoff in inches.
C = Basin capacity in cubic feet/lineal foot of log.

Basins were created behind each log by scraping soil against the log to seal it. Earthen end sills and baffles complete the structure. To contain 1.0 inch (25 mm) of runoff typically requires spacing of less than 20 feet (9.6 m) between courses. Contour placement is vital, and eliminating long, uninterrupted flow paths by "brick coursing" provides additional effectiveness. The treatments detain storm runoff on site, thereby eliminating transport of eroded soils. If the design capacity is exceeded, the structure provides some secondary benefit by reducing slope length, which interrupts concentrated flows and sediment movement. Because of their small size, the effective life of properly installed treatments is only a few years at most. Undesigned and underdesigned treatments with wide spacing and lacking runoff storage capacity can effectively concentrate runoff and cause damage that might

186

USDA Forest Service Gen. Tech. Rep. RMRS-GTR-42-vol. 4. 2005

conceivably be greater than no treatment. In high rainfall areas of the West Coast, contour log basins may be infeasible. In these cases, contour logs are placed in the same manner as above, but the exception is that they will provide only secondary benefits. It should be kept in mind that these structures are intended to detain runoff. If they immediately fill with sediment, they were likely underdesigned.

Shallow, rocky soils that are uneven are problematic for anchoring, so care must be taken to ensure that logs are adequately secured to the slope. Overly rocky and steep slopes should be avoided because benefits gained from contour-felling treatment can be easily offset by the extra implementation time required and the limited capacity to detain runoff or provide stabilization of small amounts of soil. Gentler slopes and finer textured soils (except clayey soils) lead to better installation and greater runoff control efficiency. In highly erosive soils derived from parent material such as granitics or glacial till, so much sediment can be mobilized that it might overwhelm small contour-felled logs. Availability of adequate numbers of straight trees must be considered when choosing this treatment.

Straw Wattles—Straw wattles main purpose is to break up slope length and reduce flow velocities of concentrated flow. Straw wattles are 9 to 10 inches (23 to 25 cm) in diameter and made of nylon mesh tubes filled with straw. They are permeable barriers used to detain surface runoff long enough to reduce flow velocity and provide for sediment storage. With end sills, baffles, and on the proper design spacing, straw wattles can provide runoff detention.

Straw wattles have been used in small, first order, drainages or on side slopes for detaining small amounts of sediment. They should never be placed in main or active drainages. Straw wattles function similarly to contour-felled logs. The sediment holding capacity can be increased by turning 2 feet at each end of the wattle upslope. Straw wattles are a good alternative in burned areas where logs are absent, poorly shaped, or scarce. Straw wattles are relatively inexpensive, but they can be disturbed by grazing animals or decompose or catch fire. Although the wattle netting is photodegradable, there are concerns that it persists long enough to pose hazards for small animals.

Contour Trenching and Terraces—Full-scale contour trenches are designed to totally detain the runoff from a design storm on site. The treatment must progress from the top of the slope downward as each trench course is dependent on the next one upslope. Smaller "outside" trenches can be constructed on slopes less than 30 percent. For slopes greater than 30 percent an "inside" trench must be built. This requires building a "full bench" platform for bulldozers

to operate on. In subsequent passes, the trench is tipped into the slope, forming a basin. On the final pass, bulldozers back out and push up baffles that segment the trench and allow flows to equalize into other cells. The formula for digging trenches is:

$$S = RO/12 \times C$$

Where: S = spacing of trench courses down slope measured horizontally in feet.

RO = Storm runoff in inches.
  C = Basin capacity in cubic feet/lineal foot of trench.

The practical upper limit of capacity is about 3 inches (76 mm) of runoff. Contour trenches require a minimum of 4 feet (1.2 m) of soil above bedrock for adequate construction. They work best in gravelly loams and have been applied in granitic soils and clay soils with less success (Schmidt Personal Communication 2004). Granitic soils do not maintain a structural shape well because of their coarseness and difficulty to get regenerated with cover. Clay soils can become plastic with the addition of water, and in landslide-prone topography, contour trenches can activate localized mass failures. Contour trenching has proven to be effective in a number of localities in the past, but concerns about visual effects and cultural heritage values have limited their use in the past three decades.

More recently, smaller scale contour trenches have been used to break up the slope surface, to slow runoff, to allow infiltration, and to trap sediment. These trenches or terraces are often used in conjunction with other treatments such as seeding. They can be constructed with machinery (deeper trenches) or by hand (generally shallow). Width and depth vary with design storm, spacing, soil type, and slope. When installed with heavy equipment, trenches may result in considerable soil disturbance that can create immediate erosion problems. In addition, erosion problems can occur many years after installation when runoff cuts through the trench embankment. Trenches have high visual impact when used in open areas. Shallow hand trenches tend to disappear with time as they are filled with sediment and covered by vegetation. On the other hand, large trenches installed several decades ago are still visible on the landscape. Because contour trenching and terraces are ground-disturbing activities, cultural clearances are required, and these may significantly slow the installation process.

Scarification and Ripping—Scarification and ripping are mechanical soil treatments aimed at improving infiltration rates in water repellent soils. Tractors and ATVs can be used to pull shallow harrows on slopes of 20 percent or less. Hand scarification uses steel rakes (McLeods). These treatments may increase the amount of macropore space in soils by the physical

USDA Forest Service Gen. Tech. Rep. RMRS-GTR-42-vol. 4. 2005

187

breakup of dense or water repellent soils, and thus increase the amount of rainfall that infiltrates into the soil. In addition, scarification can provide a seedbed for planting that improves germination rates. Shallow soils, rock outcrops, steep slopes, incised drainages, fine-textured soils, and high tree density create significant problems for scarification and ripping. These treatments work best where there is good soil depth, the soils are coarse textured, slopes are less than 30 percent, and woody vegetation density is low.

Silt Fences—Silt fences are installed to trap sediment in swales, small ephemeral drainages, or along hillslopes where they provide temporary sediment storage. Given the labor-intensive installation, they are used as treatment only when other methods would not be effective. They work best on gentler slopes, such as swales, but can be effective on steeper, rocky slopes where log erosion barriers would not achieve good ground contact. Silt fences are also installed to monitor sediment movement as part of effectiveness monitoring and can last several years before UV breakdown of the fabric (Robichaud and Brown 2002).

Geotextiles and Geowebbing—Polymer textiles and webbings are used to cover ground and control erosion in high-risk areas, such as extremely steep slopes, above roads or structures, or along streambanks. This material is often used in conjunction with seeding. Geotextiles come in different grades with ultraviolet inhibitors that determine how long they will last in the field. Geotextiles must be anchored securely to remain effective, especially along streambanks. The complete cover provided by some geotextiles can reduce native plant establishment.

Sand, Soil, or Gravel Bags—Sand, soil, or gravel bags are used on hillslopes or in small channels or to trap sediment and interrupt water flow. Various seed mixes or willow wands may be added to the bags to help establish vegetation. The bags are often placed in staggered rows like contour-felled logs in areas where there are no trees available. Rows of bags break water flow and promote infiltration. They store sediment temporarily, then break down and release it. They are not appropriate for use in V-shaped channels.

Temporary Fencing—Temporary fencing is used to keep grazing livestock and off-highway vehicles (OHVs) out of burned areas and riparian zones during the recovery period. Resprouting onsite vegetation and seeded species attract grazing animals and require protection to be successful.

Slash Spreading—Slash spreading covers the ground with organic material, interrupting rain impact and trapping soil. It is a common practice after timber sales, but it can also be used on burned slopes where dead vegetation is present. Slash is often used to rehabilitate firebreaks and dozer firelines. It is also used in moderately burned areas where there is more material available to spread, or below an intensely burned slope or area of water repellent soils. To be effective, slash needs to be cut so it makes good contact with the ground.

Needle Cast—Needle cast commonly occurs after low and moderate severity burns in coniferous forests. The dead needles that fall to the ground provide surface cover that functions as a naturally occurring mulch. Pannkuk and Robichaud (2003) indicate that 50 percent ground cover can reduce interrill erosion by 60 to 80 percent and rill erosion 20 to 40 percent. Although needle cast is not an applied treatment, its presence may reduce or eliminate the need for other treatments.

Hillslope Treatment Effectiveness

Increasing infiltration of rainfall and preventing soil from leaving the hillslope are considered the most effective methods to slow runoff, reduce flood peaks, retain site productivity, and reduce downstream sedimentation. Many of these hillslope treatments may be appropriate in critical areas of high risk. Monitoring of treatment effectiveness is needed to determine which treatments will work in specific settings as well as their cost effectiveness (General Accounting Office 2003, Robichaud and others 2003).

BAER Expert Rating—Hillslope treatments are implemented to keep soil in place and comprise the greatest effort in most BAER projects (fig. 10.2). Mulching and geotextiles were rated the most effective hillslope treatments because they provide immediate ground cover to reduce raindrop impact and overland flow as well as to hold soil in place. Mulching was rated "excellent" in 67 percent of the evaluations, and nobody considered it a "poor" treatment. Aerial seeding, the most frequently used BAER treatment, was rated about equally across the spectrum from "excellent" to "poor." Nearly 82 percent of the evaluations placed ground seeding effectiveness in the "good" category. Evaluations of seeding plus fertilizer covered the spectrum from "excellent" to "poor," although most responses were "fair" or "poor." After seeding, contour-felled logs are the next most commonly applied hillslope treatment. The rating for contour-felled logs was "excellent" or "good" in 66 percent of the evaluations. The remainder of the hillslope treatments received only three evaluations each, so no conclusions are offered beyond the fact that they were generally rated "excellent," "good," or "fair," and none were evaluated as "poor."

Research and Monitoring Results

Broadcast Seeding—Robichaud and others (2000) reviewed published studies of seeding effectiveness after 1 and 2 years on 34 burned sites across the Western United States. Erosion was not measured at 16 sites (47 percent); only plant cover was determined. At another 15 of the seeding sites (44 percent), seeding did not significantly reduce erosion when compared to sites that were burned but not seeded. Soil erosion was reduced by seeding on only three sites (16, 31, and 80 percent less sediment). Of 23 monitoring reports that contained some quantitative data on broadcast seeding, 15 (65 percent) did not determine soil loss, and one reported no difference in erosion due to seeding. Three reports noted reductions in sediment yield of 30 to 36 percent (2.6 to 6.2 tons/acre or 5.8 to 13.8 Mg/ha). Four monitoring reports documented increases of sediment yield on seeded areas of 118 to 386 percent of that of burned and untreated areas.

In the studies and monitoring reports examined by Robichaud and others (2000), a wide variety of grass species, mixes, and application rates were used, making generalizations difficult. However, in some of the reported studies it is noted that grass seeding does not assure increased plant cover (or any associated erosion reduction) during the first critical year after fire. In the mid-1900s, southern California foresters were urged to caution the public not to expect significant first-year sediment control from postfire seeding (Gleason 1947). Krammes (1960), in southern California, found that as much as 90 percent of first-year postfire hillslope sediment movement can occur as dry ravel before the first germination-stimulating rains even occur. Amaranthus (1989) measured the most first-year sediment movement on his Oregon study site during several storms in December, before the seeded ryegrass had produced much cover. In the reported studies, erosion was decreased by seeding in only one out of eight first-year studies (12.5 percent). However, several studies showed a trend toward lower sediment movement on seeded plots that was not statistically significant (Amaranthus 1989, Wohlgemuth and others 1998). One report suggested that measures other than seeding should be used in places where first-year control of sediment movement is critical (Ruby 1997).

Better cover and, consequently, greater erosion control may occur by the second postfire year. Amaranthus (1989) reported that in the second year after fire, seeded sites had greater total cover (plant and litter) than unseeded 42 percent of the time. Seeded species are expected to be of greatest value during the second and third rainy seasons (Esplin and Shackleford 1978), when plant litter produced by the first year's growth covers the soil. However, after the Bobcat Fire in the Colorado Front Range, Wagenbrenner (2003) found that seeding had no significant effect on sediment yields at the hillslope scale in either the first or second years. In addition, seeding had no significant effect on percent of vegetative cover compared to untreated areas (Wagenbrenner 2003).

Seeding is often most successful where it may be needed least—on gentle slopes and in riparian areas. Janicki (1989) found that two-thirds of plots with more than 30 percent annual ryegrass cover were on slopes of less than 35 percent. He also noted that grass plants concentrated in drainage bottoms indicating that seed washed off the slopes during the first two storm events. Concentration of seeded species at the base of slopes was also observed by Loftin and others (1998).

Little evidence suggests that fertilizer applied with seeded grass is effective in increasing cover or reducing erosion after fire. Several studies found no significant effect of fertilizer on plant cover or erosion (Cline and Brooks 1979, Esplin and Shackleford 1980, Tyrrel 1981).

Retention of soil onsite for productivity maintenance is an important rehabilitation treatment objective, but almost no evidence indicates whether seeding is effective in meeting this goal. Although some nutrients are inevitably lost in a fire, natural processes tend to replenish the soil over time (DeBano and others 1998).

Mulch—Straw mulch applied at a rate of 0.9 ton/acre (2 Mg/ha) significantly reduced sediment yield on burned pine-shrub forest in Spain over an 18-month period with 46 rainfall events (Bautista and others 1996). Kay (1983) tested straw mulch laid down at rates of 0.5, 1.0, 1.5, and 4 tons/acre (1.1, 2.2, 3.4, and 9.0 Mg/ha) against jute excelsior, and paper for erosion control. Straw was the most cost-effective mulch, superior in protection to hydraulic mulches and comparable to expensive fabrics. Excelsior was less effective but better than paper strip synthetic yarn. The best erosion control came from jute applied over 1.5 tons/acre (3.4 Mg/ha) straw. Miles and others (1989) studied the use of wheat straw mulch on the 1987 South Fork of the Trinity River Fire, Shasta-Trinity National Forest in California. Wheat straw mulch was applied to fill slopes adjacent to perennial streams, firelines, and areas of extreme erosion hazard. Mulch applied at rates of 1.0 to 2.0 tons/acre (2.2 to 4.5 Mg/ha) on large areas, reduced erosion significantly 4.6 to 8.0 yard^3/acre (11 to 19 m^3/ha). They considered mulching highly effective in controlling erosion. Edwards and others (1995) examined the effects of straw mulching at rates of 0.9, 1.8, 2.6, and 3.6 tons/acre (2, 4, 6, and 8 Mg/ ha) on 5 to 9 percent slopes. They reported a significant reduction in soil loss at 0.9 ton/acre (2 Mg/ ha) mulch, but increases in mulch thickness provided no additional reduction in soil loss. When comparing all the treatments used after the 2000 Cerro Grande Fire in New Mexico, mulching provided the best

USDA Forest Service Gen. Tech. Rep. RMRS-GTR-42-vol. 4. 2005

189

rehabilitation results. Although precipitation during the two study years was below normal, the plots treated with aerial seed and straw mulch yielded 70 percent less sediment than the no-treatment plots in the first year and 95 percent less in the second year. Ground cover transects showed that aerial seeding without added straw mulch provided no appreciable increase in ground cover relative to untreated plots (Dean 2001). In a 2-year postfire study in the Colorado Front Range, Wagenbrenner (2003) found that mulching reduced the erosion significantly from storms with return periods of up to 2 years. In the first study year, sediment yields from a high intensity (1.9 inches/hour, 48 mm/hour) rainfall event overwhelmed the silt fence sediment traps on both the treated and untreated study sites. However, in the second year, sediment yields from mulched hillslope sites were significantly less than the sediment yields from untreated slopes and the slopes that were seeded without mulch (Wagenbrenner 2003).

Contour Structures—Contour structures provide immediate benefits after installation in that they trap sediment during the first postfire year, which usually has the highest erosion rates. The ability of contour structures to reduce runoff and rilling, increase infiltration, and decrease downstream time-to-peak (slowing velocities) has not been documented, even though these are reasons often given for doing contour felling. If contour structures slow or eliminate runoff, sediment movement may not occur.

Logs were contour-felled on 22 acres (9 ha) of the 1979 Bridge Creek Fire, Deschutes National Forest in Oregon (McCammon and Hughes 1980). Trees 6 to 12 inches (150 to 300 mm) d.b.h. were placed and secured on slopes up to 50 percent at intervals of 10 to 20 feet (3 to 6 m). Logs were staked and holes underneath were filled. After the first storm event, about 63 percent of the contour-felled logs were judged effective in trapping sediment. The remainder were either partially effective or did not receive flow. Nearly 60 percent of the storage space behind contour-felled logs was full to capacity, 30 percent was half-full, and 10 percent had insignificant deposition. Common failures included flow under the log and not placing the logs on contour (more than 25 degrees off contour caused trap efficiency to decrease to 20 percent). More than 1,600 yard^3 (1,225 m^3) of material was estimated to be trapped behind contour-felled logs on the treated area, or about 73 yard^3/acre (135 m^3/ha). Less than 1 yard^3 (1 m^3) of sediment was deposited in the intake pond for a municipal water supply below.

The few monitoring studies done on contour-felled log treatments did not evaluate runoff, infiltration, or sediment movement changes after treatment installation; they only reported sediment storage. For example, Miles and others (1989) monitored contour-felling on the 1987 South Fork Trinity River Fires, Shasta-Trinity National Forest in California. The treatment was applied to 200 acres (80 ha) within 50,000 acres (20,240 ha) of a burned area. Trees less than 10 inches (250 mm) d.b.h. spaced 15 to 20 feet (4.5 to 6 m) apart were felled at rate of 80 to 100 trees/acre (200 to 250 trees/ha). The contour-felled logs trapped 0 to 0.07 yard^3 (0 to 0.05 m^3) of soil per log, retaining 1.6 to 6.7 yard^3/acre (3 to 13 m^3/ha) of soil onsite. Miles and others (1989) considered sediment trapping efficiency low, and the cost high for this treatment. Sediment deposition below treated areas was not measured. McCammon and Hughes (1980), on the other hand, estimated storage at just over 10 times the amounts reported by Miles and others (1989) using a higher density of logs. Depending on log barrier density and erosion rates, this treatment could trap 5 to 47 percent annual sediment production from high severity burn areas. This wide range of effectiveness indicates the need for proper estimation techniques of the erosion potential and for properly designing contour-felled log installations in terms of log numbers and spacing. If 60 percent or more of the expected sediment production can be trapped, then contour-felled logs are probably cost effective.

Dean (2001) found that plots treated with contour-felled logs as well as aerial seed and straw mulch yielded 77 percent less sediment in the first year and 96 percent in the second year as compared to untreated areas; however, these results were not significantly different from the straw mulch with seed treatment alone. Recent postfire rehabilitation monitoring efforts for six paired watersheds have indicated that contour-felled logs can be effective for low to moderate rainfall intensity storm events. However, during high intensity rainfall events, their effectiveness is greatly reduced. The effectiveness of contour-felled logs decreases over time. Once the sediment storage area behind the log is filled, the barrier can no longer trap sediment that is moving downslope (Robichaud 2000, Wagenbrenner 2003).

Contour Trenching—Contour trenches have been used as a rehabilitation treatment to reduce erosion, increase infiltration, trap sediment, and permit revegetation of fire-damaged watersheds. Although they do increase infiltration rates, the amounts are dependent on soils and geology (DeByle 1970b). Contour trenches can significantly improve revegetation by trapping more snow, but they do not affect water yield to any appreciable extent (Doty 1970, 1972). This treatment can be effective in altering the hydrologic response from short duration, high intensity storms typical of summer thunderstorms. However, this hillslope treatment does not significantly change peakflows resulting from low intensity, long duration rainfall events (DeByle 1970a).

190

USDA Forest Service Gen. Tech. Rep. RMRS-GTR-42-vol. 4. 2005

Doty (1971) noted that contour trenching in the sagebrush (*Artemisia* spp.) portion (upper 15 percent with the harshest sites) of a watershed in central Utah did not significantly change streamflow and stormflow patterns. Doty did not discuss changes in sediment yields. There were no observable changes in seasonal flows or the total flow volumes. However, the storm peakflows were substantially reduced. Costales and Costales (1984) reported on the use of contour trenching on recently burned steep slopes (40 to 50 percent) with clay loam soils in pine stands of the Philippines. Contour trenching reduced sediment yield by 81 percent of comparable, burned, but untreated areas.

Other Treatments—Straw wattles may detain surface runoff, reduce velocities, store sediment, and provide a seedbed for germination. They are a good alternative in burned areas where logs are absent, poorly shaped, or scarce. Although the wattle netting is photodegradable, there are concerns that it persists long enough to pose hazards for small animals.

Cattle exclusion with temporary fencing can be important for the first two postfire years.

Ripping and scarification is effective on roads, trails, and firebreaks with slopes less than 35 percent. Slash spreading is effective if good ground contact is maintained.

Channel Treatments and Results

Channel Treatments

In general, channel treatments need to be coupled with hillslope treatments to be really effective. Channel treatments are implemented to modify sediment and water movement in ephemeral or small-order channels to prevent flooding and debris torrents that may affect downstream values at risk. Some in-channel structures are placed and secured to slow water flow and allow sediment to settle out; sediment will later be released gradually as the structure decays. Channel clearing is done to remove large objects that might become mobilized in a flood. Much less information has been published on channel treatments than on hillslope methods.

Straw Bale Check Dams—These structures are used to prevent or reduce sediment inputs into perennial streams during the first winter or rainy season following a wildfire. Straw bales function by decreasing water velocity and detaining sediment-laden surface runoff long enough for coarser sediments to drop out and be deposited behind the check dams. The decreased water velocity also reduces downcutting in ephemeral channels. Straw bale check dams are temporary in-channel grade control structures constructed of commercially available straw or hay bales. They are inexpensive, easy to install, and effective at trapping sediment but eventually deteriorate due to climatic conditions, streamflows, or cattle and wildlife disturbance. Straw bale check dams tend to fail in large storms. Failure can occur if the dams are poorly installed or put in locations where they cannot contain runoff. Straw bales are often used where materials are not available on site to construct check dams.

Log Check Dams—Log check dams are another type of temporary in-channel grade control structure similar in function to straw bale check dams. They are used to prevent or reduce sediment inputs into perennial streams during the first winter or rainy season following a wildfire. Log dams are constructed of more durable material than straw bale dams, usually small diameter fire-killed tree stems that are available nearby. Log check dams function by decreasing water velocity and detaining sediment-laden surface runoff long enough for coarser sediments to deposit behind check dams. Decreased water velocity also reduces downcutting in ephemeral channels. Log check dams require more effort and skill to install, but will last longer than straw bale check dams.

Rock Dams and Rock Cage Gabions—Also known as rock fence check dams, these structures are used in intermittent or small perennial channels to replace large woody debris that may have been burned out during a wildfire. The rock cage dams provide a degree of grade stability and reduce flow velocities long enough to trap coarse sediments. Properly designed and installed rock check dams and rock cage (gabion) dams are semipermanent structures capable of halting gully development and reducing sediment yields by controlling channel grade and stopping head cutting of gullies. The rock cage dams must be properly sited, keyed in, and anchored to stay in place during runoff events. The dam cages should be filled with angular rock that will interlock, preventing rock from mobilizing and pounding itself apart in the cages. Downslope energy dissipaters are recommended because they reduce the risk of the rock cage dams being undercut. Construction of these structures is dependent on the availability of adequate amounts and sizes of rocks. Rock cage dams usually need to be cleaned out periodically if they are to maintain their effectiveness.

Straw Wattle Dams—Straw wattle dams work on the same principle as straw bale check dams. They trap sediment on side slopes and in the upper ends of ephemeral drainages by reducing channel gradient. Straw wattles are easy to place in contact with the soil—a distinct advantage over rigid barriers like logs—and provide a low risk barrier to soil movement. The closer together straw wattles are placed in steep terrain, the more effective they are in detaining

USDA Forest Service Gen. Tech. Rep. RMRS-GTR-42-vol. 4. 2005

191

sediment. Wattles can be used quite effectively in combination with straw bale check dams. However, they should not be placed in the channels of first order or greater drainages and swales because of high failure rates. They are most effective on hillslopes.

Log Grade Stabilizers—The purpose of log grade stabilizers is much the same as log dams, except that the emphasis is on stabilizing the channel gradient rather than trapping sediment. Numerous small log grade stabilizers are preferable to a few larger ones. In some locations, there might not be adequate, straight, woody material left after a fire to build log grade stabilizers with onsite resources.

Rock Grade Stabilizers— Rock grade stabilizers function the same as log grade stabilizers, except that they are made of rock. The emphasis is on stabilizing the channel gradient rather than trapping sediment although some sediment will be trapped by these structures. Effectiveness is impacted by (1) the use of rocks that are large enough to resist transport during runoff events and (2) placement of screening to collect and hold organic debris or sediment on the upstream side of the grade stabilizer.

Channel Debris Clearing—Channel clearing is the removal of logs, organic debris, or sediment deposits to prevent them from being mobilized in debris flows or flood events. This treatment has been done to prevent creation of channel debris dams, which might result in flash floods or increase flood heights or peakflows. Organic debris can lead to culvert failure by blocking inlets or reducing channel flow capacity. Excessive sediments in stream channels can compromise in-channel storage capacity and the function of debris basins.

Streambank Armoring and Channel Armoring—Streambank and channel armoring is done to prevent erosion of channel banks and bottoms during runoff events. In some hydrologic systems, streambanks are a major source of sediment. Factors that contribute to the success of these treatments include proper sized materials, use of geotextile fabric, avoiding overly steep areas, and the use of energy dissipaters.

In-Channel Felling—This rehabilitation channel treatment is designed to replace woody material in drainage bottoms that have been consumed by wildfire. It is intended to trap organic debris and temporarily detain or slow down storm runoff. Woody material felled into channels will ultimately alter channel gradient and may cause sediment deposition and channel aggradation. This treatment is in conflict with channel removal of woody debris as its objectives are totally different.

Debris Basins— Debris basins are constructed in stream systems that, under normal conditions, carry high sediment loads. They are intended to control runoff and reduce deterioration of water quality and threats to human life and property. Debris basins are considered a last resort because they are extremely expensive to construct and require a commitment to annual maintenance until they are abandoned. In order for debris basins to function, they must be able to trap at least 50 percent and preferably 70 to 80 percent of 100-year flows. A spillway needs to be constructed in the debris basin to safely release flow in excess of the design storage capacity. The downstream channel should be lined to prevent scour (fig. 10.11. In some instances, excavated pits in ephemeral channels have been used as debris basins. These must be large enough to trap 50 to 90 percent of flood flow. Maintenance is a key factor in effectiveness of this treatment. Although protection is immediate, maintaining debris basins is a long-term commitment.

Channel Treatment Effectiveness

BAER Expert Ratings—Effectiveness ratings for straw bale check dams and log grade stabilizers ranged relatively evenly from "excellent" to "poor" (table 10.2). While most interviewees (71 percent) thought that channel debris clearing effectiveness fell into the "good" category, 29 percent rated it "poor." Log dams and straw wattle dams were rated "excellent" or "good" in effectiveness and better than rock grade stabilizers. No one considered the effectiveness of these rehabilitation treatments to be "poor."

Figure 10.11—Rocky Mountain Research Station Engineer Joe Wagenbrenner surveys a channel scour, following the Hayman Fire, 2002, Pike-San Isabel National Forest near Deckers, CO. (Photo by A. Covert).

Table 10.2—Rehabilitation treatment effectiveness ratings from individual fires as provided by BAER program specialists. Total responses are listed as percentages in four classes. Only treatments that received three or more evaluations are included (From Robichaud and others 2000).

BAER treatment	Fires	Excellent	Good	Fair	Poor
	Number	- - - - - - - - - - - - - - - - - *Percent* - - - - - - - - - - - - - - - - -			
Hillslope treatment					
Aerial seeding	83	24.1	27.7	27.7	20.5
Contour felling	35	28.6	37.1	14.3	20.0
Mulching	12	66.8	16.6	16.6	0.0
Ground seeding	11	9.1	81.8	9.1	0.0
Silt fence	8	37.5	62.5	0.0	0.0
Seeding and fertilizer	4	25.0	0.0	50.0	25.0
Rock grade stabilizers	3	0.0	33.3	67.7	0.0
Contour trenching	3	67.7	33.3	0.0	0.0
Temporary fencing	3	0.0	67.7	33.3	0.0
Straw wattles	3	33.3	33.3	33.3	0.0
Tilling/ripping	3	33.3	33.3	33.3	0.0
Channel treatments					
Straw bale check dams	10	30.0	30.0	30.0	10.0
Log grade stabilizers	10	30.0	30.0	10.0	30.0
Channel debris clearing	7	0.0	71.4	0.0	28.6
Log dams	5	40.0	60.0	0.0	0.0
Rock grade stabilizers	3	0.0	33.3	67.7	0.0
Straw wattle dams	3	33.3	67.7	0.0	0.0
Road treatments					
Culvert upgrading	6	6.7	66.6	0.0	16.7
Trash racks	4	50.0	0.0	25.0	25.0

Channel Treatment Research and Monitoring Results—Here we look at various results of interest to researchers and land managers.

Straw bale check dams: Straw bale check dams are designed to reduce sediment inputs into streams. They often fill in the first few storms, so their effectiveness diminishes quickly, and they can blow out during high flows. Thus, their usefulness is short-lived. Miles and others (1989) reported on the results of installing 1,300 straw bale check dams after the 1987 South Fork Trinity River Fires, Shasta-Trinity National Forest in California. Most dams were constructed with five bales. About 13 percent of the straw bale check dams failed due to piping under or between bales or undercutting of the central bale. Each dam stored an average 1.1 yard^3 (0.8 m^3) of sediment. Miles and others (1989) reported that filter fabric on the upside of each dam and a spillway apron would have increased effectiveness. They considered straw bale check dams easy to install and highly effective when they did not fail.

Collins and Johnston (1995) evaluated the effectiveness of straw bales on sediment retention after the Oakland Hills fire. About 5,000 bales were installed in 440 straw bale check dams and 100 hillslope barriers. Three months after installation, 43 to 46 percent of the check dams were functioning. This decreased to 37 to 43 percent by 4.5 months, at which time 9 percent were side cut, 22 percent were undercut, 30 percent had moved, 24 percent were filled, 12 percent were unfilled, and 3 percent were filled but cut. Sediment storage amounted to 55 yard^3 (42 m^3) behind all the straw bale check dams and another 122 yard^3 (93 m^3) on an alluvial fan.

Goldman and others (1986) recommended that the drainage area for straw bale check dams be kept to less than 20 acres (8 ha). Bales usually last less than 3 months, flow should not be greater than 11feet^3/sec (0.3 m^3/sec), and bales should be removed when sediment depth upstream is one-half of bale height. More damage can result from failed barriers than if no barrier were installed (Goldman and others 1986).

Log check dams: Log dams can trap sediment by decreasing velocities and allowing coarse sediment to drop out. However, if these structures fail, they usually aggravate erosion problems. Logs 12 to 18 inches (300 to 450 mm) diameter were used to build 14 log check dams that retained from 1.5 to 93 yard^3 (mean 29 yard^3) (1.1 to 71 m^3; mean 22 m^3) of sediment after the 1987 South Fork Trinity River Fires on the Shasta-Trinity National Forest in California

USDA Forest Service Gen. Tech. Rep. RMRS-GTR-42-vol. 4. 2005

193

(Miles and others 1989). While log check dams have a high effectiveness rating and 15 to 30 years of life expectancy (Miles and others 1989), they are costly to install.

Rock dams and rock cage gabions: Heede (1970, 1976) reported that these structures can reduce sediment yields by 60 percent or more. Although these cross-channel structures are relatively expensive, they can be used in conjunction with vegetation treatments to reduce erosion by 80 percent and suspended sediment concentrations by 95 percent (Heede 1981). While vegetation treatments, such as grassed waterways, augment rock check dams and are less expensive, their maintenance costs are considerably greater.

On mild gradients, these structures work well. Some failures occurred on steeper slopes when high velocity flows are greater than 3 feet^3/sec (1 m^3/sec). This is a common theme for all channel treatments. Most of the failures occur where treatments are imposed on steep gradient sections of ephemeral or first to second order perennial channels. Rock cage dams often last long enough and trap enough fine sediments to provide microsites for woody riparian vegetation to get reestablished.

Check dams constructed in Taiwan watersheds with annual sediment yields of 10 to 30 yard^3/acre (19 to 57 m^3/ha) filled within 2 to 3 years. Sediment yield rates decreased upstream of the check dams but were offset by increased scouring downstream (Chiun-Ming 1985).

Other channel treatments: No published information was found on the effectiveness of straw wattle dams, log grade stabilizers, rock grade stabilizers, in-channel debris basins, in-channel debris clearing, streambank armoring, or other channel rehabilitation treatments. However, several considerations have been related to the effectiveness of these treatments.

Log and rock grade stabilizers emphasize channel stabilization rather than storing sediment. They tend to work for low and moderate flows, not high flows.

Channel clearing (removing logs and other organic debris) was rated "good" 71 percent of the time because it prevents logs from being mobilized in debris flow or floods. Nonetheless, use of this treatment has declined since the early 1990s (and in-channel felling has increased) because in-stream woody debris has been clearly linked to improved fish habitat. Despite the lack of effectiveness data for rock cage dams, they do provide grade stability and reduce velocities enough to drop out coarse sediment. Debris basins are designed to store runoff and sediment and are often used to prevent downstream flooding and sedimentation in the Southwestern United States. They are usually designed to trap 50 to 70 percent of the expected flows.

Road and Trail Treatments and Results

Road and Trail Treatments

Road rehabilitation treatments consist of a variety of practices aimed at increasing the water and sediment processing capabilities of roads and road structures, such as culverts and bridges, in order to prevent large cut-and-fill failures and the movement of sediment downstream. The functionality of the road drainage system is not affected by fire, but the increased stream and storm flows in a burned-over watershed can exceed the functionality of that system. Road treatments are not designed to retain water and sediment but rather to manage water's erosive forces and avoid damage to the road structures.

Rolling Dips/Waterbars/Cross Drain/Culvert Overflow/Bypass—These treatments are designed to provide drainage relief for road sections or water in the inside ditch to the downhill side of roads especially when the existing culvert is expected to be overwhelmed. Rolling dips are easily constructed with road grader, dozer, or backhoe. Rolling dips or waterbars need to be deep enough to contain the expected flow and their location carefully assessed to prevent damages to other portions of the road prism. Waterbars can be made out of rocks or logs, but they are not as effective as earthen bars placed diagonally across roads to divert runoff away from the road surface. Armoring of dip and fillslope at the outlet is often needed to prevent incision and gullying.

Culvert Upgrades—Culvert improvements increase the flow capacity, which may prevent road damage. Upgraded culverts need to be sized and installed (approaches, exits, and slope) to handle expected increased flows. Flexible down spouts and culvert extensions often are needed to keep exiting water from highly erodible slopes.

Culvert Inlet/Outlet Armoring/Risers—These treatments reduce scouring around the culvert entrance and exit. They allow heavy particles to settle out of sediment-laden water and reduce the chance of debris plugging the culvert. Culvert risers allow for sediment accumulation while allowing water to flow through the culvert. Sometimes culvert risers can clog and may be difficult to clean.

Culvert Removal—This procedure is a planned removal of undersized culverts that would probably fail due to increased flows. After culvert removal, armoring the stream crossing will allow for continued use of the road. If the road is not needed, culvert removal is done in conjunction with road obliteration.

194

USDA Forest Service Gen. Tech. Rep. RMRS-GTR-42-vol. 4. 2005

Trash Racks—Trash racks are installed to prevent debris from clogging culverts or down-stream structures. These structures are generally built out of logs but occasionally from milled lumber or metal, and they are anchored to the sides and bottom of the channel. Trash racks are sized to handle expected flows and protect downstream structures. Most trash rack designs allow debris to ride up and to the side of the cage. Trash racks generally perform better in smaller drainages and need to be cleared after each storm to be effective.

Storm Patrols—Patrols during storms provide immediate response for assessment of flood risk, clearing of blocked culvert entrances and drainage ditches, and closing areas that are at risk for floods, landslides, and so forth. This treatment can include early warning systems, such as radio-activated rain gauges or stream gauge alarms, that signal potential flooding conditions.

Ditch Cleaning and Ditch Armoring—Cleaning and armoring provides adequate water flow capacity and prevents downcutting of ditches. Without this treatment, high water levels can overtop roadways leading to gully development in roadbeds.

Armoring Ford Crossing—Armored crossings provide low-cost access across stream channels that are generally capable of producing large flows that flood the road surface. Large riprap is placed upstream and downstream of actual road crossing areas. Armored crossings are most often used for gravel roads.

Outsloping—Outsloping prevents concentration of flow on road surfaces that produces rilling, gullying, and rutting. This is one of the few rehabilitation treatments that have both immediate and long-term facility and resource benefits. Given that roads are a major source of sediment in forests, road improvements that reduce erosion from roads are beneficial. Sometimes after regrading, compaction does not occur due to low traffic volume, which may produce some short-term erosion. Traffic should be curtailed during wet road conditions to prevent rutting and road subgrade damages.

Trail Work—The purpose of rehabilitation treatments on trails is to provide adequate drainage and stability so trails do not contribute to concentrated flows or become sources of sediment. This treatment is labor intensive, as all the work must be done by hand with materials that can be hand carried or brought in on ATVs. Water bars need to be installed correctly at proper slopes and depths to be effective.

Other Treatments—A variety of other minor treatments are available as solutions to specific problems.

They include wetting agents to reduce water repellency on high erosion hazard areas, gully plugs to prevent headcutting in meadows, flood signing installation to warn residents and visitors of flooding potential, and removal of loose rocks above roadways that were held in place by roots, forest debris, and duff consumed by fire.

Road Treatment Effectiveness

Road treatments are designed to move water to desired locations and prevent washout of roads. There is little quantitative research evaluating and comparing road treatment effectiveness. A recent computer model, X-DRAIN, can provide sediment estimates for various spacings of cross drains (Elliot and others 1998), and the computer model, WEPP-Road, provides sedimentation estimates for various road configurations and mitigation treatments (Elliot and others 1999). Thus, effectiveness of various spacings of rolling dips, waterbars, cross drains, and culvert bypasses can be compared.

BAER Expert Ratings—Only two road treatments—culvert upgrading and trash racks—received more than three effectiveness evaluations. The responses covered the range from "excellent" to "poor," although 73 percent of the BAER experts rated culvert upgrading "excellent" or "good" in effectiveness (table 10.2). Evaluations of trash racks were evenly split as "excellent," "fair," or "poor."

Research and Monitoring Results—Furniss and others (1998) developed an excellent analysis of factors contributing to road failures at culverted stream crossings. These locations are important because 80 to 90 percent of fluvial hillslope erosion in wildlands can be traced to road fill failures and diversions of road-stream crossings that are unrelated to wildfires (Best and others 1995). Because it is impossible to design and build all stream crossings to withstand extreme stormflows, they recommended increasing crossing capacity and designing to minimize the consequences of culvert exceedence as the best approaches for forest road stream crossings.

Comprehensive discussions of road-related treatments and their effectiveness can be found in Packer and Christensen (1977), Goldman and others (1986), and Burroughs and King (1989). Recently the USDA Forest Service's San Dimas Technology and Development Program developed a Water/Road Interaction Technologies Series (Copstead 1997) that covers design standards, improvement techniques, and evaluates some surface drainage treatments for reducing sedimentation.

USDA Forest Service Gen. Tech. Rep. RMRS-GTR-42-vol. 4. 2005

195

Summary, Conclusions, and Recommendations _____

Spending on postfire emergency watershed rehabilitation has steadily increased since about 1990. An evaluation of USDA Forest Service burned area emergency rehabilitation treatment effectiveness was completed jointly by the USDA Forest Service Research and Development and National Forest System staffs. The resulting study by Robichaud and others (2000) analyzed BAER treatment attributes and conditions that led to success or failure in achieving BAER goals after wildfires in the continental Western United States. The study found that spending on rehabilitation had risen sharply during the previous decade because the perceived threat of debris flows and floods had increased where fires were closer to the wildland-urban interface. Existing literature on treatment effectiveness is limited, thus making treatment comparisons difficult; however, the amount of protection provided by any treatment is limited—especially during short-duration, high-intensity rainfall events.

Relatively little monitoring of postfire rehabilitation treatments had been conducted between 1970 and 2000. During that time there were at least 321 fires that received BAER treatment, which cost the Forest Service around $110 million. Some level of monitoring occurred on about 33 percent of the fires that received BAER treatments. Since 2000, the number, size, and severity of fires in the Western United States have dramatically increased, with a concurrent increase in BAER spending to about $80 million annually. However, monitoring efforts continue to be short-term and inconsistent. Analysis of the literature, Burned Area Report forms, interview comments, monitoring reports, treatment effectiveness ratings, and the authors' continuing work in the area of postfire rehabilitation have led to the conclusions and recommendations discussed in this section.

Recommendations: Models and Predictions

- Rainfall intensity as well as rainfall amount and duration affect the success of rehabilitation treatment. Rehabilitation treatments are least ineffective in short-duration, high-intensity rainstorms (that is, convective thunderstorms), particularly in the first 2 years after burning.
- Quantitative data are needed to guide future responses to postfire rehabilitation and to build, test, and refine predictive models for different burned forest environments. Accurate climate, runoff, and erosion models for burned forest environments is dependent on

(1) improved mapping of burn severity and better characterization of postfire soil water repellency, (2) improved prediction of runoff responses at different spatial scales, including short-duration high-intensity thunderstorms, (3) quantitative data for the relative magnitudes and consequences of hillslope verses channel erosion, and (4) refined sediment deposition and routing models for various drainages.

Recommendations: Postfire Rehabilitation Treatment

- Rehabilitation should be done only if the risk to life and property is high since significant resources have to be invested to ensure improvement over natural recovery. In most watersheds, it is best not to do any treatments. If treatments are necessary, then it is more effective to detain runoff and reduce erosion on site (hillslope treatment) than to collect it downstream (channel treatment).
- Seeding treatment may not be needed as often as previously thought. Seeding has a low probability of reducing erosion the first wet season after a fire when erosion rates are highest. Thus, it is often necessary to do other treatments in critical areas.
- Mulching can be an effective treatment because it provides immediate protection from raindrop impact and overland flow. Mulching rates that provide at least 70 percent ground cover are desirable in critical areas.
- Contour-felled log structures have limited benefit as an effective treatment compared to other hillslope treatments if: (a) the density and size of the felled logs are matched to the expected erosion, (b) the logs, basins, and sills are properly located, and (c) the treated area is not likely to be subjected to short-duration high-intensity storms. This is considered to be true for areas where runoff and erosion rates are expected to be high. These treatments provide storm-by-storm protection during the first year postfire where runoff and erosion rates are highest. In areas that lack available trees, straw wattles can provide an alternative. However, the overall effectiveness of properly designed and constructed contour-felled logs and straw wattles needs adequate study and documentation in the scientific literature.
- Channel treatments, such as straw bale check-dams, should be viewed as secondary mitigation treatments. Sediment has already been transported from the hillslopes and

will eventually be released though the stream system unless it is physically removed from the channel. Most channel treatments hold sediment temporarily so that the release is desynchronized from the storm flow event.

- To reduce the threat of road failure, road treatments such as rolling dips, water bars, and relief culverts, properly spaced, provide a reasonable method to move water past the road prism. Storm patrol attempts to keep culverts clear and close areas as needed. This approach shows promise as a cost effective technique to reduce road failure due to culvert blockage.

- The development and launch of the Web-based database of past and current BAER projects should be expedited so that future decisions are based on the best data available. This database will include treatment design criteria and specifications, contract implementation specifications, example Burned Area Report calculations, and monitoring techniques. This database needs to be kept current as new information is obtained.

Recommendations: Effectiveness monitoring

- The need for improved effectiveness monitoring has gained momentum. In April 2003, the Government Accounting Office published a report entitled *Wildland Fires: Better Information Needed on Effectiveness of Emergency Stabilization and Rehabilitation Treatments,* which clearly stated that neither the USDA Forest Service nor the Department of the Interior's Bureau of Land Management could determine whether emergency stabilization and rehabilitation treatments were achieving their intended results. Although treatment monitoring is required, there is no agreed-on protocol for how and what to collect and analyze for determining effectiveness. Such protocol needs to be pursued.

- Effectiveness monitoring needs to be initiated as quickly as possible after treatments are applied, as the first storms typically pose the greatest risk to downstream resources. Effectiveness monitoring should include quantifying reductions in erosion, sedimentation, and/or downstream flooding. It may also include measurement of changes in infiltration, soil productivity, ecosystem recovery, and water quality parameters. Burned but untreated areas must be available to provide a control, or baseline, from which to assess both short- and long-term effectiveness of treatments as well as ecosystem response to the fire and natural recovery rates. Recently published techniques are now available that may aid in the development of monitoring protocols (Robichaud and Brown 2002).

- Funding for effectiveness monitoring must be part of the BAER treatment funding request and extend for at least 5 years after the fire. This necessitates a change in BAER funding protocol, which currently is limited to 2 years.

- Policy and funding mechanisms should be established to take advantage of the overlap between research and monitoring by supporting activities that can accomplish the goals of both programs. This should include testing of new rehabilitation technologies as they become available.

USDA Forest Service Gen. Tech. Rep. RMRS-GTR-42-vol. 4. 2005

197

Notes

Malcolm J. Zwolinski
Daniel G. Neary
Kevin C. Ryan

Chapter 11:
Information Sources

Introduction

New research and development result in continually improved information on the effects of fire on soils and water. Nevertheless, as the wildfire seasons of the recent past point out, there is need for additional work in this area. Also, expansion of populations in the Western United States have placed greater demands on forested watersheds to provide stable supplies of water for municipalities. And more physical resources are now at risk from postfire streamflow events.

This volume updates the information available on impacts of fire on soils and water, to a given point in time. Barring more frequent updates of this volume (originally published in 1979; Wells and others 1979, Tiedemann and others 1979), future information retrieval must be dynamic and current.

This chapter outlines additional sources of information, particularly those that are likely to be easily updated and accessible. We have attempted to identify some of the more common places where fire personnel may search for general and specific fire effects and associated fire environment information. Examples shown are meant to be illustrative and not inclusive of all sources.

USDA Forest Service Gen. Tech. Rep. RMRS-GTR-42-vol. 4. 2005

199

Databases

U.S. Fire Administration

The U.S. Fire Administration Federal Fire Links database is an online, searchable database containing links to Web sites in a variety of categories that are related to fire and emergency services. Location:

http://www.usfa.fema.gov/applications/fflinks/

Current Wildland Fire Information

Up-to-date information on current wildland fire situations, statistics, and other information is maintained by the National Interagency Fire Center's Current Wildland Fire Information site. It can be found at:

http://www.nifc.gov/information.html

Fire Effects Information System

The Fire Effects Information System (FEIS) provides up-to-date information about fire effects on plants and animals. The FEIS database contains nearly 900 plant species, 100 animal species, and 16 Kuchler plant communities found in North America. Each synopsis emphasizes fire and how it affects each species. Synopses are documented with complete bibliographies. Several Federal agencies provide maintenance support and updating of the database. It is located at:

http://www.fs.fed.us/database/feis/

National Climatic Data Center

The National Climatic Data Center (NCDC) database contains the largest active archive of weather data. NCDC provides numerous climate publications and provides access to data from all NOAA Data Centers through the National Virtual Data System (NVDS). Its Web site is located at:

http://lwf.ncdc.noaa.gov/oa/ncdc.html

PLANTS

The PLANTS Database is a single source of standardized information about vascular plants, mosses, and lichens of the United States and its Territories. PLANTS includes names, checklists, identification information, distributional data, references, and so forth. The database will have threatened and endangered plant status for States.Web site is located at:

http://plants.usda.gov/home_page.html

Fire Ecology Database

The E. V. Komarek Fire Ecology Database contains a broad range of fire-related information. Literature on control of wildfires and applications of prescribed burning is included. Citations include reference books, chapters in books, journal articles, conference papers, State and Federal documents. The database contains more than 10,000 citations and is updated on a continuous basis, with both current and historical information. The Tall Timbers Fire Ecology Thesaurus, a guide for doing keyword searches, is also available to be downloaded. This Web site is located at:

http://www.ttrs.org/fedbintro.htm

Wildland Fire Assessment System

The Wildland Fire Assessment System (WFAS) generates daily national maps of selected fire weather and fire danger components of the National Fire Danger Rating System (NFDRS). It was developed by the USDA Forest Service Fire Sciences Laboratory in Missoula, MT. Maps available include Fire Danger, Fire Weather Observations and Next Day Forecasts, Dead Fuel Moisture, Live Fuel Moisture-Greenness, Drought, Lower Atmosphere Stability Index, and Lightning Ignition Efficiency. Its location is:

http://www.fs.fed.us/land/wfas/welcome.htm

National FIA Database Systems

The National Forest Inventory and Analysis (FIA) Database Retrieval System produces tables and maps for geographic areas of interest based on the national forest inventory conducted by the USDA Forest Service. The Timber Product Output (TPO) Database Retrieval System describes for each county the round-wood products harvested, logging residues left behind, and wood/bark residues generated by wood-using mills. The Web site is:

http://ncrsz.fs.fed.us/4801/fiadb/rpa_tpo/wc_rpa_tpo.asp

Web Sites

A number of sites on the World Wide Web contain information on fire effects. These include research, fire coordination centers, and other related sites.

USDA Forest Service, Rocky Mountain Research Station Wildland Fire Research Program, Missoula, Montana

Fire Behavior Research Work Unit RMRS-4401:

http://www.fs.fed.us/rm/main/labs/miss_fire/rmrs4401.html

Fire Effects Research Work Unit RMRS-4403:

http://www.fs.fed.us/rm/main/labs/miss_fire/rmrs4403.html

Fire Chemistry Research Work Unit RMRS-4404:

http://www.fs.fed.us/rm/main/labs/miss_fire/rmrs4404.html

USDA Forest Service, Rocky Mountain Research Station Watershed, Wildlife, and Wildland-Urban Interface Fire Research Program, Flagstaff, Arizona

Watershed Research Unit RMRS-4302

http://www.rmrs.nau.edu/lab/4302/

Wildland-Urban Interface Research Unit RMRS-4156

http://www.rmrs.nau.edu/lab/4156/

Wildlife Research Unit

http://www.rmrs.nau.edu/lab/4251/

USDA Forest Service, Pacific Northwest Research Station Pacific Wildlands Fire Science Laboratory, Seattle, Washington

http://www.fs.fed.us.nw/pwfsl/

USDA Forest Service, Pacific Southwest Research Station, Fire Science Laboratory, Riverside, California

http://www.fs.fed.us/psw/rfl/

Fire and Fire Surrogate Program

USDA Forest Service Fire and Fire Surrogate Treatments for Ecosystem Restoration:

http://www.fs.fed.us/ffs/

National Fire Coordination Centers

National Interagency Coordination Center:
http://www.nifc.gov/news/nicc.html

Alaska Coordination Center:
http://fire.ak.blm.gov

Eastern Area Coordination Center:

http://www.fs.fed.us/eacc

Eastern Great Basin Coordination Center:

http://gacc.nifc.gov/egbc/

Northern California Geographic Area Coordination Center:

http://gacc.nifc.gov/oncc/

Northern Rockies Coordination Center:

http://gacc.nifc.gov/nrcc/

Northwest Interagency Coordination Center:

http://www.nwccweb.us/

Rocky Mountain Coordination Center:

http://www.fs.fed.us/r2/fire/rmacc.html

Southwest Coordination Center:

http://gacc.nifc.gov/swcc/

Southern Area Coordination Center:

http://gacc.nifc.gov/sacc/

Southern California Coordination Center:

http://gacc.nifc.gov/oscc/

Western Great Basin Coordination Center:

http://gacc.nifc.gov/wgbc/

USDA Forest Service Fire & Aviation Management Web site:

http://www.fs.fed.us/fire/

Bureau of Land Management Office of Fire and Aviation:

    http://www.fire.blm.gov/

National Park Service Fire Management Program Center:

    http:data2.itc.gov/fire/index.cfm

U.S. Fish and Wildlife Service Fire Management:

    http://fire.r9.fws.gov/

Other Web Sites

Joint Fire Sciences web site:

    http://jfsp.nifc.gov

Natural Resources Canada, Forest Fire in Canada

    http://fire.cfs.nrcan.gc.ca/index_e.php

Mexico Incendios Forestales Info

    http://www.incendiosforestales.info/

Federal Emergency Management Agency:

    http://www.fema.gov/

Smokey Bear Web site:

    http://smokeybear.com/

USDA-ARS-National Sedimentation Laboratory Web site:

    http://ars.usda.gov/main/site_main.htm?modecode=64_08_05_00

Firewise Communities Web site:

    http://www.firewise.org/

U.S. Geological Survey Fire Research Program

    http://www.usgs.gov/themes/wildfire.html

Laboratory of Tree-Ring Research, University of Arizona:

    http://www.ltrr.arizona.edu/

Wildfire News:

    http://www.wildfirenews.com

Smokejumpers:

    http://fs.fed.us/fire/people/smokejumpers/
    http://fire.blm.gov/smokejumper/

A-10 Warthog Air Tankers:

    http://FireHogs.com

Fire Weather (National Weather Service, Boise):

    http://www.boi.noaa.gov/

Textbooks

The following is a list of textbooks that are cited in various chapters of this or the other volumes in this series that deal with fire effects on vegetation, soils, and water:

1. Biswell, Harold H. 1989. *Prescribed Burning in California Wildlands Vegetation Management.* University of California Press. 255 pp.

2. Bond, William J., and Brian W. vanWilgen. 1996. *Fire and Plants.* Chapman & Hall, London. 263 pp.

3. Bradstock, Ross A., Jann E. Williams, and A. Malcolm Gill, editors. 2002. *Flammable Australia, the Fire Regimes and Biodiversity of a Continent.* Cambridge University Press. 462 pp.

4. Collins, Scott L., and Linda L. Wallace, editors. 1990. *Fire in North American Tallgrass Prairies.* University of Oklahoma Press. 175 pp.

5. DeBano, Leonard F., Daniel G. Neary, and Peter F. Ffolliott. 1998. *Fire's Effects on Ecosystems.* John Wiley & Sons, New York. 333 pp.

6. Kozlowski, T. T., and C. E. Ahlgren, editors. 1974. *Fire and Ecosystems.* Academic Press, Inc., New York. 542 pp.

7. Pyne, Stephen J. 1982. *Fire in America—A Cultural History of Wildland and Rural Fire.* University of Washington Press. 654 pp.

8. Pyne, Stephen J. 2001. *Fire—A Brief History.* University of Washington Press. 204 pp.

9. Pyne, Stephen J., Patricia L. Andrews, and Richard D. Laven. 1996. *Introduction to Wildland Fire.* 2nd Edition. John Wiley & Sons, New York. 769 pp.

10. Tere, William C. 1994. *Firefighter's Handbook on Wildland Firefighting— Strategy, Tactics and Safety.* Deer Valley Press, Rescue, CA. 313 pp.

11. Whelan, Robert J. 1995. *The Ecology of Fire.* Cambridge University Press. 346 pp.

12. Wright, Henry A., and Arthur W. Bailey. 1982. *Fire Ecology, United States and Southern Canada.* John Wiley & Sons, New York. 501 pp.

Journals and Magazines

Journals and magazines constitute another traditional source of fire effects information. Some of the useful ones include:

Fire Management Today (formerly *Fire Management Notes*). Superintendent of Documents, P.O. Box 371954, Pittsburgh, PA 15250-7954. Also available at:

http://www.fs.fed.us/fire/fmt/index.htm

Wildfire. Official publication of the International Association of Wildland Fire, 4025 Fair Ridge Drive, Suite 300, Fairfax, VA 22033-2868. Also available at:

http://www.iawfonline.org/

Wildland Firefighter. Wildland Firefighter, P.O. Box 130, Brownsville, OR 97327. Also available at:

http://wildlandfire.com

International Journal of Wildland Fire. CSIRO Publishing, P.O. Box 1139, Collingwood, Victoria 3066, Australia. Also available at:

http://www.publish.csiro.au/nid/114.htm

Journal of Forestry, Forest Science and The Forestry Source. Society of American Foresters, 5400 Grosvenor Lane, Bethesda, MD. 20814-2198. Also includes its regional journals. Also available at:

http://safnet.org/periodicals/

Forest Ecology and Management. Elsevier Science, Regional Sales Office, Customer Support Department, P.O. Box 945, New York, NY 10159-0945. Also available at:

http://www.elsevier.com/locate/issn/foreco

Journal of Range Management. Society for Range Management, 445 Union Blvd, Suite 230, Lakewood, CO 80228. Also available at:

http://uvalde.tamu.edu/jrm/jrmhome.htm

Canadian Journal of Forest Research. NRC Research Press, National Research Council of Canada, Ottawa, ON K1A 0R6. Also available at:

http://www.cif-ifc.org/engusu/e-cjfr-spec-sub.shtml

Environmental Science and Technology. American Chemical Society, 1155 16th St., N.W., Washington, DC 20036. Also available at:

http://pubs.acs.org/journals/esthag/index.html

Journal of Environmental Quality. American Society of Agronomy, 677 South Segoe Road, Madison, WI 53711. Also available at:

http://jeq.scijournals.org/

Ecology and *Ecological Applications.* Ecological Society of America, Suite 400, 1707 H Street NW, Washington, DC 20006. Also available at the following Web site:

http://www.esapubs.org/publications/

Journal of the American Water Resources Association. 4 West Federal Street, P.O. Box 1626, Middleburg VA 20118-1626. Also available at:

http://www.awra.org/publicationindex.htm

Journal of Wildlife Management. The Wildlife Society, 5410 Grosvenor Lane, Bethesda, MD 20814-2197. Also available at:

http://www.wildlife.org/publications/

Other Sources _____

Other information sources for fire effects on soils and water include USDA Forest Service Research Station reports, bulletins, notes and other publications:

North Central Forest Experiment Station, 1992 Folwell Avenue, St. Paul, MN 55108

http://www.ncrs.fs.fed.us/

USDA Forest Service Gen. Tech. Rep. RMRS-GTR-42-vol. 4. 2005

205

Northeastern Research Station, 11 Campus Boulevard, Newton Square, PA 19073

http://www.fs.fed.us/ne/

Pacific Northwest Research Station, P.O. Box 3890, Portland, OR 97208-3890

http://www.fs.fed.us/pnw/

Pacific Southwest Research Station, P.O. Box 245, Berkeley, CA 94701-0245

http://www.psw.fs.fed.us/psw/

Rocky Mountain Research Station, 2150 Centre Avenue, Fort Collins, CO. 80526

http://www.fs.fed.us/rm/

Southern Research Station, 200 Weaver Boulevard, P.O. Box 2680, Asheville, NC 28802

http://www.srs.fs.fed.us/

National Fire Plan is a cooperative effort of the USDA Forest Service, Department of the Interior, and the National Association of State Foresters for managing the impacts of wildfires on communities and the environment. It publishes annual attainment and information reports as well as maintaining a Web site:

http://www.fireplan.gov/

The National Incident Information Center Morning Fire Report, issued during periods of high fire activity, provides up-to-date fire activity and ecosystem impacts information at:

http://www.fs.fed.us/news/fire/

Daniel G. Neary
Kevin C. Ryan
Leonard F. DeBano

Chapter 12: Summary and Research Needs

Volume Objective

The objective of this volume is to provide an overview of the state-of-the-art understanding of the effects of fire on soils and water in wildland ecosystems. Our challenge was to provide a meaningful summary for North American fire effects on these resources despite enormous variations produced by climate, topography, fuel loadings, and fire regimes. This volume is meant to be an information guide to assist land managers with fire management planning and public education, and a reference on fire effects processes, pertinent publications, and other information sources. Although it contains far more information and detailed site-specific effects of fire on soils and water than the original 1979 Rainbow volumes, it is not designed to be a comprehensive research-level treatise or compendium. That challenge is left to several textbooks (Chandler and others 1991, Agee 1993, Pyne and others 1996, DeBano and others 1998).

Soil Physical Properties Summary

The physical processes occurring during fires are complex and include both heat transfer and the associated change in soil physical characteristics. The most important soil physical characteristic affected by fire is soil structure because the organic matter component can be lost at relatively low temperatures. The loss of soil structure increases the bulk density of the soil and reduces its porosity, thereby reducing soil productivity and making the soil more vulnerable to postfire runoff and erosion. Although heat is transferred in the soil by several mechanisms, its movement by vaporization and condensation is the most important. The result of heat transfer in the soil is an increase in soil temperature that affects the physical, chemical, and biological properties of the soil. When organic substances are moved downward in the soil by vaporization and condensation, they can cause a water-repellent soil condition that further accentuates

USDA Forest Service Gen. Tech. Rep. RMRS-GTR-42-vol. 4. 2005

207

postfire runoff and extensive networks of surface rill erosion or erosion by raindrop splash. The magnitude of change in soil physical properties depends on the temperature threshold of the soil properties and the severity of the fire. The greatest change in soil physical properties occurs when smoldering fires burn for long periods.

Soil Chemistry Summary

The most basic soil chemical property affected by soil heating during fires is organic matter. Organic matter not only plays a key role in the chemistry of the soil, but it also affects the physical properties (see chapter 2) and the biological properties (see chapter 4) of soils as well. Soil organic matter plays a key role in nutrient cycling, cation exchange, and water retention in soils. When organic matter is combusted, the stored nutrients are either volatilized or are changed into highly available forms that can be taken up readily by microbial organisms and vegetation. Those available nutrients not immobilized are easily lost by leaching or surface runoff and erosion. Nitrogen is the most important nutrient affected by fire, and it is easily volatilized and lost from the site at relatively low temperatures. The amount of change in organic matter and nitrogen is directly related to the magnitude of soil heating and the severity of the fire. High- and moderate-severity fires cause the greatest losses. Nitrogen loss by volatilization during fires is of particular concern on low-fertility sites because N can only be replaced by N-fixing organisms. Cations are not easily volatilized and usually remain on the site in a highly available form. An abundance of cations can be found in the thick ash layers (or ash-bed) remaining on the soil surface following high-severity fires.

Soil Biology Summary

Soil microorganisms are complex. Community members range in activity from those merely trying to survive, to others responsible for biochemical reactions that are among the most elegant and intricate known. How they respond to fire will depend on numerous factors, including fire intensity and severity, site characteristics, and preburn community composition. Some generalities can be made, however. First, most studies have shown strong resilience by microbial communities to fire. Recolonization to preburn levels is common, with the amount of time required for recovery generally varying in proportion to fire severity. Second, the effect of fire is greatest in the forest floor (litter and duff). Prescriptions that consume major fuels but protect forest floor, humus layers, and soil humus are recommended.

Fire and Streamflow Regimes Summary

Fires affect water cycle processes to a greater or lesser extent depending on severity. Fires can produce some substantial effects on the streamflow regime of both small streams and rivers, affecting annual and seasonal water yield, peakflows and floods, baseflows, and timing of flows. Adequate baseflows are necessary to support the continued existence of many wildlife populations. Water yields are important because many forest, scrubland, and grassland watersheds function as municipal water supplies. Peakflows and floods are of great concern because of their potential impacts on human safety and property. Next to the physical destruction of a fire itself, postfire floods are the most damaging aspect of fire in the wildland environment. It is important that resource specialists and managers become aware of the potential of fires to increase peakflows.

Following wildfires, flood peakflows can increase dramatically, severely affecting stream physical conditions, aquatic habitat, aquatic biota, cultural resources, and human health and safety. Often, increased flood peakflows of up to 100 times those previously recorded, well beyond observed ranges of variability in managed watersheds, have been measured after wildfires. Potentials exist for peak flood flows to jump to 2,300 times prewildfire levels. Managers must be aware of these potential watershed responses in order to adequately and safely manage their lands and other resources in the postwildfire environment.

Water Quality Summary

When a wildland fire occurs, the principal concerns for change in water quality are: (1) the introduction of sediment; (2) the potential for increasing nitrates, especially if the foliage being burned is in an area of chronic atmospheric deposition; (3) the possible introduction of heavy metals from soils and geologic sources within the burned area; and (4) the introduction of fire retardant chemicals into streams that can reach levels toxic to aquatic organisms.

The magnitude of the effects of fire on water quality is primarily driven by fire severity, and not necessarily by fire intensity. Fire severity is a qualitative term describing the amount of fuel consumed, while fire intensity is a quantitative measure of the rate of heat release (see chapter 1). In other words, the more severe the fire the greater the amount of fuel consumed and nutrients released and the more susceptible the site is to erosion of soil and nutrients into the stream where it could potentially affect water quality. Wildfires usually are more severe than prescribed fires. As a result, they are more likely to produce significant effects on water quality. On the other hand, prescribed fires are designed to be less severe and would be expected to produce less effect on water quality. Use of prescribed fire allows the manager the opportunity to control the severity of the fire and to avoid creating large areas burned at high severity.

The degree of fire severity is also related to the vegetation type. For example, in grasslands the differences between prescribed fire and wildfire are probably small. In forested environments, the magnitude of the effects of fire on water quality will probably be much lower after a prescribed fire than after a wildfire because of the larger amount of fuel consumed in a wildfire. Canopy-consuming wildfires would be expected to be of the most concern to managers because of the loss of canopy coupled with the destruction of soil aggregates. These losses present the worst-case scenario in terms of water quality. The differences between wild and prescribed fire in shrublands are probably intermediate between those seen in grass and forest environments.

Another important determinant of the magnitude of the effects of fire on water quality is slope. Steepness of the slope has a significant influence on movement of soil and nutrients into stream channels where it can affect water quality. Wright and others (1976) found that as slope increased in a prescribed fire, erosion from slopes is accelerated. If at all possible, the vegetative canopy on steep, erodible slopes needs to be maintained, particularly if adequate streamside buffer strips do not exist to trap the large amounts of sediment and nutrients that can be transported quickly into the stream channel. It is important to maintain streamside buffer strips whenever possible, especially when developing prescribed fire plans. These buffer strips will capture much of the sediment and nutrients from burned upslope areas.

Nitrogen is of concern to water quality. If soils on a particular site are close to N saturation, it is possible to exceed maximum contamination levels of NO_3-N (10 ppm or 10 mg/L) after a severe fire. Such areas should not have N-containing fertilizer applied after the fire. Chapter 3 contains more discussion of N. Fire retardants typically contain large amounts of N, and they can cause water quality problems where drops are made close to streams.

The propensity for a site to develop water repellency after fire must be considered (see chapter 2). Water-repellent soils do not allow precipitation to penetrate down into the soil and therefore are conducive to erosion. Severe fires on such sites can put large amounts of sediment and nutrients into surface water.

Finally, heavy rain on recently burned land can seriously degrade water quality. Severe erosion and runoff are not limited to wildfire sites alone. But if postfire storms deliver large amounts of precipitation or short-duration, high-intensity rainfalls, accelerated erosion and runoff can occur even after a carefully planned prescribed fire. Conversely, if below-average precipitation occurs after a wildfire, there may not be a substantial increase in erosion and runoff and no effect on water quality.

Fire managers can influence the effects of fire on water quality by careful planning before prescribed burning. Limiting fire severity, avoiding burning on steep slopes, and limiting burning on potentially water-repellent soils will reduce the magnitude of the effects of fire on water quality.

Aquatic Biota Summary

The effects of wildland fire on fish are mostly indirect in nature. There are some documented instances of fires killing fish directly. The largest problems arise from the longer term impact on habitat. This includes changes in stream temperature due to plant understory and overstory removal, ash-laden slurry flows, increases in flood peakflows, and sedimentation due to increased landscape erosion. Most information on the effects of wildfire on fishes has been generated since about 1990. Limited observational information exists on the immediate, direct effect of fire on fishes. Most information is indirect and demonstrates the effects of ash flows, changes in hydrologic regimes, and increases in suspended sediment on fishes. These impacts are marked, ranging from 70 percent to total loss of fishes. The effects of fire retardants on fishes are observational and not well documented at present. Further, all information is from forested biomes as opposed to grasslands.

Anthropogenic influences, largely land use activities over the past century, cumulatively influence fire effects on fishes. Fire suppression alone has affected vegetation densities on the landscape and the severity and extent of wildfire and, in turn, its effects on aquatic ecosystems and fishes. Most all studies of fire effects on fishes are short term (less than 5 years) and local in nature. A landscape approach to analyses has not been made to date. Fish can recover rapidly from population reductions or loss but can be markedly limited or precluded by loss of stream connectivity imposed by human-induced barriers. Fisheries management postfire should be based on species and fisheries and their management status. Managers should be vigilant of opportunities to restore native fishes in event of removal of introduced, nonnative, translocated species.

Although the impact of fire on fishes appears to be marked, a larger impact may loom in the future. Recent advances in atmospheric, marine, and terrestrial

USDA Forest Service Gen. Tech. Rep. RMRS-GTR-42-vol. 4. 2005

209

ecosystem science have resulted in the correlation of ocean temperature oscillations, tree ring data, and drought. Based on current information on fire effects on fishes, combined with that of climate change, drought, and recent insect infestations, the greatest impacts on fish may lie ahead. The information emerging in new climate change analyses suggests that the Southeastern and parts of the Western United States have a high probability of continuing and future drought, with potential impact of wildfire on fishes in the Southwest. Based on some of these overarching global indicators, the rest of the story is starting to unfold.

The effects of fires on reptile and amphibian populations are also covered in volume 1 of this series (Smith 2000). A review by Russell and others (1999) concluded that there are few reports of fire-caused injury to herpetofauna in general much less aquatic and wetland species.

Similar to fishes, the recorded effects of fire on aquatic macroinvertebrates are indirect as opposed to direct. Response varies from minimal to no changes in abundance, diversity, or richness, to considerable and significant changes. Abundance of macroinvertebrates may actually increase in fire-affected streams, but diversity generally is reduced. These differences are undoubtedly related to landscape variability, burn size and severity, stream size, nature and timing of postfire flooding events, and postfire time. Temporally, changes in macroinvertebrate indices in the first 5 years postfire can be different from ensuing years, and long-term (10 to 30 years) effects have been suggested.

The effects of fires on bird populations are covered in volume 1 of this series (Smith 2000). Aquatic areas and wetlands often provide refugia during fires. However, wetlands such as cienegas, marshes, cypress swamps, spruce, and larch swamps do burn under the right conditions. The impacts of fires on individual birds and populations in wetlands would then depend upon the season, uniformity, and severity of burning (Smith 2000).

Aquatic and wetland dwelling mammals are usually not adversely impacted by fires due to animal mobility and the lower frequency of fires in these areas. Lyon and others (2000) discuss factors such as fire uniformity, size, duration, and severity that affect mammals. However, most of their discussion relates to terrestrial mammals, not aquatic and wetland ones. Aquatic and wetland habitats also provide safety zones for mammals during fires.

Wetlands Summary

While the connection between wetlands and fire appears incongruous, fire plays an integral role in the creation and persistence of wetland species and ecosystems. Wetland systems have come under various classification systems, reflecting both the current understanding of wetland processes and the missions of the individual land management agencies. We examined some of the complex water inflows and outflows that are balanced by the interrelationships between biotic and abiotic factors. The presence and movement of water are dominant factors controlling the interactions of fire with wetland dynamics, nutrient and energy flow, soil chemistry, organic matter decomposition, and plant and animal community composition.

In wetland soils, the effects produced by surface and ground fires are related to the intensity and duration of fire at either the soil surface or the interface between burning and nonburning organic soil materials. In general, surface fires can be characterized as short duration, variable severity ones, while ground fires can be characterized as having longer duration and lower severity. However, the latter type of fires can produce profound physical and chemical changes in wetlands.

Models Summary

A number of older modeling technologies are commonly used in estimating fire effects during and after fire (FOFEM, WATSED, WEPP, RUSLE, and others). Newer models include DELTA-Q and FOREST, and others are under construction. These process-based models provide managers with additional tools to estimate the magnitude of fire effects on soil and water produced by land disturbance. FOFEM was developed to meet the needs of resource managers, planners, and analysts in predicting and planning for fire effects. Quantitative predictions of fire effects are needed for planning prescribed fires that best accomplish resource needs, for impact assessment, and for long-range planning and policy development. FOFEM was developed to meet this information need. The WATSED technology was developed for watershed analysis. The RUSLE model was developed for agriculture and rangeland hillslopes and has been extended to forest lands. The WEPP model was designed as an improvement over RUSLE that can either be run as a stand-alone computer model by specialists, or accessed through a special Internet interface designed for forest applications, including wild fires.

All of these models have limitations that must be understood by fire managers or watershed specialists before they are applied. The models are only as good as the data used to create and validate them. Some processes such as extreme flow and erosion events are not simulated very well because of the lack of good data or the complexity of the processes. However, they do provide useful tools to estimate landscape changes to disturbances such as fire. Potential users should make use of the extensive documentation of these models and consult with the developers to ensure the most appropriate usage of the models.

Watershed Rehabilitation Summary

Spending on postfire emergency watershed rehabilitation has steadily increased since 1990. An evaluation of USDA Forest Service burned area emergency rehabilitation treatment effectiveness was completed jointly by the USDA Forest Service Research and Development and National Forest System staffs. Robichaud and others (2000) collected and analyzed information on past use of BAER treatments in order to determine attributes and conditions that led to treatment success or failure in achieving BAER goals after wildfires in the continental Western United States. The study found that spending on rehabilitation has risen sharply recently because the perceived threat of debris flows and floods has increased where fires are closer to the wildland-urban interface. Existing literature on treatment effectiveness is limited, thus making treatment comparisons difficult; however, the amount of protection provided by any treatment is limited—especially during short-duration, high-intensity rainfall events.

Relatively little monitoring of postfire rehabilitation treatments has been conducted in the last three decades. In the three decades prior to 2000, there were at least 321 fires that received BAER treatment, which cost the Forest Service around $110 million. Some level of monitoring occurred on about 33 percent of these project fires. In the early 2000s, the number, size, and severity of Western United States fires dramatically increased, with a concurrent increase in BAER spending of about $50 million annually. Monitoring efforts continue to be short term and inconsistent. Analysis of the literature, Burned Area Report forms, interview comments, monitoring reports, treatment effectiveness ratings, and the authors' continuing work in the area of postfire rehabilitation have led to the conclusions and recommendations.

Information Sources Summary

This volume updates the information available on fire's impacts on soils and water. It is just one source of information prepared to a given point in time. Barring more frequent updates of this volume (originally published in 1979; Wells and others 1979, Tiedemann and others 1979), some mechanisms for future information retrieval need to be continually dynamic and current. The World Wide Web has produced a quantum leap in the availability of information now at the fire manager's finger tips, with past and current information on the Web growing daily. The future problem will be synthesizing the mountains of information now available. It is hoped that this volume provides some of that synthesis.

Research Needs

This volume points out information gaps and research needs throughout its chapters. To complicate matters more, some regions of the country are experiencing larger and more severe fires that are producing ecosystem effects not studied before. Some of the key research needs are:

- Beyond the initial vegetation and watershed condition impacts of the physical process of fire, suppression activities can add further levels of disturbance to both soils and water. These disturbances need to be evaluated along with those produced by fire.
- The burning of concentrated fuels (such as slash, large woody debris) can cause substantial damage to the soil resource even though these long-term effects are limited to only a small proportion of the landscape where the fuels are piled. The physical, chemical, and biological effects of concentrated fuel burning need to be better understood.
- Improved understanding of the effects of fire on soil chemical properties is needed for managing fire on all ecosystems, and particularly in fire-dependent systems.
- Soil microorganisms are complex. Community members range in activity from those merely trying to survive to others responsible for biochemical reactions among the most elegant and intricate known. How they respond to fire will depend on numerous factors, including fire intensity and severity, site characteristics, and preburn community composition. Understanding of these responses is needed.
- Peakflows and floods are of great concern because of their potential impacts on human safety and property. Next to the physical destruction of a fire itself, postfire floods are the most damaging aspect of fire in the wildland environment. It is important that research focus on improving the ability of resource specialists and managers to understand the potential of fires to increase peakflows and to improve predictions of floods.
- Heavy rain on recently burned land can seriously degrade water quality. Severe erosion

USDA Forest Service Gen. Tech. Rep. RMRS-GTR-42-vol. 4. 2005

211

and runoff are not limited to wildfire sites alone. But if postfire storms deliver large amounts of precipitation or short-duration, high-intensity rainfalls, accelerated erosion and runoff can occur even after a carefully planned prescribed fire. The effects of fire on municipal watersheds are not well documented, and the ability to predict magnitude and duration of water quality change is limited.

- Limited observational information exists on the immediate, direct effect of fire on fishes. Most information is indirect and demonstrates the effects of ash flows, changes in hydrologic regimes, and increases in suspended sediment on fishes. These impacts are marked, ranging from 70 percent to total loss of fishes. Also, the effects of fire retardants on fishes is observational and not well documented. Further, most information is from forested biomes, so research needs to be expanded to grasslands.

- Anthropogenic influences, largely land use activities over the past century, cumulatively influence fire effects on fishes. Fire suppression alone has affected vegetation densities on the landscape and the severity and extent of wildfire and, in turn, its effects on aquatic ecosystems and fishes. Most all studies of fire effects on fishes are short term (less than 5 years) and local in nature. A landscape approach to analyses and research needs to be made.

- Riparian areas are particularly important because they provide buffer strips that trap sediment and nutrients that are released when surrounding watersheds are burned. The width of these buffer strips is critical for minimizing sediment and nutrient movement into the streams. But little information exists on effective buffer strips and sizes.

- Although we have several models, all of them have limitations that must be understood by fire managers or watershed specialists before they are applied. The models are only as good as the data used to create and validate them. Some processes such as extreme flow and erosion events are not simulated well because of the lack of good data or the complexity of the processes.

- The need for improved effectiveness monitoring has gained momentum. In April 2003, the Government Accounting Office of Congress published a report (GAO 2003) that clearly stated that neither the U.S. Department of Agriculture's Forest Service nor the Department of the Interior's Bureau of Land Management could determine whether emergency stabilization and rehabilitation treatments were achieving their intended results. Although treatment monitoring is required, there is no agreed-on protocol for how and what to collect and analyze for determining effectiveness.

- Rainfall intensity as well as rainfall amount and duration affect the success of rehabilitation treatment. Rehabilitation treatments are least effective in short-duration, high-intensity rainstorms (that is, convective thunderstorms), particularly in the first 2 years after burning. Information is limited in this area.

- Quantitative data are needed to guide future responses to postfire rehabilitation and to build, test, and refine predictive models for different burned forest environments. Accurate climate, runoff, and erosion models for burned forest environments are dependent on (1) improved mapping of burn severity and better characterization of postfire soil water repellency; (2) improved prediction of runoff responses at different spatial scales, including short-duration, high-intensity thunderstorms; (3) quantitative data for the relative magnitudes and consequences of hillslope verses channel erosion; and (4) refined sediment deposition and routing models for various drainages.

- Consumption of organic soil horizons has only been quantified in a limited number of vegetation types. Although moisture vs. consumption patterns have emerged there is still a need to quantify consumption and soil heating in a wider range of vegetation types. Likewise additional research is needed to further elucidate the physical mechanisms of organic soil consumption such that robust models based on combustion and heat transfer relationships can eventually replace the empirical studies leading to wider application.

- Salvage logging after wildfires is becoming a more common occurrence. Research in the past has examined the soil and water impacts of logging and wildfire separately, but rarely together. There is a large need for research on logging and its associated roading activities at various time intervals post-wildfire, and for different forest ecosystems and physiographic regions.

212

USDA Forest Service Gen. Tech. Rep. RMRS-GTR-42-vol. 4. 2005

References

Acea, M.J.; Carballas, T. 1996. Changes in physiological groups of microorganisms in soil following wildfire. FEMS Microbiology Ecology. 20: 33–39.

Acea, M.J.; Carballas, T. 1999. Microbial fluctuations after soil heating and organic amendment. Bioresource Technology. 67: 65–71.

Adams, R.; Simmons. 1999. Ecological effects of fire fighting foams and retardants. In: Lunt, Ian; Green, David G.; Lord, Brian, (eds.). Proceedings, Australian bushfire conference; 1999 July; Albury, New South Wales, Australia. Available on line: http://life.csu.edu.au/bushfire99. [May 13, 2005].

Agee, J.K. 1973. Prescribed fire effects on physical and hydrologic properties of mixed-conifer forest floor and soil. Report 143. Davis: University of California Resources Center. 57 p.

Agee, J.K. 1993. Fire ecology of Pacific Northwest forests. Washington, DC: Island Press. 493 p.

Ahlgren, I.F. 1974. The effect of fire on soil organisms. In: Kozlowski, T.T.; Ahlgren, C.E., eds. Fire and ecosystems. New York: Academic Press: 47–72.

Ahlgren, I.F.; Ahlgren, C.E. 1965. Effects of prescribed burning on soil microorganisms in a Minnesota jack pine forest. Ecology. 46: 304–310.

Albin, D.P. 1979. Fire and stream ecology in some Yellowstone tributaries. California Fish and Game. 65: 216–238.

Albini, F.A. 1975. An attempt (and failure) to correlate duff removal and slash fire heat. Gen. Tech. Rep. INT-24. Ogden, UT: U.S. Department of Agriculture, Forest Service, Intermountain Forest and Range Experiment Station. 16 p.

Albini, F.A. 1976. Estimating wildfire behavior and effects. Gen. Tech. Rep. INT-30. Ogden, UT: U.S. Department of Agriculture, Forest Service, Intermountain Forest and Range Experiment Station. 92 p.

Albini, F.A.; Reinhardt, E.D. 1995. Modeling ignition and burning rate of large woody natural fuels. International Journal of Wildland Fire. 5(2): 81–91.

Albini, F.; Ruhul Amin, M.; Hungerford, R.D.; Frandsen, W.H.; Ryan, K.C. 1996. Models for fire-driven heat and moisture transport in soils. Gen. Tech. Rep. INT-GTR-335. Ogden, UT: U.S. Department of Agriculture, Forest Service, Intermountain Forest and Range Experiment Station. 16 p.

Aldon, E.F. 1960. Research in the ponderosa pine type. In: Progress report on watershed management research in Arizona. Fort Collins, CO: U.S. Department of Agriculture, Forest Service, Rocky Mountain Forest and Range Experiment Station: 17–24.

Alexander, M.E. 1982. Calculating and interpreting forest fire intensities. Canadian Journal of Botany. 60(4): 349–357.

Allen, L. 1998. Grazing and fire management. In: Tellman, B.; Finch, D.M.; Edminster, C.; Hamre, R., (eds.). The future of arid grasslands: identifying issues, seeking solutions. Proc. RMRS-P-3. Fort Collins, CO: U.S. Department of Agriculture, Forest Service, Rocky Mountain Research Station: 97–100.

Almendros, G.; Gonzalez-Vila, F.J.; Martin, F. 1990. Fire-induced transformation of soil organic matter from an oak forest: an experimental approach to the effects of fire on humic substances. Soil Science. 149: 158–168.

Amaranthus, M.P. 1989. Effect of grass seeding and fertilizing on surface erosion in two intensely burned sites in southwest Oregon. In: Berg, N.H., (tech. coord.). Proceedings of the symposium on fire and watershed management; 1988 October 26–28; Sacramento, CA. Gen. Tech. Rep. PSW-109. Berkeley, CA: U.S. Department of Agriculture, Forest Service, Pacific Southwest Forest and Range Experiment Station: 148–149.

Amaranthus, M.P.; Trappe, R.J.; Molina, R.J. 1989. Long-term forest productivity and the living soil. In: Gessel, S.P.; Lacate, D.S.; Weetman, G.F.; Powers, R.F., (eds.). Sustained productivity of forest soils; proceedings 7th North American forest soils conference. Faculty of Forestry Publications. Vancouver: University of British Columbia: 36–52.

Andersen, A.N.; Müller, W.J. 2000. Arthopod responses to experimental fire regimes in an Australian tropical savannah: ordinal-level analysis. Austral Ecology. 25: 199–209.

Anderson, D.C.; Harper, K.T.; Rushforth, S.R. 1982. Recovery of cryptogamic soil crusts from grazing on Utah winter ranges. Journal of Range Management. 35: 355–359.

Anderson, E.W. 1987. Riparian area definition—a viewpoint. Rangelands. 9: 70.

Anderson, H.E. 1969. Heat transfer and fire spread. Res. Pap. INT-69. Ogden, UT: U.S. Department of Agriculture, Forest Service, Intermountain Forest and Range Experiment Station. 20 p.

Anderson, H.E. 1982. Aids to determining fuel models for estimating fire behavior. Gen. Tech. Rep. INT-122. Ogden, UT: U.S. Department of Agriculture, Forest Service, Intermountain Forest and Range Experiment Station. 20 p.

Anderson, H.W. 1974. Sediment deposition in resiviors associated with rual roads, forest fires, and catchment attributes. In: Proceedings symposium on man's effect on erosion and sedimentation, UNESCO [1974 September 9–12; Paris.] International Association Hydrological Science Publications. 113: 87–95.

Anderson, H.W. 1976. Fire effects on water supply, floods, and sedimentation. In: Proceedings, of the Tall Timbers fire ecology conference, Pacific Northwest; 1974 October 16–17; Portland, OR. Tallahassee, FL: Tall Timbers Research Station. 15: 249–260.

Anderson, H.W.; Coleman, G.B.; Zinke, P.J. 1959. Summer slides and winter scour—dry-wet erosion in southern California mountains. Res. Pap. PSW-36. Berkeley, CA: U.S. Department of Agriculture, Forest Service, Pacific Southwest Forest and Range Experiment Station. 12 p.

Anderson, H.W.; Hoover, M.D.; Reinhart, K.G. 1976. Forests and water: effects of forest management on floods, sedimentation, and water supply. Gen. Tech. Rep. PSW-18. Berkeley, CA: U.S. Department of Agriculture, Forest Service, Pacific Southwest Forest and Range Experiment Station. 115 p.

Anderson, J.M. 1991. The effects of climate change on decomposition processes in grassland and coniferous forest. Ecological Applications. 1: 326–347.

Andrews, P.L. 1986. BEHAVE: Fire behavior prediction and fuel modeling system: BURN subsystem, Part 1. Gen. Tech. Rep. INT-194. U.S. Department of Agriculture, Forest Service, Intermountain Forest and Range Experiment Station. 130 p.

Angino, E.E.; Magunson, L.M.; Waugh, T.C. 1974. Mineralogy of suspended sediment and concentrations of Fe, Mn, Ni, Cu, and Pb in water and Fe, Mn, and Pb in suspended load of selected Kansas streams. Water Resources Research. 10: 1187–1191.

Armentano, T.V.; Menges, E.S. 1986. Patterns of change in the carbon balance of organic soil wetlands of the temperate zone. Journal of Ecology. 74: 755–774.

Arno, S.F. 2000. Fire in western forest ecosystems. In: Brown, J.K.; Smith, J.K., (eds.). Wildland fire in ecosystems: effects of fire on flora. Gen. Tech. Rep. RMRS-GTR-42-vol. 2. Ogden, UT: U.S. Department of Agriculture, Forest Service, Rocky Mountain Research Station: 97–120.

Artsybashev, E.S. 1983. Forest fires and their control. [Translated from Russian] New Delhi: Amerind Publishing Co. 160 p.

Atlas, R.M.; Horowitz, A.; Krichevsky, M.; Bej, A.K. 1991. Response of microbial populations to environmental disturbance. Microbial Ecology. 22: 249–256.

Aubertin, G.M.; Patric, J.H. 1974. Water quality water after clearcutting a small watershed in West Virginia. Journal of Environmental Quality. 3: 243–249.

Auclair, A.N.D. 1977. Factors affecting tissue nutrient concentrations in a Carex meadow. Oecologia. 28: 233–246.

Autobee, R. 1993. The Salt River project. Bureau of Reclamation History Program. Denver, CO: U.S. Department of the Interior, Bureau of Reclamation. 41 p.

Baath, E.; Frostegard, A.; Pennanen, T.; Fritze, H. 1995. Microbial community structure and pH response in relation to soil organic matter quality in wood-ash fertilized, clear-cut or burned coniferous forest soils. Soil Biology and Biochemistry. 27: 229–240.

Bacchus, S.T. 1995. Ground levels are critical to the success of prescribed burns. In: Cerulean, S.I.; Engstrom, R.T., (eds.) Fire in wetlands: a management prespective. Proceedings of the Tall Timbers fire ecology conference. Tallahassee, FL: Tall Timbers Research Station. 19: 117–133.

Bailey, R.G. 1995. Descriptions of ecoregions of the United States. Misc. Publ. 1391. Washington, DC: U.S. Department of Agriculture, Forest Service. 108 p.

Bailey, R.W. 1948. Reducing runoff and siltation through forest and range management. Journal of Soil and Water Conservation. 3: 24–31.

Bailey, R.W.; Copeland, O.L. 1961. Vegetation and engineering structures in flood and erosion control. Paper 11-1. In: Proceedings, 13th Congress; 1961 September; Vienna, Austria. International Union of Forest Research Organization. 23 p.

Baker, M.B., Jr. 1986. Effects of ponderosa pine treatments on water yields in Arizona. Water Resources Research. 22: 67–73.

Baker, M.B., Jr. 1990. Hydrologic and water quality effects of fire. In: Krammes, J.S., (tech. coord.). Effects of fire management of southwestern natural resources. Gen. Tech. Rep. RM-191. Fort Collins, CO: U.S. Department of Agriculture, Forest Service, Rocky Mountain Forest and Range Experiment Station: 31–42.

Baker, M.B., Jr., comp. 1999. History of watershed management in the central Arizona highlands. Gen. Tech. Rep. RMRS-GTR-29. Fort Collins, CO: U.S. Department of Agriculture, Forest Service, Rocky Mountain Forest and Range Experiment Station. 56 p.

Baker, M.B., Jr.; DeBano, L.F.; Ffolliott, P.F.; Gottfried, G.J. 1998. Riparian-watershed linkages in the Southwest. In: Potts, D.E., (ed.). Rangeland management and water resources. Proceedings of the American Water Resources Association specialty conference; Hendon, VA: 347–357.

Barnett, D. 1989. Fire effects on Coast Range soils of Oregon and Washington and management implications. Soils R-6 Tech. Rep. Portland, OR: U.S. Department of Agriculture, Forest Service, Pacific Northwest Region. 89 p.

Barro, S.C.; Conard, S.G. 1987. Use of ryegrass seeding as an emergency revegetation measure in chaparral ecosystems. Gen. Tech. Rep. PSW-102. Berkeley, CA: U.S. Department of Agriculture, Forest Service, Pacific Southwest Forest and Range Experiment Station. 12 p.

Barro, S.C.; Wohlemuth, P.M.; Campbell, A.G. 1989. Postfire interactions between riparian vegetation and channel morphology and the implications for stream channel rehabilitation. In: Abell, D.L., (tech. coord.). Proceedings of the California riparian conference: Protection, management, and restoration for the 1990s. Gen. Tech. Rep. PSW-100. Berkeley, CA: U.S. Department of Agriculture, Forest Service, Pacific Southwest Forest and Range Experiment Station: 51–53.

Bautista, S.; Bellot, J.; Vallejo, V.R. 1996. Mulching treatment for post-fire soil conservation in a semiarid ecosystem. Arid Soil Research and Rehabilitation. 10: 235–242.

Beasley, R.S. 1979. Intensive site preparation and sediment loss on steep watersheds in the Gulf Coastal Plain. Soil Science Society of America Journal. 43: 412–417.

Beasley, R.S.; Granillo, A.B.; Zillmer, V. 1986. Sediment losses from forest management: mechanical vs. chemical site preparation after cutting. Journal of Environmental Quality. 17: 219–225.

Bell, R.L.; Binkley, D. 1989. Soil nitrogen mineralization and immobilization in response to periodic prescribed fire in a loblolly pine plantation. Canadian Journal of Forestry Research. 19: 816–820.

Bellgard, S.E.; Whelan, R.J.; Muston, R.M. 1994. The impact of wildfire on vesicular-arbucscular mycorrhizal fungi and their potential to influence the re-establishment of post-fire plant communities. Mycorrhiza. 4: 139–146.

Belnap, J. 1994. Potential role of cryptobiotic soil crust in semiarid rangelands. In: Monsen, S.B.; Kitchen, S.G., (eds.). Proceedings, ecology and management of annual rangelands. Gen. Tech. Rep. INT-GTR-313. Ogden, UT: U.S. Department of Agriculture, Forest Service, Intermountain Forest and Range Experiment Station: 179–185.

Belnap, J.; Kaltenecker, J.H.; Rosentreter, R.; Williams, J; Leonard, S.; Eldridge, D. 2001. Biological soil crusts: ecology and management. Report No. BLM/ID/ST-01/001+1730. Denver, CO. U.S. Department of the Interior, National Science and Technology Center. 110 p.

Benda, L.E.; Cundy, T.W. 1990. Predicting deposition of debris flows in mountain channels. Canadian Geotechnical Journal. 27: 409–417.

Bennett, K.A. 1982. Effects of slash burning on surface soil erosion rates in the Oregon Coast Range. Corvallis: Oregon State University. 70 p. Thesis.

Berndt, H.W. 1971. Early effects of forest fire on streamflow characteristics. Res. Note PNW-148. Portland, OR: U.S. Department of Agriculture, Forest Service, Pacific Northwest Forest and Range Experiment Station. 9 p.

Beschta, R.L. 1980. Turbidity and suspended sediment relationships. In: Proceedings symposium on watershed management. St. Anthony, MN: American Society of Civil Engineers: 271–281.

Beschta, R.L. 1990. Effects of fire on water quantity and quality. In: Walstad, J.D.; Radosevich, S.R.; Sandberg, D.V. (eds.). Natural and prescribed fire in Pacific Northwest forests. Corvallis: Oregon State University Press: 219–231.

Best, D.W.; Kelsey, D.K.; Hagans, D.K.; Alpert, M. 1995. Role of fluvial hillslope erosion and road construction in the sediment budget of Garret Creek, Humboldt County, California. In: Nolan, K.M.; Kelsey, H.M.; Marron, D.C. (eds.). Geomorphic processes and aquatic habitat in the Redwood Creek Basin, Northwestern California. Prof. Paper 1454. Washington, DC: U.S. Geological Survey: M1–M9.

Beyers, J.L.; Conard, S.G.; Wakeman, C.D. 1994. Impacts of an introduced grass, seeded for erosion control, on post-fire community composition and species diversity in southern California chaparral. In: Proceedings of the 12th international conference on fire and forest meteorology; 1993 October 26–28; Jekyll Island, GA. Bethesda, MD: Society of American Foresters: 594–601.

Bhadauria, T.; Ramakrishnan, P.S.; Srivastava, K.N. 2000. Diversity and distribution of endemic and exotic earthworms in natural and regenerating ecosystems in the central Himalayas, India. Soil Biology and Biochemistry. 32: 2045–2054.

Biggio, E.R.; Cannon, S.H. 2001. Compilation of post-wildfire runoff data from the Western United States. Open-file Report 2001-474. U.S. Geological Survey. 24 p.

Binkley, D.; Richter, D.; Davis, M.B.; Caldwell, B. 1992. Soil chemistry in a loblolly/longleaf pine forest with interval burning. Ecological Applicatons. 2: 157–164.

Bird, M.I.; Veenendaal, E.M.; Moyo, C.; Lloyd, J.; Frost P. 2000. Effect of fire and soil texture on soil carbon in a sub-humid savanna, Matopos, Zimbabwe. Geoderma. 9: 71–90.

Bissett, J.; Parkinson, D. 1980. Long-term effects of fire on the composition and activity of the soil microflora of a subalpine, coniferous forest. Canadian Journal of Botany. 58: 1704–1721.

Biswell, H.H. 1973. Prescribed fire effects on water repellency, infiltration, and retention in mixed conifer litter. Water Resource Report. Davis: University of California: 25–40.

Biswell, H.H.; Schultz, A.M. 1965. Surface runoff and erosion as related to prescribed burning. Journal of Forestry. 55: 372–373.

Black, C.A. 1968. Soil plant relationships. New York: John Wiley & Sons, Inc. 792 p.

Boerner, R.E.J. 1982. Fire and nutrient cycling in a temperate ecosystem. Bioscience. 32: 187–192.

Boerner, R.E.J.; Decker, K.L.M.; Sutherland, E.K. 2000. Prescribed burning effects on soil enzyme activity in a southern Ohio hardwood forest: a landscape-scale analysis. Soil Biology and Biochemistry. 32: 899–908.

Bohn, C. 1986. Biological importance of streambank stability. Rangelands. 8: 55–56.

Bolin, S.B.; Ward, T.J. 1987. Recovery of a New Mexico drainage basin from a forest fire. In: Proceedings of the symposium on forest hydrology and watershed management. Publ. 167. Washington, DC: International Association of Hydrological Sciences: 191–198.

Bollen, W.B. 1974. Soil microbes. In: Cramer, O. (ed.). Environmental effects of forest residues management in the Pacific Northwest. A state-of-knowledge compendium. Gen. Tech. Rep. PNW-24. Portland, OR: U.S. Department of Agriculture, Forest Service, Pacific Northwest Forest and Range Experiment Station: B-1-B-30.

Bond, W.J.; Van Wilgen, B.W. 1996. Fire and plants. London: Chapman & Hall. 263 p.

214

USDA Forest Service Gen. Tech. Rep. RMRS-GTR-42-vol. 4. 2005

Borchers, J.G.; Perry, D.A. 1990. Effects of prescribed fire on soil organisms. In: Walstad, J.D.; Radosevich, S.R.; Sandberg, D.V. (eds.). Natural and prescribed fire in Pacific Northwest forests. Corvallis: Oregon State University Press: 143–157.

Bosch, J.M.; Hewlett, J.D. 1982. A review of catchment experiments to determine the effect of vegetation changes on water yield and evapotranspiration. Journal of Hydrology. 55: 3–23.

Bozek, M.A.; Young, M.K. 1994. Fish mortality resulting from delayed effects of fire in the greater Yellowstone ecosystem. Great Basin Naturalist. 54(1): 91–95.

Branson, W.J.; Miller, R.F.; McQueen, I.S. 1976. Moisture relationships in twelve northern shrub communities near Grand Junction, Colorado. Ecology. 57: 1104–1124.

Bridgham, S.D.; Chein Lu, P.; Richardson, J.L.; Updegraff, K. 2001. Soils of northern peatlands: histisols and gelisols. In: Richardson, J.L.; Vepraskas, M.J. (eds.). Wetland soils—genesis, hydrology, landscapes, and classification. Lewis Publishers: 343–370.

Brooks, K.N.; Ffolliott, P.F.; Gregersen, H.M.; DeBano, L.F. 2003. Hydrology and the management of watersheds. 3rd Edition. Ames: Iowa State Press. 704 p.

Brown, A.A.; Davis, K.P. 1973. Forest fire: control and use. New York: McGraw-Hill Book Company. 584 p.

Brown, G.W.; Krygier, J.T. 1970. Effects of clearcutting on stream temperature. Water Resources Research. 6(5): 1189–1198.

Brown, G.W. 1980. Forestry and water quality. Corvallis: Oregon State University Bookstores. 124 p.

Brown, G.W.; Gahler, A.R.; Marston, R.B. 1973. Nutrient losses after clear-cut logging and slash burning in the Oregon Coast Range. Water Resources Research. 9: 1450–1453.

Brown, H.E.; Thompson, J.R. 1965. Summer water used by aspen, spruce, and grassland in western Colorado. Journal of Forestry. 63: 756–760.

Brown, J.K. 1970. Physical fuel properties of ponderosa pine forest floors and cheatgrass. Res. Pap. INT-74. Ogden, UT: U.S. Department of Agriculture Forest Service, Forest Service, Intermountain Forest and Range Experiment Station. 16 p.

Brown, J.K. 1981. Bulk densities of nonuniform surface fuels and their application to fire modeling. Forest Science. 27(4): 667–683.

Brown, J.K. 2000. Introduction and fire regimes. In: Brown, J.K.; Smith, J.K. (eds.). Wildland fire in ecosystems—effects of fire on flora. Gen. Tech. Rep. GTR-RMRS-42-Vol. 2. Ogden, UT: U.S. Department of Agriculture, Forest Service: 1–8.

Brown, J.K.; Smith, J.K. 2000. Wildland fire in ecosystems: effects of fire on floral. Gen. Tech. Rep. RMRS-GTR-42-Vol. 2. Fort Collins, CO: U.S. Department of Agriculture, Forest Service, Rocky Mountain Research Station. 257 p.

Brown, J.K.; Marsden, M.A.; Ryan, K.C.; Reinhardt, E.D. 1985. Predicting duff and woody fuel consumed by prescribed fire in the northern Rocky Mountains. Res. Pap. INT-337. Ogden, UT: U.S. Department of Agriculture, Forest Service, Intermountain Forest and Range Experiment Station. 23 p.

Brown, T.C.; Binckley, D. 1994. Effect of management on water quality in North American forests. Gen. Tech. Rep. RM-248. Fort Collins, CO: U.S. Department of Agriculture, Forest Service, Rocky Mountain Forest and Range Experiment Station. 27 p.

Bryant, W.L.; Brinson, M.M.; Hook, P.B.; Jones, M.N. 1991. Response of vegetation in a brackish marsh to simulated burning and to nitrogen and phosphorus enrichment. In: Ecology of a nontidal brackish marsh in coastal North Carolina. Open File Report 91-03. Slidell, LA: U.S. Department of the Interior, Fish and Wildlife Service, National Wetlands Research Center: 283–306.

Bullard, W.E. 1954. A review of soil freezing by snow cover, plant cover, and soil conditions in Northwestern United States. Region 6 Soils Report. Portland, OR: U.S. Department of Agriculture, Forest Service, Pacific Northwest Region. 12 p.

Burgan, R.E.; Rothermel, R.C. 1984. BEHAVE: Fire behavior prediction and fuel modeling system—fuel subsystem. Gen. Tech. Rep. GTR INT-167. Ogden, UT: U.S. Department of Agriculture, Forest Service, Intermountain Forest and Range Experiment Station. 126 p.

Burroughs, E.R., Jr.; King, J.G. 1989. Reduction in soil erosion of forest roads. Gen. Tech. Rep. INT-264. Ogden, UT: U.S. Department of Agriculture, Forest Service, Intermountain Research Station. 21 p.

Byram, G.M. 1959. Combustion of forest fuels. In: K.P. Davis (ed.). Forest fire: control and use. New York: McGraw-Hill: 61–123.

California Department of Forestry. 1998. California forest practices rules. Sacramento: California Department of Forestry and Fire Protection. 257 p.

Callaham, M.A., Jr.; Hendrix, P.F.; Phillips, R.J. 2003. Occurrence of an exotic earthworm (Amynthas agrestis) in undisturbed soils of the southern Appalachian Mountains, USA. Pedobiologia. 47(5/6): 466–470.

Callison, J.; Brotherson, J.D.; Brown, J.E. 1985. The effects of fire on the blackbrush (Coleogyne ramosissima) community of southwestern Utah. Journal of Range Management. 38: 535–538.

Campbell, G.S.; Jungbauer, J.D., Jr.; Bidlake, W.R.; Hungerford, R.D. 1994. Predicting the effect of temperature on soil thermal conductivity. Soil Science. 158(5): 307–313.

Campbell, G.S.; Jungbauer, J.D., Jr.; Bristow, K.L.; Hungerford, R.D. 1995. Soil temperature and water content beneath a surface fire. Soil Science. 159(6): 363–374.

Campbell, G.S.; Jungbauer, J.D., Jr.; Bristow, K.L.; Bidlake, W.R. 1992. Simulation of heat and water flow in soil under high temperature (fire) conditions. Pullman: Washington State University, Department of Agronomy and Soils. Unpublished report to U.S. Department of Agriculture, Forest Service, Intermountain Forest and Range Experiment Station, Intermountain Fire Sciences Laboratory, Missoula, MT.

Campbell, R.E., Baker, M.B., Jr.; Ffolliott, P.F.; Larson, F.R.; Avery, C.C. 1977. Wildfire effects on a ponderosa pine ecosystem: an Arizona case study. Res. Pap. RM-191. Fort Collins, CO: U.S. Department of Agriculture, Forest Service, Rocky Mountain Forest and Range Experiment Station. 12 p.

Cannon, S.H. 2001. Debris-flow generation from recently burned watersheds. Environmental & Engineering Geoscience. 7(4): 321–241.

Carballas, M.; Acea, M.J.; Cabaneiro, A.; Trasar, C.; Villar, M.C.; Diaz-Ravina, M.; Fernandez, I.; Prieto, A.; Saa, A.; Vazquez, F.J.; Zehner, R.; Carballas, T. 1993. Organic matter, nitrogen, phosphorus and microbial population evolution in forest humiferous acid soils after wildfires. In: Trabaud, L.; Prodon, P. (eds.). Fire in Mediterranean ecosystems. Ecosystem Research Report 5. Brussels, Belgium: Commission of the European Countries: 379–385.

Carreira, J.A.; Niell, F.X. 1992. Plant nutrient changes in a semi-arid Mediterranean shrubland after fire. Journal of Vegetation Science. 3: 457–466.

Carter, C. D. and J. N. Rinne. In Press. Short-term effects of the Picture Fire on fishes and aquatic habitat. Hydrology and Water Resources in Arizona and the Southwest. 35:

Chambers, D.P.; Attiwill, P.M. 1994. The ash-bed effect in Eucalyptus regnans forest: chemical, physical and microbiological changes in soil after heating or partial sterilization. Australian Journal of Botany. 42: 739–749.

Chandler, C.P; Cheney, P.; Thomas, P; Trabaud, L.; Williams, D. 1991. Fire in forestry - Volume I: Forest fire behavior and effects. New York: John Wiley & Sons, Inc. 450 p.

Chiun-Ming, L. 1985. Impact of check dams on steep mountain channels in northeastern Taiwan. In: El-Swaify, S.A.; Moldenhauer, W.C.; Lo, A., (eds.). Soil erosion and conservation: 540–548.

Christ, J.H. 1934. Reseeding burned-over lands in northern Idaho. Ag. Exp. Sta. Bull. 201. Moscow: University of Idaho. 27 p.

Christensen, N.L. 1973. Fire and the nitrogen cycle in California chaparral. Science. 181: 66–68.

Christensen, N.L. 1981. Fire regimes in southeastern ecosystems. In: Mooney, H.A.; Bonnicksen, T.M.; Christensen, N.L.; Lotan J.E.; Reiners, R.A. (tech. coords.). Fire Regimes and ecosystem properties: proceedings of the conference; 1978 December 11–15; Honolulu, HI. Gen. Tech. Rep. WO-26. Washington, DC: U.S. Department of Agriculture, Forest Service: 112–136.

Christensen, N.L. 1985. Shrubland fire regimes and their evolutionary consequences. In: Pickett, S.T.A.; White, P.S. (eds.). The ecology of natural disturbances and patch dynamics. New York: Academic Press: 85–100.

Christensen, N.L.; Mueller, C.H. 1975. Effects of fire on factors controlling plant growth in Adenostoma chaparral. Ecological Monographs. 45: 29–55.

USDA Forest Service Gen. Tech. Rep. RMRS-GTR-42-vol. 4. 2005

215

Christensen, N.L.; Wilbur, R.B.; McLean, J. S. 1988. Soil-Vegetation correlations in the pocosins of Croatan National Forest, North Carolina. Biol. Rep. 88(28). Washington, DC: U.S. Department of the Interior, Fish and Wildlife Service. 97 p.

Clark, J.S. 1998. Effects of long-term water balances on fire regimes, north-western Minnesota. Journal of Ecology. 77: 989–1004.

Clary, W.P.; Baker, M.B., Jr.; O'Connell, P.F.; Johnsen, T.N., Jr.; Campbell, R.E. 1974. Effects of pinyon-juniper removal on natural resource products and users in Arizona. Res. Pap. RM-128. Fort Collins, CO: U.S. Department of Agriculture, Forest Service, Rocky Mountain Forest and Range Experiment Station. 28 p.

Clayton, J.L. 1976. Nutrient gains to adjacent ecosystems during a forest fire: an evaluation. Forest Science. 22: 162–166.

Clayton, J.L.; Kennedy, D.A. 1985. Nutrient losses from timber harvest in the Idaho batholith. Soil Science Society American Journal. 49(4): 1041–1049.

Cline, G.G.; Brooks, W.M. 1979. Effect of light seed and fertilizer application in steep landscapes with infertile soils after fire. Northern Region Soil, Air, Water Notes. Missoula, MT: U.S. Department of Agriculture, Forest Service, Northern Region: 79–86.

Cline, R.G.; Haupt, H.F.; Campbell, G.S. 1977. Potential water yield response following clearcut harvesting on north and south slopes in Northern Idaho. Res. Pap. INT-191. Ogden, UT: U.S. Department of Agriculture, Forest Service, Intermountain Forest and Range Experiment Station. 16 p.

Clinton, B.D.; Vose, J.M.; Swank, W.T. 1996. Shifts in aboveground and forest floor carbon and nitrogen pools after felling and burning in the southern Appalachians. Forest Science. 42: 431–441.

Cohen, A.D.; Casagrande, D.J.; Andrejko, M.J.; Best, G.R. 1984. The Okefenokee Swamp: Its natural history, geology, and geochemistry. Los Alamos, NM: Wetland Surveys. 709 p.

Coleman, D.C.; Crossley, D.A., Jr. 1996. Fundamentals of soil ecology. New York: Academic Press. 205 p.

Collins, L.M.; Johnston, C.E. 1995. Effectiveness of straw bale dams for erosion control in the Oakland Hills following the fire of 1991. In: Keeley, J.E.; Scott, T. (eds.). Brushfires in California wildlands: ecology and resources management. Fairfield, WA: International Association of Wildland Fire: 171–183.

Conard, S. G.; Ivanova, G.A. 1997. Wildfire in Russian boreal forests—potential impacts of fire regime characteristics on emissions and global carbon balance estimates. Environmental Pollution. 98: 305–313.

Conrad, C.E.; Poulton, C.E. 1966. Effect of wildfire on Idaho fescue and bluebunch wheatgrass. Journal of Range Management. 19: 138–141.

Cope, M.J.; Chaloner, W.G. 1985. WildFire: an interaction of biological and physical processes. In: Tiffney, B.H. (ed.). Geologic factors and the evolution of plants. New Haven, CT: Yale University Press: 257–277

Copeland, O.L. 1961. Watershed management and reservoir life. Journal of American Water Works Association. 53(5): 569–578.

Copeland, O.L. 1968. Forest Service Research in erosion control in the Western United States. Annual meeting, American Society of Agricultural Engineers; 1968 June 18–21; Logan, UT. Paper 68-227. St. Joseph, MI: American Society of Agricultural Engineers. 11 p.

Copley, T.L.; Forrest, L.A.; McColl, A.G.; Bell, F.G. 1944. Investigations in erosion control and reclamation of eroded lands at the Central Piedmont Conservation Station. Statesville, NC, 1930–1940. Tech. Bull. 873. Washington, DC: U.S. Department of Agriculture, Soil Conservation Service. 66 p.

Copstead, R. 1997. The water/road interaction technology series: an introduction. No.9777 1805-SDTDC. San Dimas, CA: U.S. Department of Agriculture, Forest Service, San Dimas Technology Center. 4 p.

Costales, E.F., Jr.; Costales, A.B. 1984. Determination and evaluation of some emergency measures for the quick rehabilitation of newly burned watershed areas in the pine forest. Sylvtrop Philippine Forestry Research. 9: 33–53.

Coults, J.R.H. 1945. Effect of veld burning on the base exchange capacity of a soil. South Africa Journal of Science. 41: 218–224.

Countryman, C.M. 1975. The nature of heat—Its role in wildland fire—Part 1. Unnumbered Publication. Berkeley, CA: U.S. Department of Agriculture, Forest Service, Pacific Southwest Forest and Range Experiment Station. 8 p.

Countryman, C.M. 1976a. Radiation. Heat—Its role in wildland fire—Part 4. Unnumbered Publication. Berkeley, CA: U.S. Department of Agriculture, Forest Service, Pacific Southwest Forest and Range Experiment Station. 8 p.

Countryman, C.M. 1976b. Radiation. Heat—Its role in wildland fire—Part 5. Unnumbered Publication. Berkeley, CA: U.S. Department of Agriculture, Forest Service, Pacific Southwest Forest and Range Experiment Station. 12 p.

Covington, W.W.; Moore, M.M. 1992. Restoration of presettlement tree densities and natural fire regimes in ponderosa pine ecosystems. Bulletin Ecology Society of America. 73(2 suppl): 142–148.

Covington, W.W.; Sackett, S.S. 1986. Effect of periodic burning on soil nitrogen concentrations in ponderosa pine. Soil Science Society of America Journal. 50: 452–457.

Covington, W.W.; Sackett, S.S. 1992. Soil mineral changes following prescribed burning in ponderosa pine. Forest Ecology and Management. 54: 175–191.

Covington, W.W.; Everette, R.L.; Steele, R.; Irwin, L.L.; Daer, T.A.; Auclaire, A.N.D. 1994. Historical and anticipated changes in forest ecosystems in the inland west of the United States. Journal of Sustainable Forestry. 2: 13–63.

Cowardin, L.M.; Carter, V.; Golet, F.C.; LaRoe, E.T. 1979. Classification of wetlands and deepwater habitats of the United States. Pub. FWS/OBS-79/31. Washington, DC: U.S. Department of the Interior, Fish and Wildlife Service. 103 p.

Craft, C.B. 1999. Biology of wetland soils. In: Richardson, J.L.; Vepraskas, M.J (eds.). Wetland soils-genesis, hydrology, landscapes, and classification. Lewis Publishers: 107–135.

Croft, A.R.; Monninger, L.V. 1953. Evapotranspiration and other water losses on some aspen forest types in relation to water available for stream flow. Transactions, American Geophysical Union. 34: 563–574.

Croft, A.R.; Marston, R.B. 1950. Summer rainfall characteristics in northern Utah. Transactions, American Geophysical Union. 31(1): 83–95.

Crouse, R. P. 1961. First-year effects of land treatment on dry-season streamflow after a fire in southern California. Res. Note 191. Berkeley, CA: U.S. Department of Agriculture, Forest Service, Pacific Southwest Forest and Range Experiment Station. 4 p.

Crow, T. R.; Baker, M.E.; Burton, V. Barnes. 2000. Chapter 3: Diversity in riparian landscapes. In: Verry, E.S.; Hornbeck, J.W.; Dolloff, C.A. (eds.). Riparian management in forests of the continental eastern United States. New York: Lewis Publishers: 43–66.

Cummins, K.R. 1978. Ecology and distribution of aquatic insects. In: Merritt, R.W.; Cummins, K.W. (eds.). An introduction to the aquatic insects of North America. Dubuque, IA: Kendall-Hunt: 29–31.

Curtis, J.T. 1959. The vegetation of Wisconsin—An ordination of plant communities. Madison: The University of Wisconsin Press. 657 p.

Daniel, C.C., III. 1981. NAME OF ARTICLE?? In: Richardson, C.J., Matthews, M.L.; Anderson, S.A. (eds.) Pocosin wetlands: An integrated analysis of coastal plain freshwater bogs in North Carolina. Stroudsburg, PA: Hutchinson Ross Publishing Company; New York, NY: Academic Press: 69–108.

Daniel, H.A.; Elwell, H.M.; Cox, M.B. 1943. Investigations in erosion control and the reclamation of eroded land at the Red Plains Conservation Experiment Station, Guthrie, OK, 1930–1940. Tech. Bull. U.S. Department of Agriculture, Soil Conservation Service. 837 p.

Davis, A.M., 1979. Wetland succession, fire and the pollen record: a midwestern example. The American Midland Naturalist. 102(1) 86–102.

Davis, E.A. 1984. Conversion of Arizona chaparral increases water yield and nitrate loss. Water Resources Research. 20: 1643–1649.

Davis, E.A. 1987. Chaparral conversion and streamflow: Nitrate increase is balanced mainly by a decrease in bicarbonate. Water Resources Research. 23: 215–224.

Davis, J.B. 1984. Burning another empire. Fire Management Notes. 45(4): 12–17.

Davis, J.D. 1977. Southern California reservoir sedimentation. Preprint. American Society of Civil Engineers Fall Convention and Exhibit; San Francisco, CA.

De Las Haras, J.; Herranz, J.M.; Martinez, J.J. 1993. Influence of bryophyte pioneer communities on edaphic changes in soils of Mediterranean ecosystems damaged by fire (SE Spain). In: Trabaud, L.; Prodon, P. (eds.). Fire in Mediterranean ecosystems. Ecosystem Res. Rep. 5. Brussels, Belgium: Commission of the European Countries: 387–393.

Dean, AE. 2001. Evaluating effectiveness of watershed conservation treatments applied after the Cerro Grande Fire, Los Alamos, New Mexico. Tucson: University of Arizona. 116 p. Thesis.

DeBano, L.F. 1969. Observations on water-repellent soils in Western United States. In: DeBano, L.F.; Letey, J. (eds.). Proceedings of a symposium on water repellent soils. 1968 May 6–10; Riverside, CA Davis: University of California: 17–29.

DeBano, L.F. 1974. Chaparral soils. In: Proceedings of a symposium on living with the chaparral. San Francisco, CA: Sierra Club: 19–26.

DeBano, L.F. 1981. Water repellent soils: a state-of-the-art. Gen. Tech. Rep. PSW-46. Berkeley, CA: U.S. Department of Agriculture, Forest Service, Pacific Southwest Forest and Range Experiment Station. 21 p.

DeBano, L.F. 1990. Effects of fire on soil resource in Arizona chaparral. In: Krammes, J.S. (tech. coord.). Effects of fire management of southwestern natural resources. Gen. Tech. Rep. RM-191. Fort Collins, CO: U.S. Department of Agriculture, Forest Service, Rocky Mountain Forest and Range Experiment Station: 65–77.

DeBano, L.F. 1991. The effect of fire on soil. In: Harvey. A. E.; Neuenschwander, L.F.(eds.). Management and productivity of western-montane forest soils. Gen. Tech. Rep. INT-280. Ogden, UT: U.S. Department of Agriculture, Forest Service, Intermountain Forest and Range Experiment Station: 32–50.

DeBano, L.F. 2000a. Water repellency in soils: a historical overview. Journal of Hydrology. 231-232: 4–32.

DeBano, L.F. 2000b. The role of fire and soil heating on water repellency in wildland environments: a review. Journal of Hydrology. 231-232: 195–206.

DeBano, L.F.; Conrad, C.E. 1976. Nutrients lost in debris and runoff water from a burned chaparral watershed. In: Proceedings of the third Federal interagency sedimentation conference;1976 March; Denver, CO. Washington, DC: Water Resources Council: 3: 13–27

DeBano, L.F.; Conrad, C.E. 1978. Effects of fire on nutrients in a chaparral ecosystem. Ecology. 59: 489–497.

DeBano, L.F.; Klopatek, J.M. 1988. Phosphorus dynamics of pinyon-juniper soils following simulated burning. Soil Science Society of American Journal. 52: 271–277.

DeBano, L.F.; Krammes, J.S. 1966. Water repellent soils and their relation to wildfire temperatures. Bulletin of the International Association of Scientific Hydrology. 11(2): 14–19.

DeBano, L. F.; Neary, D.G. 1996. Effects of fire on riparian systems. In: Ffolliott, P. F.; DeBano, L.F.; Baker, M.B., Jr., Gottfried, G.J.; Solis-Garza, G.; Edminster, C.B.; Neary, D.G.; Allen, L.S.; Hamre, R.H. (tech. coords.) Effects of fire on Madrean Province ecosystems: a symposium proceedings. Gen. Tech. Rep. RM-GTR-289. Fort Collins, CO: U.S. Department of Agriculture, Forest Service, Rocky Mountain Forest and Range Experiment Station: 69–76.

DeBano, L.F.; Eberlein, G.E.; Dunn, P.H. 1979. Effects of burning on chaparral soils: I. Soil nitrogen. Soil Science Society of American Journal. 43: 504–509.

DeBano, L.F.; Ffolliott, P.F.; Baker, M.B., Jr. 1996. Fire severity effects on water resources. In: Ffolliott, P.F.; DeBano, L.F.; Baker, M.B., Jr.; Gottfried, G.J.; Solis-Garza, G.; Edminster, C.B.; Allen, L.S.; Hamre, R.H. (tech. coords.). Effects of fire on Madrean Province ecosystems: a symposium proceedings. Gen. Tech. Rep. RM-GTR-289. Fort Collins, CO: U.S. Department of Agriculture, Forest Service, Rocky Mountain Forest and Range Experiment Station: 77–84.

DeBano, L.F.; Neary, D.G.; Ffolliott, P.F. 1998. Fire's effects on ecosystems. New York: John Wiley & Sons, Inc. 333 p.

DeBano, L.F.; Rice, R.M.; Conrad, C.E. 1979. Soil heating in chaparral fires: effects on soil properties, plant nutrients, erosion and runoff. Res. Pap. PSW-145. Berkeley, CA: U.S. Department of Agriculture, Forest Service, Pacific Southwest Forest and Range Experiment Station. 21 p.

DeBano, L.F.; Savage, S.M.; Hamilton, D.A. 1976. The transfer of heat and hydrophobic substances during burning. Soil Science Society of America Journal. 40: 779–782.

DeBell, D.S.; Ralston, C.W. 1970. Release of nitrogen by burning light forest fuels. Soil Science Society of America Proceedings. 34: 936–938.

DeByle, N.V. 1970a. Do contour trenches reduce wet-mantle flood peaks? Res. Note INT-108. Ogden, UT: U.S. Department of Agriculture, Forest Service, Intermountain Forest and Range Experiment Station. 8 p.

DeByle, N.V. 1970b. Infiltration in contour trenches in the Sierra Nevada. Res. Note INT-115. Ogden, UT: U.S.Department of Agriculture, Forest Service, Intermountain Forest and Range Experiment Station. 5 p.

DeByle, N.V. 1981. Clearcutting and fire in the larch/Douglas-fir forests of western Montana—a multifaceted research summary. Gen. Tech. Rep. INT-99. Ogden, UT: U.S. Department of Agriculture, Forest Service, Intermountain Forest and Range Experiment Station. 73 p.

DeByle, N.V.; Packer, P.E. 1972. Plant nutrient and soil losses in overland flow from burned forest clearcuts. In: Watersheds in transition symposium; 1972 June; Fort Collins, CO American Water Resources Association Proceedings Series. 14: 296–307.

Deeming, J.E.; Burgan, R.E.; Cohen, J.D. 1977. The national fire danger rating system, 1978. Gen. Tech. Rep. INT-39. Ogden, UT: U.S.Department of Agriculture, Forest Service, Intermountain Forest and Range Experiment Station. 63 p.

DeGraff, J.V. 1982. Final evaluation of felled trees as a sediment retaining measure, Rock Creek Fire, Kings RD, Sierra National Forest, Fresno, CA: Special Report. Fresno, CA: U.S. Department of Agriculture, Forest Service, Sierra National Forest. 15 p.

Deka, H.K.; Mishra, R.R. 1983. The effect of slash burning on soil microflora. Plant and Soil. 73: 167–175.

Deka, H.K.; Mishra, R.R.; Sharma, G.D. 1990. Effect of fuel burning on VA mycorrhizal fungi and their influence on the growth of early plant colonizing species. Acta Botanica Indica. 18: 184–189.

Diaz-Ravina, M.; Prieto, A.; Baath, E. 1996. Bacterial activity in a forest soil after soil heating and organic amendments measured by the thymidine and leucine incorporation techniques. Soil Biology and Biochemistry. 28: 419–426.

Dickinson, M.B.; Johnson, E.A. 2001. Fire effects on trees. In: Johnson, E.A.; Miyanishi, K. (eds.). Forest fires, behavior and ecological effects. San Francisco, CA: Academic Press: 477–525.

Dimitrakopoulos, A.P.; Martin, R.E.; Papamichos, N.T. 1994. A simulation model of soil heating during wildland fires. In: Sala, M.; Rubio, J.L. (eds.). Soil erosion as a consequence of forest fires. Logrono, Spain: Geoforma Ediciones: 207–216.

Doerr, S.H.; Shakesby, R.A.; Walsh, R.P.D. 2000. Soil water repellency: its causes, characteristics and hydro-geomorphological significance. Earth Science Reviews. 51(1-4): 33–65.

Dortignac, E.J. 1956. Wateshed resources and problems of the upper Rio Grande Basin. Special Station Paper. Fort Collins, CO: U.S. Department of Agriculture, Forest Service, Rocky Mountain Forest and Range Experiment Station. 107 p.

Doty, R.D. 1970. Influence of contour trenching on snow accumulation. Journal of Soil and Water Conservation. 25(3): 102–104.

Doty, R.D. 1971. Contour trenching effects on streamflow from a Utah watershed. Res. Pap. INT-95. Ogden, UT: U.S.Department of Agriculture, Forest Service, Intermountain Forest and Range Experiment Station. 19 p.

Doty, R.D. 1972. Soil water distribution on a contour-trenched area. Res. Note INT-163. Ogden, UT: U.S. Department of Agriculture, Forest Service, Intermountain Forest and Range Experiment Station. 6 p.

Douglass, J.E.; Godwin, R.C. 1980. Runoff and soil erosion from site preparation practices. In: U.S. forestry and water quality: what course in the 80's. Richmond, VA. Washington, DC: Water Pollution Control Federation: 51–73.

Douglass, J.E.; Van Lear, D.H. 1983. Prescribed burning and water quality of ephemeral streams in the Piedmont of South Carolina. Forest Science. 29(1): 181–189.

Duchesne, L.C.; Hawkes, B.C. 2000. Fire in northern ecosystems. In: Brown, J.K.; Smith, J.K. (eds.). Wildland fire in ecosystems—Effects of fire on flora. Gen. Tech. Rep. RMRS-GTR-42-vol. 2. Ogden, UT: U.S. Department of Agriculture, Forest Service, Rocky Mountain Research Station: 35–52.

Dumontet, S.; Dinel, H.; Scopa, A.; Mazzatura, A.; Saracino, A. 1996. Post-fire soil microbial biomass and nutrient content of a pine forest soil from a dunal Mediterranean environment. Soil Biology and Biochemistry. 28: 1467–1475.

Dunn, P.F.; DeBano, L.F. 1977. Fire's effect on biological and chemical properties of chaparral soils. In: Mooney, H.A.; Conrad; C.E. (tech. coords.). Proceedings of the symposium on the environmental consequences of fire and fuel management in Mediterranean ecosystems. Gen. Tech. Rep. WO-3: Washington, DC: U.S. Department of Agriculture, Forest Service: 75–84.

Dunn, P.H.; Barro, S.C.; Poth, M. 1985. Soil moisture affects survival of microorganisms in heated chaparral soil. Soil Biology and Biochemistry. 17: 143–148.

Dunn, P.H.; Barro, S.C.; Wells, W.G., III.; Poth, A.; Wohlgemuth, P.M.; Colver, C.G. 1988. The San Dimas Experimental Forest: 50 years of research. Gen. Tech. Rep. PSW-104. Berkeley, CA: U.S. Department of Agriculture, Forest Service, Pacific Southwest Forest and Range Experiment Station. 49 p.

Dunn, P.H.; DeBano, L.F.; Eberlein, G.E. 1979. Effects of burning on chaparral soils: II. Soil microbes and nitrogen mineralization. Soil Science Society of America Journal. 43: 509–514.

Dunne, T.; Leopold, L.B. 1978. Water in environmental planning. San Francisco, CA: W.H. Freeman and Company. 818 p.

Dyrness, C.T.; Norum, R.A. 1983. The effects of experimental fires on black spruce forest floors in interior Alaska. Canadian Journal of Forest Research. 13: 879–893.

Dyrness, C.T.; Van Cleve, K.; Levison, J.D. 1989. The effect of wildfire on soil chemistry in four forest types in interior Alaska. Canadan Jounal of Forest Research. 19: 1389–1396.

Edwards, L.; Burney, J.; DeHaan, R. 1995. Researching the effects of mulching on cool-period soil erosion in Prince Edward Island. Canadian Journal of Soil and Water Conservation. 50: 184–187.

Eivazi, F.; Bayan, M.R. 1996. Effects of long-term prescribed burning on the activity of select soil enzymes in an oak-hickory forest. Canadian Journal of Forest Research. 26: 1799–1804.

Elliot, W.J.; Hall, D.E. 1997. Water Erosion Prediction Project (WEPP) forest applications. Gen. Tech. Rep. INT-GTR-365. Ogden, UT: U.S. Department of Agriculture, Forest Service, Intermountain Research Station. 11 p.

Elliot, W.J.; Graves, S.M.; Hall, D.E.; Moll, J.E. 1998. The X-DRAIN cross drain spacing and sediment yields. No.9877 1801-SDTDC. San Dimas, CA: U.S. Department of Agriculture, Forest Service, San Dimas Technology Center. 24 p.

Elliot, W.J.; Hall, D.E.; Scheele, D.L. 1999. WEPP-Road: WEPP interface for predicting forest road runoff, erosion and sediment delivery. [Online] Available: http://forest.moscowfsl.wsu.edu/fswep/docs/wepproaddoc.html.

Elliot, W.J.; Scheele, D.L.; Hall, D.E. 2000. The Forest Service WEPP interfaces. American Society of Agricultural Engineers Summer Meeting, 2000. Paper No. 005021. St. Joseph, MI: American Society of Agricultural Engineers. 9 p.

Esplin, D.H.; Shackleford, J.R. 1978. Cachuma burn reseeding evaluation: coated vs. uncoated seed, first growing season. Unpublished report. Goleta, CA: U.S. Department of Agriculture, Forest Service, Los Padres National Forest. 24 p.

Esplin, D.H.; Shackleford, J.R. 1980. Cachuma burn revisited. Unpublished report on file at: U.S. Department of Agriculture, Forest Service, Los Padres National Forest, San Diego, CA 5 p.

Evans, R.D.; Johansen, J.R. 1999. Microbiotic crusts and ecosystem processes. Critical Reviews in Plant Sciences. 18: 183–225.

Everett, R.L.; Java-Sharpe, B.J.; Scherer, G.R.; Wilt, F.M.; Ottmar, R.D. 1995. Co-occurrence of hydrophobicity and allelopathy in sand pits under burned slash. Soil Science Society of America Journal. 59: 1176–1183.

Ewel, K.C. 1990. Swamps. In: Myers, R.L.; Ewel, J.J. (eds.). Ecosystems of Florida. Orlando: University of Central Florida Press: 281–323.

Farmer, E.E.; Fletcher, J.E. 1972. Some intra-storm characteristics of high-intensity rainfall burst. In: Distribution of precipitation in mountainous areas; proceedings, Geilo Symposium, Norway; 1972 July 31–August 5. Geneva, Switzerland: World Meteorological Organization. 2: 525–531.

Faulkner, S.P.; de la Cruz, A.A. 1982. Nutrient mobilization following winter fires in an irregularly flooded marsh. Journal of Environmental Quality. 11: 129–133

Faust, R. 1998. Lesson plan: Fork fire soil loss validation monitoring. Unit VIII, Long-term recovery and monitoring. In: Burned area emergency rehabilitation (BAER) techniques. San Francisco: U.S. Department of Agriculture, Forest Service, Pacific Southwest Region.

Feller, M.C. 1998. The influence of fire severity, not fire intensity, on understory vegetation biomass in British Columbia. In: Proceedings, 13th conference on fire and forest meteorology; 1996 October 27–31; Lorne, Australia. International Journal of Wildland Fire: 335-348.

Feller, M.C.; Kimmins, J.P. 1984. Effects of clearcutting and slash burning on streamwater chemistry and watershed nutrient budgets in southwestern British Columbia. Water Resources Research. 20: 29–40.

Ferguson, E.R. 1957. Prescribed burning in shortleaf-loblolly pine on rolling uplands in east Texas. Fire Control Notes 18. Washington, DC: U.S. Department of Agriculture, Forest Service:130-132.

Fernandez, I.; Cabaneiro, A.; Carballas, T. 1997. Organic matter changes immediately after a wildfire in an Atlantic forest soil and comparison with laboratory soil heating. Soil Biology and Biochemistry. 29: 1–11.

Ferrell, W.R. 1959. Report on debris reduction studies for mountain watersheds. Los Angeles, CA: Los Angeles County Flood Control District, Dams and Conservation Branch. 164 p.

Ffolliott, P.F.; Baker, M.B. Jr. 2000. Snowpack hydrology in the southwestern United States: contributions to watershed management. In: Ffolliott, P.F., M.B. Baker, Jr., C.B. Edminster; Dillion, M.C.; Mora, K.L. (tech. coords.). Land stewardship in the 21st century: the contributions of watershed management. Proc. RMRS-P-13. Fort Collins, CO: U.S. Department of Agriculture, Forest Service, Rocky Mountain Forest and Range Experiment Station: 274–276.

Ffolliott, P.F.; Brooks, M.B., Jr. 1996. Process studies in forestry hydrology: a worldwide review. In: V.P. Singh; Kumer, B. (eds.). Surface-water hydrology. The Netherlands: Kluwer Academic Publishers: 1–18.

Ffolliott, P.F.; Neary, D.G. 2003. Impacts of a historical wildfire on hydrologic processes: a case study in Arizona. Proceedings of the American Water Resources Association international congress on watershed management for water supply systems; 2003 June 29–July 2. New York, NY. Middleburg, VA: American Water Resources Association. 10 p.

Ffolliott, P.F.; Thorud, D.B. 1977. Water yield improvement by vegetation management. Water Resources Bulletin. 13: 563–571.

Ffolliott, P.F.; Arriaga, L.; Mercado Guido, C. 1996. Use of fire in the future: benefits, concerns, constraints. In: Ffolliott, P.F.; DeBano, L.F.; Baker, M.B., Jr.; Gottfried, G.J.; Solis-Garza, G.; Edminster, C.B.; Neary, D.G.; Allen, L.S.; Hamre, R.H. (tech. coords.). Effects of fire on Madrean Providence ecosystems: a symposium proceedings. Gen. Tech. Rep. RM-GTR-289. Fort Colllins, CO: U.S. Department of Agriculture, Forest Service, Rocky Mountain Forest and Range Experiment Station: 217–222.

Ffolliott, P.F.; Gottfried, G.J.; Baker, M.B., Jr. 1989. Water yield from forest snowpack management: Research findings in Arizona and New Mexico. Water Resources Research 25: 1999–2007.

Finney, M.A. 1998. FARSITE: Fire area simulator—model development and evaluation. Res. Pap. RMRS-RP-4. Ogden UT: US Department of Agriculture, Forest Service, Rocky Mountain Research Station. 47p.

Fisher, S.G.; Minckley, W.L. 1978. Chemical characteristics of a desert stream in flash flood. Journal of Arid Environments. 1: 25–33.

Flanagan, D.C.; Livingston, S.J (eds.). 1995. WEPP user summary. NSERL Report No. 11, W. Lafayette, IN: Natural Resources Conservation Service, National Soil Erosion Research Laboratory. 131 p.

Flanagan, D.C.; Whittemore, D.A.; Livingston, S.J.; Ascough, J.C., II; Savabi, M. 1994. Interface for the water erosion prediction project model. Symposium, American Society of Agricultural Engineers; 1994 June 20–23; Kansas City, MO; St. Joseph, MI: American Society of Agricultural Engineers. 16 p.

Flinn, M.A.; Wein, R.W. 1977. Depth of underground plant organs and theoretical survival during fire. Canadian Journal of Botany. 55: 2550–2554.

218

USDA Forest Service Gen. Tech. Rep. RMRS-GTR-42-vol. 4. 2005

Foster D.R. 1983. The history and pattern of fire in the boreal forest of southeastern Labrador. Canadian Journal of Botany. 61: 2459–2471.

Fowler, W.B.; Helvey, J.D. 1978. Changes in the thermal regimes after prescribed burning and selected tree removal (Grass Camp 1975). Res. Pap. PNW-235. Portland, OR: U.S. Department of Agriculture, Forest Service, Pacific Northwest Forest and Range Experiment Station. 17 p.

Fox, T.R.; Burger, J.A.; Kreh, R.E. 1983. Impact of site preparation on nutrient dynamics and stream water quality on a Piedmont site. American Society of Agronomy Abstracts. 207 p.

Frandsen, W.H. 1987. The influence of moisture and mineral soil on the combustion of smoldering forest duff. Canadian Journal of Forest Research. 17: 1540–1544.

Frandsen, W.H. 1991a. Burning rate of smoldering peat. Northwest Science. 64(4): 166–172.

Frandsen, W.H. 1991b. Heat evolved from smoldering peat. International Journal of Wildland Fire. 1: 197–204.

Frandsen, W.H. 1997. Ignition probability of organic soils. Canadian Journal of Forest Restoration. 27: 1471–1477.

Frandsen, W.H.; Ryan, K.C. 1986. Soil moisture reduces belowground heat flux and soil temperature under a burning fuel pile. Canadian Journal of Forest Research. 16: 244–248.

Fredriksen, R.L. 1971. Comparative chemical water quality—natural and disturbed streams following logging and slash burning. In: Proceedings of a symposium on forest land uses and stream environment; 1970 October 19–21; Corvalis, OR, Corvalis: Oregon State University, Continuing Eduction Publications: 125–137.

Fredricksen, R.L.; Moore, D.G.; Norris, L.A. 1975. Impact of timber harvest, fertilization, and herbicide treatment on stream water quality in the Douglas-fir regions. In: Bernier, B.; Winget, C.H. (eds.). Forest soils and forest land management. Proceedings of the 4th North American forest soils conference, Laval University, Quebec City, Canada: 283–313.

Fritze, H.; Pennanen, T.; Pietikainen, J. 1993. Recovery of soil microbial biomass and activity from prescribed burning. Canadian Journal of Forest Research. 23: 1286–1290.

Fritze, H.; Smolander, A.; Levula, T.; Kitunene, V.; Malkonen, E. 1994. Wood-ash fertilization and fire treatments in a scots pine forest stand: effects on the organic layer, microbial biomass, and microbial activity. Biology and Fertility of Soils. 17: 57–63.

Fritze, H; Pennanen, T.; Kitunen, V. 1998. Characterization of dissolved organic carbon from burned humus and its effects on microbial activity and community structure. Soil Biology and Biochemistry. 30: 687–693.

Frost, C. 1995. Presettlement fire regimes in southeastern marshes, peatlands and swamps. Proceedings, Tall Timbers fire ecology conference; 1993 November 3-6. Tallahassee, FL: Tall Timbers Research Station. 19: 39–60.

Frost, C.C. 1998. Presettlement fire frequency regimes of the United States: a first approximation. In: Pruden, T.L.; Brennan, L.(eds.). Fire in ecosystem management: shifting paradigm from suppression to prescription. Proceedings, Tall Timbers fire ecology conference; 1996 May 7–10. Tallahassee, FL: Tall Timbers Research Station: 20:70–81.

Fuhrer, E. 1981. Interception measurements in beech stands. Ereszeti Kutatasek. 74: 125–137.

Furniss, M.J.; Ledwith, T.S.; Love, M.A.; McFadin, B.C.; Flanagan, S.A. 1998. Response of road-stream crossings to large flood events in Washington, Oregon, and Northern California. 9877 1806-SDTDC. San Dimas, CA: U.S. Department of Agriculture, Forest Service, San Dimas Technology Development Center. 14 p.

Fyles, J.W.; Fyles, I.H.; Beese, W.J.; Feller, M.C. 1991. Forest floor characteristics and soil nitrogen availability on slash-burned sites in coastal British Columbia. Canadian Journal of Forest Research. 21: 1516–1522.

Gartner, J.E.; Bigio, E.R.; Cannon, S.H. 2004. Compilation of post wildfire runoff-event data from the Western United States. Open-File Report 04-1085. Denver, CO: U.S. Department of the Interior, U.S. Geological Survey. 22 p.

General Accounting Office. 2003. Wildland fires: better information needed on effectiveness of emergency stabilization and rehabilitation treatments. GAO-03-430. Washington, DC: United States General Accounting Office. 55 p. plus appendices.

Georgia Forestry Association. 1995. Best management practices for forested wetlands in Georgia. Atlanta: Georgia Forestry Association Wetlands Committee. 26 p.

Giardina, C.P.; Sanford, R.L.; Dockersmith, I.C. 2000. Changes in soil phosphorus and nitrogen during slash-and-burn clearing of a dry tropical forest. Soil Science Society of America Journal. 64: 399–405.

Gimenez, A.; Pastor, E.; Zarate, L.; Planas, E.; Arnaldos, J. 2004. Long-term forest fire retardants: a review of quality, effectiveness, application and environmental considerations. Intenational Journal of Wildland Fire. 13(1): 1–15.

Gibbons, D.R.; Salo, E.O. 1973. An annotated bibliography of the effects of logging on fish of the Western United States and Canada. Portland, OR: U.S. Departement of Agriculture, Forest Service, Pacific Northwest Forest and Range Experiment Station. 145 p.

Gifford, G.F. 1973. Loss of particulate organic materials from semiarid watersheds as a result of extreme hydrologic events. Water Resources Research. 9: 1443–1449.

Gifford, G.F.; Buckhouse, J.C.; Busby, F.E. 1976. Hydrologic impact of burning and grazing on a chained pinyon-juniper site in southeastern Utah. Publ. PRJNR 012-1. Ogden: Utah Water Resources Laboratory. 22 p.

Gillon, D.; Gomendy, V.; Houssard, C.; Marechal, J.; Valette, J.C. 1995. Combustion and nutrient losses during laboratory burns. International Journal of Wildland Fire. 5: 1–12.

Giovannini, G.; Lucchesi, S. 1983. Effect of fire on hydrophobic and cementing substances of soil aggregates. Soil Science. 136: 231–236.

Giovannini, G.; Lucchesi, S.; Cervelli, S. 1983. Water-repellent substances and aggregate stability in hydrophobic soil. Soil Science. 135: 110–113.

Giovannini, G.; Lucchesi, S., Giachetti, M. 1987. The naural evolution of a burned soil: a three year investigation. Soil Science. 143: 220–226.

Glaser, P.H.; Wheeler, G.A.; Gorham E.; Wright, H.E. 1981. The patterned mires of the Red Lake peatland, Northern Minnesota: vegetation, water chemistry and landforms. Journal of Ecology. 69: 575–599.

Gleason, C.H. 1947. Guide for mustard sowing in burned watersheds of southern California. San Francisco, CA: U.S.Department of Agriculture, Forest Service, California Region. 46 p.

Glendening, G.E.; Pase, C.P.; Ingebo, P. 1961. Preliminary hydrologic effects of wildfire in chaparral. In: Proceedings, 5th annual Arizona Watershed Symposium; 1961 September 21; Phoenix, AZ. Tucson: University of Arizona: 12–15.

Goldman, S.J.; Jackson, K.; Bursztynsky, T.A. 1986. Erosion and sediment control handbook. San Francisco, CA: McGraw-Hill. 360 p.

Gosz, J.R.; White, C.S.; Ffolliott, P.F. 1980. Nutrient and heavy metal transport capacities of sediment in the southwestern United States. Water Resources Bulletin. 16: 927–933.

Gottfried, G.J. 1991. Moderate timber harvesting increases water yields from an Arizona mixed conifer watershed. Water Resources Bulletin. 27: 537–547.

Gottfried, G.J.; DeBano, L.F. 1990. Streamflow and water quality responses to preharvest prescribed burning in an undisturbed ponderosa pine watershed. In: Krammes, J.S. (tech. cord.). Effects of fire management of southwestern natural resources. Gen. Tech. Rep. RM-191. Fort Collins, CO: U.S. Department of Agriculture, Forest Service, Rocky Mountain Forest and Range Experiment Station: 222–228.

Gottfried, G.J.; Neary, D.G.; Baker, M.B., Jr.; Ffolliott, P.F. 2003. Impacts of wildfires on hydrologic processes in forested ecosystems: two case studies. In: Renard, K.G.; McElroy, S.A.; Gburek, W.J.; Canfield, H.E.; Scott, R.L. (eds.). First interagency conference on research in the watersheds; 2003 October 27–30. Washington, DC: U.S. Department of Agriculture, Agricultural Research Service: 668–673.

Graham, R.T.; Harvey, A.E.; Jurgensen, M.F.; Jain, T.B.; Tonn, J.R.; Page-Dumroese, D.S. 1994. Managing coarse woody debris in forests of the Rocky Mountains. Res. Pap. INT-RP-477. Ogden, UT: U.S. Department of Agriculture, Forest Service, Intermountain Forest and Range Experiment Station. 12 p.

USDA Forest Service Gen. Tech. Rep. RMRS-GTR-42-vol. 4. 2005

219

Greene, R.S.B.; Chartres, C.J.; Hodgkinson, K.C. 1990. The effects of fire on the soil in a degraded semi-arid woodland: I. Cryptogam cover and physical and micromorphological properties. Australian Journal of Soil Research. 28: 755–777.

Gresswell, R.E. 1999. Fire and aquatic ecosystems in forested biomes of Northern America. Transactions of the American Fisheries Society. 128: 193–221.

Grier, C.C. 1975. Wildfire effects on nutrient distribution and leaching in a coniferous ecosystem. Canadian Journal of Forestry Research.5: 599–607.

Griffith, R.W. 1989. Memo, silt fence monitoring. Stanislaus National Forest. Unpublished report on file at: U.S.Department of Agriculture, Forest Service, Stanislaus National Forest, Placerville, CA. 4 p.

Groeschl, D.A.; Johnson, J.E.; Smith, D.W. 1990. Forest soil characteristics following wildfire in the Shenandoah National Park, Virginia. In: Nodvin, Stephen C.; Waldrop, Thomas A. (eds.). Fire and the environment: ecological and cultural perspective: proceedings of an international symposium; 1990 Marcy 20–24; Knoxville, TN. Gen. Tech. Rep. SE-69. Asheville, NC: U.S. Department of Agriculture, Forest Service.Southeastern Forest Experiment Station: 129–137.

Groeschl, D.A.; Johnson, J.E.; Smith D.W. 1992. Early vegetative response to wildfire in a table mountain-pitch pine forest. International Journal of Wildland Fire. 2: 177–184.

Groeschl, D.A.; Johnson, J.E.; Smith, D.W. 1993. Wildfire effects on forest floor and surface soil in a table mountain pine-pitch pine forest. International Journal of Wildland Fire. 3: 149–154.

Grove, T.S.; O'Connell, A.M.; Dimmock, G.M. 1986. Nutrient changes in surface soils after an intense fire in jarrah (Eucalyptus marginata Donn ex Sm.) forest. Australian Journal of Ecology. 11: 303–317.

Hall, J.D.; Brown, G.W.; Lantz, R.L. 1987. The Alsea watershed study: a retrospective. In: Salo, E.O.; Cundy, T.W. (eds.). Streamside management: Forestry and fishery interactions. Contr. No. 57. Seattle: University of Washington. 399–416.

Hardy, C.C.; Menakis, J.P.; Long, D.G.; Brown, J.K. 1998. Mapping historic fire regimes for the Western United States: Integrating remote sensing and biophysical data. In: Greer, J.D. (ed.). Proceedings of the 7th Forest Service Remote Sensing Applications Conference; 1998 April 6-10;. Nassau Bay, TX. Bethesda, MD: American Society for Photogrammetry and Remote Sensing: 288–300.

Hardy, C.C.; Schmidt, K.M.; Menakis, J.P.; Sampson, R.N. 2001. Spatial data for national fire planning and fuel management. International Journal of Wildland Fire. 10(3 and 4): 353–372.

Hare, R.C. 1961. Heat effects on living plants. Occasional Paper 183. New Orleans, LA: U.S. Department of Agriculture, Forest Service, Southern Forest Experiment Station. 32 p.

Harr, R.D. 1976. Forest practices and streamflow in western Oregon. Gen. Tech. Rep. PNW-49. Portland, OR: U.S. Department of Agriculture, Forest Service, Pacific Northwest Forest and Range Experiment Station. 18 p.

Hartford, R.A. 1989. Smoldering combustion limits in peat as influenced by moisture, mineral content, and organic bulk density. In: MacIver, D.C.; Auld, H.; Whitewood, R. (eds.). Proceedings,10th conference on fire and forest meteorology; 1989 April 17–21; Ottawa, Ontario. Chalk River, Ontario: Forestry Canada, Petawawa National Forestry Institute: 282–286.

Hartford, R.A.; Frandsen, W.H. 1992. When it's hot, it's hot ... or maybe it's not (surface flaming may not portend extensive soil heating). International Journal of Wildland Fire. 2: 139–144.

Harvey, A.E. 1994. Integrated roles for insects, diseases and decomposers in fire dominated forests of the Inland Western United States: past, present and future forest health. Journal of Sustainable Forestry. 2: 211–220.

Harvey, A.E.; Jurgensen, M.F.; Graham, R.T. 1989. Fire-soil interactions governing site productivity in the northern Rocky Mountains. In: Baumgartner, D.M.; Bruer, D.W.; Zamora, B.A.; Neuenschwander, L.F.; Wakinoto, R.H. (comps. and eds.). Prescribed fire in the Intermountain region: forest site preparation and range improvements. Symposium proceedings. Pullman: Washington State University, Cooperative Extension Service: 9-18.

Harvey, A.E.; Jurgensen, M.F.; Larsen, M.J. 1980a. Clearcut harvesting and ectomycorrhizae: survival of activity on residual roots and influence on a bordering forest stand in western Montana. Canadian Journal of Forest Research. 10: 300–303.

Harvey, A.E.; Jurgensen, M.F.; Larsen, M.J. 1981. Organic reserves: importance to ectomycorrhizae in forest soils of western Montana. Forest Science. 27: 442–445.

Harvey, A.E.; Larsen, M.J.; Jurgensen, M.F. 1976. Distribution of ectomycorrhizae in a mature Douglas-fir/larch forest soil in western Montana. Forest Science. 22: 393–398.

Harvey, A.E.; Larsen, M.J.; Jurgensen, M.F. 1980b. Partial cut harvesting and ectomycorrhizae: early effects in Douglas-fir-larch forests of western Montana. Canadian Journal of Forest Research. 10: 436–440.

Hatch, A.B. 1960. Ash-bed effects in western Australian forest soils. Bulletin 64. Forestry Department Western Australia: 1–19.

Hauer, F.R.; Spencer, C.N. 1998. Phosphorus and nitrogen dynamics in streams associated with wildfire: a study of immediate and longterm effects. International Journal of Wildland Fire. 8(4): 183–198.

Hauser and Spence 1998 [CHAPTER 6 Table 6.2] not in reference list. Probably Hauer and Spencer 1998???

Haworth, K.; McPherson, G.R. 1991. Effect of Quercus emporyi on precipitation distribution. Journal of the Arizona-Nevada Academy of Science; thirty-fifth annual meeting; 1991 April 20; Flagstaff, AZ. Proceedings Supplement. 26: 21.

Haynes, R.J. 1986. Origin, distribution, and cycling of nitrogen in terrestrial ecosystems. In: Hayes, R.J. (ed.). Mineral in the plant-soils systems. New York: Academic Press: 1–51.

Heede, B.H. 1960. A study of early gully-control structures in the Colorado Front Range. Station Paper 55. Fort Collins, CO: U.S. Department of Agriculture, Forest Service, Rocky Mountain Forest and Range Experiment Station. 42 p.

Heede, B.H. 1970. Design, construction and cost of rock check dams. Res. Pap. RM-20. Fort Collins, CO: U.S. Department of Agriculture, Forest Service, Rocky Mountain Forest and Range Experiment Station. 24 p.

Heede, B.H. 1976. Gully development and control: the status of our knowledge. Res. Pap. RM-169. Fort Collins, CO: U.S. Department of Agriculture, Forest Service, Rocky Mountain Forest and Range Experiment Station. 42 p.

Heede, B.H. 1981. Rehabilitation of disturbed watershed through vegetation treatment and physical structures. In: Proceedings, Interior West watershed management symposium; 1980 April; Spokane, WA. Pullman: Washington State University, Cooperative Extension: 257–268.

Heede, B.H.; Harvey, M.D.; Laird, J.G. 1988. Sediment delivery linkages in a chaparrel watershed following a fire. Environmental Management. 12: 349–358.

Heilman, P.E.; Gessel, S.P. 1963. Nitrogen requirements and the biological cycling of nitrogen in Douglas-fir stands in relation to effects of nitrogen fertilization. Plant and Soil. 18: 386–402.

Heinselman, M.L. 1978. Fire in wilderness ecosystems. In: Hendee, J.C.; Stankey, G.H.; Lucas, R.C. Wilderness management. Misc. Pub. No. 1365. Washington, DC: U.S. Department of Agriculture, Forest Service: 249–278.

Heinselman, M.L. 1981. Fire intensity and frequency as factors in the distribution and structure of northern ecosystems. In: Mooney, H.A., et al. (coords.). Fire regimes and ecosystem properties. Gen. Tech. Rep. WO-26. Washington, DC: U.S. Department of Agriculture, Forest Service: 7–57.

Heinselman, M.L. 1983. Fire and succession in the conifer forests of northern North America. In: West, D.C.; Shugart, H.H.; Botkin, D.B. (eds.). Forest succession, concepts and application. New York: Springer Verlag: 374–405.

Helvey, J.D. 1971. A summary of rainfall interception by certain conifers of North America. In; Monike, E.J. (ed.) Proceedings of the third international seminar for hydrology professors: Biological effects in the hydrological cycle. West Lafayette, IN: Purdue University: 103–113.

Helvey, J.D. 1973. Watershed behavior after forest fire in Washington. In: Agriculture and urban considerations in irrigation and drainage;. proceedings of the irrigation and drainage specialty conference. New York: American Society of Civil Engineers: 403–422.

220

USDA Forest Service Gen. Tech. Rep. RMRS-GTR-42-vol. 4. 2005

Helvey, J.D. 1980. Effects of a north-central Washington wildfire on runoff and sediment production. Water Resources Bulletin. 16: 627–634.

Helvey, J.D.; Patric, J.H. 1965. Canopy and litter interception by hardwoods of eastern United States. Water Resource Research. 1: 193–206.

Helvey, J.D.; Tiedemann, A.R.; Anderson, T.D. 1985. Nutrient loss by soil erosion and mass movement after wildfire. Journal of Soil and Water Conservation. 40: 168–173.

Helvey, J.D.; Tiedemann, A.R.; Fowler, W.B. 1976. Some climatic and hydrologic effects of wildfire in Washington State. Proceedings of the Tall Timber fire ecology conference. Tallahassee, FL: Tall Timbers Research Station: 15: 201–222.

Hendricks, B.A.; Johnson, J.M. 1944. Effects of fire on steep mountain slopes in central Arizona. Journal of Forestry. 42: 568–571.

Hendrickson, D. A.; Minkley, W.L. 1984. Cienegas—vanishing climax communities of the American Southwest. Desert Plants. 6: 131–175.

Hermann, S.M.; Phernetton, R.A.; Carter, A.; Gooch, T. 1991. Fire and vegetation in peatbased marshes of the coastal plain: examples from the Okefenokee and Great dismal Swamps. In: High intensity fire in wildlands: management challenges and options; proceedings of the Tall Timbers fire ecology conference. Tallahassee, FL: Tall Timbers Research Station: 217–234.

Hernandez, T.; Garcia, C.; Reinhardt, I. 1997. Short-term effect of wildfire on the chemical, biochemical, and microbiological properties of Mediterranean pine forest soils. Biology and Fertility of Soils. 25: 109–116.

Herr, D.G.; Duchesne, L.C.; Tellier, R.; McAlpine, R.S.; Peterson, R.L. 1994. Effect of prescribed burning on the ectomycorrhizal infectivity of a forest soil. International Journal of Wildland Fire. 4: 95–102.

Hewlett, J.D. 1982. Principals of hydrology. Athens: University of Georgia Press. 183 p.

Hewlett, J.D.; Hibbert, A.R. 1967. Factors affecting the response of small watersheds to precipitation in humid areas. In: Sopper, W.E.; Lull, H.W. (eds.). International symposium on forest hydrology. Oxford, England: Pergamon Press: 275–290.

Hewlett, J.D.; Helvey, J.D. 1970. The effects of clear-felling on the storm hydrograph. Water Resources Research. 6: 768–782.

Hewlett, J.D.; Troendle, C.A. 1975. Non-point and diffused water sources: a variable source area problem. In: Watershed management symposium proceedings; 1975 August 11–13; Logan, UT,. New York: American Society of Civil Engineers: 21–46.

Hibbert, A.R. 1971. Increases in streamflow after converting chaparral to grass. Water Resources Bulletin. 7: 71–80.

Hibbert, A.R. 1984. Stormflows after fire and conversion of chaparral. In: Dell, B. (ed.). Proceedings of the 4th international conference on Mediterranean ecosystems; 1984 August 13–17; Perth, Australia. Nedlands, Australia: University of Western Australia, Botany Department: 71–72.

Hibbert, A.R.; Davis, E.A.; Knipe, O.D. 1982. Water yield changes resulting from treatment of Arizona chaparral. Gen. Tech. Rep. PSW-58. Berkeley, CA: U.S. Department of Agriculture, Forest Service, Pacific Southwest Forest and Range Experiment Station: 382–389.

Hibbert, A.R; Davis, E.A.; Scholl, D.G. 1974. Chaparral conversion potential in Arizona. Part I: Water yield response and effects on other resources. Res. Pap. RM-126. Fort Collins, CO: U.S. Department of Agriculture, Forest Service, Rocky Mountain Forest and Range Experiment Station. 36 p.

Hobbs, N.T.; Schimel, D.S. 1984. Fire effects on nitrogen mineralization and fixation in mountain shrub and grassland communities. Journal of Range Management. 37: 402–404.

Hoffman, R.J.; Ferreira, R.F. 1976. A reconnaissance of the effects of a forest fire on water quality in Kings Canyon National Park, California. Open File Report 76–497. Monlo Park, CA: U.S. Department of the Interior, Geological Survey. 17 p.

Hoover, M.D.; Leaf, C.F. 1967. Process and significance of interception in Colorado subalpine forests. In: Sopper, W.E.; Lull, H.W. (eds.). Proceedings of the international symposium on forest hydrology. New York: Pergamon Press: 212–222.

Hornbeck, J.W. 1973. Stormflow from hardwood forested and cleared watersheds in New Hampshire. Water Resources Research. 9(2): 346–354.

Hornbeck, J.W.; Adams, M.B.; Corbett, E.S.; Verry, E.S.; Lynch, J.A. 1993. Long-term impacts of forest treatments on water yields: a summary for northeastern USA. Journal of Hydrology. 150: 323–344.

Hornbeck, J.W.; Martin, C.W.; Pierce, R.S.; Bormann, F.H.; Likens, G.E.; Eaton, J.S. 1987. The northern hardwood forest ecosystem: 10 years of recovery from clearcutting. Res. Pap. NE-596. Broomall, PA. U.S. Department of Agriculture, Forest Service, Northeastern Forest Experiment Station. 30 p.

Horwath, W.R.; Paul, E.A. 1994. Microbial biomass. In: Weaver, R.W. (ed.). Methods of soil analysis: Part 2, Microbiological and biochemical properties. Madison, WI: Soil Science Society of America: 753–773.

Hosking, J.S. 1938. The ignition at low temperatures of the organic matter in soils. Journal of Agricultural Science. 28: 393–400.

Hoyt, W.G.; Troxell, H.C. 1934. Forests and stream flow. Paper No. 1858. Transacions of the American Society of Civil Engineers. 111 p.

Hudson, Norman. 1981. Soil conservation. 2d ed. Ithaca, NY: Cornell University Press. 324 p.

Hulbert, L.C. 1969. Fire and litter effects in undisturbed bluestem prairie in Kansas. Ecology. 50(5): 874–877.

Humphreys, F.R.; Lambert, M.J. 1965. An examination of a forest site which has exhibited the ash-bed effect. Australian Journal of Soil Research. 3: 81–94.

Hungerford, R.D. 1990. [SHOULD THIS BE HUNGERFORD 1990a?] Describing downward heat flow for predicting fire effects. Fire effects: prescribed and wildfire. Problem analysis, Problem No. 1, Addendum 7/9/90. Missoula, MT: U.S. Department of Agriculture, Forest Service, Intermountain Research Station, Intermountain Fires Sciences Laboratory. 100 p.

Hungerford, R.D. 1990 [SHOULD THIS BE HUNGERFORD 1990b?]. Modeling the downward heat pulse from fire in soils and plant tissue. In: MacIver, D.C.; Auld, H.; Whitewood, R., (eds.). Proceedings of the 10th conference on fire and forest meteorology; Ottawa, Canada: 148–154.

Hungerford, R.D.; Ryan, K.C. 1996. Prescribed fire considerations in southern forested wetlands. In: Flynn, K.M., (ed.)., Proceedings of the southern forest wetland ecology and management conference. Clemson University, Clemson, South Carolina: 87-92.

Hungerford, R.D. Unpublished Data, Forestry Sciences Laboratory, Rocky Mountain Research Station, Missoula, MT.

Hungerford, R.D.; Frandsen W.H.; Ryan, K.C. 1995a. Heat transfer into the duff and organic soil. Final Project Report. Agreement No. 14-48-009-92-962. U.S. Department of the Interior, Fish and Wildlife Service. 48 p.

Hungerford, R.D.; Frandsen, W.H.; Ryan, K.C. 1995b. Ignition and burning characteristics of organic soils. In: Cerulean, S.I.; Engstrom, R.T. (eds.). Fire in wetlands: a management perspective. Proceedings of the Tall Timbers Fire Egology Conference. Tallahassee, FL: Tall Timbers Research Station: 19: 78–91.

Hungerford, R.D.; Harrington, M.G.; Frandsen, W.H.; Ryan, K.C.; Niehoff, G.J. 1991. The influence of fire on factors that affect site productivity. In: Harvey, A.C.; Nuenschwander, L.F. (comps.). Proceedings, Management and productivity of western montane forest soils. Gen. Tech. Rep. INT-280. Ogden, UT: U.S. Department of Agriculture, Forest Service, Intermountain Forest and Range Experiment Station: 32–50.

Hungerford, R.D.; Reardon, J. Unpublished Data, Forestry Sciences Laboratory, Rocky Mountain Research Station, Missoula, MT.

Hungerford, R.D.; Reardon, J.; Ryan, K.C. unpublished data. [Name of paper and where on file?]

Hungerford, R.D.; Ryan, K.C.; Reardon, J. 1994. Duff consumption: new insights from laboratory burning. In: Proceedings of the 12th international conference on fire and forest meteorology; 1993 October 26–28; Jekyll Island, GA. Bethesda, MD: Society of American Foresters: 594–601.

Ice, G.G. 1985. Catalog of landslide inventories for the Northwest. Tech. Bull. No. 456. Corvallis, OR: National Council of the Paper Industry for Air and Stream Improvement. 78 p.

Ice, G.G. 1996. Forest management options to control excess nutrients for the Tualatin River, Oregon. Special Report 96-04. Raleigh, NC: National Council of the Paper Industry for Air and Stream Improvement. 16 p.

USDA Forest Service Gen. Tech. Rep. RMRS-GTR-42-vol. 4. 2005

221

Ice, G.G.; Light, J.; Reiter, M. [In Press]. Use of natural temperature patterns to identify achievable stream temperature criteria for forest streams. Western Journal of Applied Forestry.

Iglesias, T.; Cala, V.; Gonzalez, J. 1997. Mineralogical and chemical modifications in soils affected by forest fire in the Mediterranean area. The Science of the Total Environment. 204: 89–96.

Ilhardt, B.L.; Verry, E.S.; Palik, G.J. 2000. Chapter 2: Defining riparian areas. In: Verry, E.S.; Hornbeck, J.W.; Dolloff, C.A. (eds.). Riparian management in forests of the continental eastern United States. New York: Lewis Publishers: 43–66.

Isichei, A.O. 1990. The role of algae and cyanobacteria in arid lands: a review. Arid Soil Research and Rehabilitation. 4: 1–17.

James, S.W. 1982. Effects of fire and soil type on earthworm populations in a tallgrass prairie. Pedobiologia. 24: 140–147.

Janicki, A. 1989. Emergency revegetation of the Stanislaus Complex fire. Unpublished report on file at:. U.S. Department of Agriculture, Forest Service, California Region, Stanislaus National Forest, Placerville, CA. 17 p.

Jasieniuk, M.A.; Johnson, E.A. 1982. Peatland vegetation organization and dynamics in the western subarctic, Northwest Territories, Canada. Canadian Journal of Botany. 60: 2581–2593.

Jeffries, D.L.; Link, S.O.; Klopatek, J.M. 1993. CO_2 fluxes of crytogamic crusts: I. Response to restoration. The New Psychologist. 125: 163–173.

Jenny, H. 1941. Factors of soil formation. New York: McGraw-Hill Book Company, Inc. 281 p.

Johansen, J.R. 1993. Cryptogamic crusts of semiarid and arid lands of North America. Journal Phycology. 29: 140–147.

Johansen, J.R.; Ashley, J.; Rayburn, W.R. 1993. The effects of range fire on soil algal crusts in semiarid shrub-steppe of the Lower Columbia Basin and their subsequent recovery. Great Basin Naturalist. 53: 73–88.

Johansen, J.R.; St. Clair, L.L.; Nebeker, G.T. 1984. Recovery patterns of cryptogamic soil crusts in desert rangelands following fie disturbance. Bryologist. 87: 238–243.

Johnson, D.W. 1992. Effects of forest management on soil carbon storage. Water, Air, Soil Pollution. 64: 83–120.

Johnson, D.W.; Cole, D.W. 1977. Anion mobility in soils: Relevance to nutrient transport from terrestrial to aquatic ecosystems. Ecological Research Series, EPA-600/3-77-068. Corvallis, OR: U.S. Environmental Protection Agency. 27 p.

Johnson, D.W.; Curtis, P.S. 2001. Effects of forest management on soil C and N storage: meta analysis. Forest Ecology Management. 140: 227–238.

Johnson, E.A. 1992. Fire and vegetation dynamics: studies from the North American boreal forest. New York: Cambridge University Press. 129 p.

Johnson, E.A.; Miyanishi, K. 2001. Forest fires, behavior, and ecological effects. San Francisco, CA: Academic Press. 594 p.

Johnson, M.G. 1978. Infiltration capacities and surface erodiblility associated with forest harvesting activities in the Oregon Cascades. Corvallis: Oregon State University. 172 p. Thesis.

Johnson, M.G.; Beschta, R.L. 1980. Logging, infiltration capacity, and surface readability in western Oregon. Journal of Forestry. 78: 334–337.

Johnson, R.S. 1970. Evaporation from bare, herbaceous, and aspen plots: a check on a former study. Water Resources Research. 6: 324–327.

Johnson, W. [Personal communication]. Bend, OR: U.S. Department of Agriculture, Deschutes National Forest.

Johnson, W.W.; Sanders, H.O. 1977. Chemical forest fire retardants: acute toxicity to five freshwater fishes and a scud. Tech. Pap. 91, Washington, DC: U.S. Department of the Interior, Fish and Wildlife Service. 7 p.

Johnston, M.; Elliott, J. 1998. The effect of fire severity on ash, and plant and soil nutrient levels following experimental burning in a boreal mixedwood stand. Canadian Journal of Soil Science. 78: 35–44.

Jones, A.T.; Ryan, K.C. (tech. cords.). In preparation. Wildland fire in ecosystems: effects of fire on cultural resources and archeology. Gen. Tech. Rep. RMRS-GTR-42, Volume 3. Fort Collins, CO: U.S. Department of Agriculture, Forest Service, Rocky Mountain Research Station.

Jones, R.D., and six coauthors. 1989. Fishery and aquatic management program in Yellowstone National Park. U.S. Fish and Wildlife Service, Technical Report for 1988, Yellowstone National Park, Wyoming.

Jones, R.D.; Botlz, G.; Carty, D.G.; Keading, L.R.; Mahony, D.L.; Olliff, S.T. 1993. Fishery and aquatic management program in Yellowstone National Park. Technical Report for 1988. Yellowstone National Park, WY: U.S. Department of the Interior, Fish and Wildlife Service. 171 p.

Jorgensen, J.R.; Wells, C.G. 1971. Apparent nitrogen fixation in soil influenced by prescribed burning. Soil Science Society of America Proceedings. 35: 806–810.

Jorgensen, J.R.; Hodges, C.S., Jr. 1970. Microbial characteristics of a forest soil after twenty years of prescribed burning. Mycologia. 62: 721–726.

Jurgensen, M.F.; Arno, S.F.; Harvey, A.E.; Larsen, M.J.; Pfister, R.D. 1979. Symbiotic and nonsymbiotic nitrogen fixation in northern Rocky Mountain forest ecosystems. In: Gordon, J.C.; Wheeler, C.R.; Perry, D.A., (eds.). Symposium proceedings on symbiotic nitrogen fixation in the management of temperate forests. Corvallis: Oregon State University: 294–308.

Jurgensen, M.F.; Graham R.T.; Larsen, M.J.; Harvey, A.E. 1992. Clear-cutting, woody residue removal, and nonsymbiotic nitrogen fixation in forest soils of the inland Pacific Northwest. Canadian Journal of Forest Research. 22: 1172–1178.

Jurgensen, M.F.; Harvey, A.E.; Jain, T.B. 1997. Impacts of timber harvesting on soil organic matter, nitrogen, productivity, and health of Inland Northwest forests. Forest Science. 43: 234–251.

Jurgensen, M.F.; Harvey, A.E.; Larsen, M.J. 1981. Effects of prescribed fire on soil nitrogen levels in a cutover Douglas-fir/ western larch forest. Res. Pap. INT-275. Ogden, UT: U.S. Department of Agriculture, Forest Service, Intermountain Forest and Range Experiment Station. 6 p.

Juste, C.; Dureau, P. 1967. Production of ammonia nitrogen by thermal decomposition of amino acids with a clay-loam soil. Series D 265C.R. Paris, France: Academy of Science: 1167–1169.

Kakabokidis, K.D. 2000. Effects of wildfire supression chemicals on people and the environment—a review. Global Nest: The International Journal. 2(2): 129–137.

Kasischke, E.S.; Stocks, B.J. 2000. Fire, climate change, and carbon cycling in the boreal forest. New York: Springer-Verlag. 461 p.

Kauffman, J.B.; Sanford, R.L., Jr.; Cummings, D.L.; Salcedo, I.H.; Sampaia, E.V.S.B. 1993. Biomass and nutrient dynamics associated with slash fires in neotropical dry forests. Ecology. 74: 140–151.

Kauffman, J.B.; Steele, M.D.; Cummings, D.; Jaramillo, V.J. 2003. Biomass dynamics associated with deforestation, fire, and conversion to cattle pasture in a Mexican tropical dry forest. Forest Ecology and Management. 176(1-3): 1–12.

Kay, B.L. 1983. Straw as an erosion control mulch. Agronomy Progress Report No.140. Davis: University of California Agricultural Experiment Station. 11 p.

Key, C.H.; Benson, N.C. 2004. Ground measure of severity, the Composite Burn Index; and remote sensing of severity, the Normalized Burn Ratio; FIREMON landscape assessment documents. National Park Service and U.S. Geological Survey National Burn Severity Mapping Project: http://burnseverity.cr.usgs.gov/methodology.asp U.S. Washington, DC: U.S. Department of the Interior, National Park Service and U.S. Geological Survey.

Kilgore, B.M. 1981. The role of fire frequency and intensity in ecosystem distribution and structure: Western forests and scrublands. Gen. Tech. Rep. WO-26, Washington, DC: U.S. Department of Agriculture, Forest Service: 58–89.

Kilmaskossu, M.S.E. 1988. Fire as a management tool to improve the renewable natural resources of Indonesia. Tucson: University of Arizona. 65 p. Thesis.

King, N.K.; Packham, D.R.; Vines, R.G. 1977. On the loss of selenium and other elements from burning forest litter. Australian Forestry Research. 7: 265–268.

Kleiner, E.F.; Harper, K.T. 1977. Soil properties in relation to cryptogamic ground cover in Canyonlands National Park. Journal of Range Management. 30: 202–205.

Klemmedson, J.O. 1994. New Mexico locust and parent material: Influence on macronutrients of forest floor and soil. Soil Science Society of America Journal. 58: 974–980.

222

USDA Forest Service Gen. Tech. Rep. RMRS-GTR-42-vol. 4. 2005

Klock, G.O.; Helvey, J.D. 1976. Soil-water trends following a wild-fire on the Entiat Experimental Forest. Proceedings of the Tall Timbers fire ecology conference, Pacific Northwest; 1974 October 16–17; Portland, OR. Tallahassee, FL: Tall Timbers Research Station. 15: 193–200.

Klock, G.O.; Tiedemann, A.R.; Lopushinsky, W. 1975. Seeding recommendations for disturbed mountain slopes in north central Washington. Res. Note PNW-244. Portland, OR: U.S. Department of Agriculture, Forest Service, Pacific Northwest Forest and Range Experiment Station. 8 p.

Klopatek, C.C.; DeBano, L.F.; Klopatek, J.M. 1988. Effects of simulated fire on vesicular-arbuscular mycorrhizae in pinyon-juniper woodland soil. Plant and Soil. 109: 245–249.

Klopatek, J.M.; Klopatek, C.C.; DeBano, L.F. 1991. Fire effects on nutrient pools of woodland floor materials and soils in a pinyon-juniper ecosystem. In: Fire and the environment: ecological and cultural perspectives: proceedings of an international symposium. Gen. Tech. Rep. SE-69. Asheville, NC: U.S. Department of Agriculture, Forest Service, Southeastern Forest Experiment Station: 154–159.

Knight, H. 1966. Loss of nitrogen from the forest floor by burning. Forestry Chronicle. 42: 149–152.

Knighton, M.D. 1977. Hydrologic response and nutrient concentrations following spring burns in an oak-hickory forest. Soil Science Society American Journal. 41: 627–632.

Knoepp, J.D.; Swank, W.T. 1993a. Effects of prescribed burning in the southern Appalachians on soil nitrogen. Canadian Journal of Forest Research. 23: 2263–2270.

Knoepp, J.D.; Swank, W.T. 1993b. Site preparation burning to improve southern Appalachian pine-hardwood stands: Nitrogen responses in soil, soil water, and streams. Canadian Journal of Forest Research. 23: 2263–2270.

Knoepp, J.D.; Swank, W.T. 1994. Long-term soil chemistry changes in aggrading forest ecosystems. Soil Science Society of America Journal. 58: 325–331.

Knoepp, J.D.; Swank, W.T. 1995. Comparison of available soil nitrogen assays in control and burned forested sites. Soil Science Society of America Journal 59: 1750–1754.

Knoepp, J.D.; Swank, W.T. 1998. Rates of nitrogen mineralization across an elevation and vegetation gradient in the southern Appalachians. Plant and Soil. 204: 235–241.

Koelling, M.; Kucera, C.L. 1965. The influence of fire on composition of central Missouri prairie. America Midland Naturalist. 72: 142–147.

Kologiski, R.L. 1977. The phytosociology of the green swamp, North Carolina. Tech. Bull. No. 250. North Carolina Agricultural Experiment Station.

Kovacic, D.A.; Swift, D.M.; Ellis, J.E.; Hakonson, T.E. 1986. Immediate effects of prescribed burning on mineral soil nitrogen in ponderosa pine of New Mexico. Soil Science. 141: 71–76.

Kovacic, D.A.; Swift, D.M.; Ellis, J.E.; Hakonson, T.E. 1986. Immediate effects of prescribed burning on mineral soil nitrogen in ponderosa pine of New Mexico. Soil Science. 141(1): 71–76.

Krabbenhoft, D.P.; Fink, L.E.; Olson, M.L.; Rawlik, P.S., II. 2001. The effect of dry down and natural fires on mercury methylation in the Florida Everglades. In: Nriagu, Jerome, (ed.). The Everglades; Proceedings of the 11th annual international conference on heavy metals in The environment; 2000 August 6–8; Ann Arbor: University of Michigan, School of Public Health. 14 p.

Kraemer, J.F.; Hermann, R.K. 1979. Broadcast burning: 25-year effects on forest soils in the western flanks of the Cascade Mountains. Forest Science. 25: 427–439.

Krammes, J.S. 1960. Erosion from mountain side slopes after fire in southern California. Res. Note PSW-171. Berkeley, CA: U.S. Department of Agriculture, Forest Service, Pacific Southwest Forest and Range Experiment Station. 8 p.

Krammes, J.S.; Rice, R.M. 1963. Effect of fire on the San Dimas Experimental Forest. In: Proceedings, 7th annual meeting, Arizona watershed symposium; 1963 September 18; Phoenix, AZ: 31–34.

Kuhry, P. 1994. The role of fire in the development of sphagnum-dominated peatlands in western boreal forests. Canadian Journal of Ecology. 82: 899–910.

Kuhry, P.; Vitt, D.H. 1996. Fossil carbon/nitrogen ratios as a measure of peat decomposition. Ecology. 77(1): 271–275.

Kutiel, P.; Naveh, Z. 1987. Soil properties beneath *Pinus halepensis* and *Quercus calliprinos* trees on burned and unburned mixed forest on Mt. Carmel, Israel. Forest Ecology and Management. 20: 11–24.

Kutiel, P.; Shaviv, A. 1992. Effects of soil type, plant composition and leaching on soil nutrients following a simulated forest fire. Forest Ecology and Management. 53: 329–343.

La Point, T.W.; Price, F.T.; Little E.E. 1996. Environmental toxicology and risk assessment, 4th Edition. Special Pubication No. 1262. West Conshohocken, PA: American Society for Testing Materials. 280 p.

Labat Anderson Inc. 1994. Human health risks assessment: chemicals used in wildland fire suppression. Contract 53-3187-9-30. Washington, DC: U.S. Department of Agriculture, Forest Service, Fire and Aviation Management.

Laderman, A.D. 1989. The ecology of Atlantic white cedar wetlands: a community profile. Biol. Rep. 85. Washington, DC: U.S. Department of the Interior, Fish and Wildlife Service. 114 p.

Landsberg, J.D.; Tiedemann, A.R. 2000. Fire management. In: Dissmeyer, G.E. (ed.). Drinking water from forests and grasslands. Gen. Tech. Rep. SRS-39. Asheville, NC: U.S. Department of Agriculture, Forest Service, Southern Forest Experiment Station: 124–138.

Landsberg, J.L.; Lavorel, S.; Stol, J. 1999. Grazing response groups among understorey plants in arid ranglands. Journal of Vegitation Science. 10: 683–696.

Lapin, M.; Barnes, B.V. 1995. Using the landscape ecosystem approach to assess species and ecosystem diversity. Conservation Biology. 9: 1148–1158.

Larson, W.E.; Pierce, F.J.; Dowdy, R.H. 1983. The threat of soil erosion to long-term crop production. Science. 219: 458–465.

Lathrop, R.G., Jr. 1994. Impacts of the 1988 wildfires on the water quality of Yellowstone and Lewis Lakes, Wyoming. International Journal of Wildland Fire. 4(3): 169–175.

Lavabre, J.D.; Gaweda, D.S.; H.A. Froehlich, H.A. 1993. Changes in the hydrological response of a small Mediterranean basin a year after fire. Journal of Hydrology. 142: 273–299.

Lavelle, P. 1988. Earthworm activities and the soil system. Biology and Fertility of Soils. 6: 237–251.

Laverty, L.; Williams, J. 2000. Protecting people and sustaining resources in fire-adapted ecosystems: a cohesive strategy. The Forest Service management response to the General Accounting Office Report GAO/RCED-99-65. Washington, DC: U.S. Department of Agriculture, Forest Service. 85 p.

Lawrence, D.E.; Minshall, G.W. 1994. Short- and long-term changes in riparian zone vegetation and stream macroinvertebrate community structure. In: Despain, D.G. (ed.). Plants and their environments: Proceedings of the first biennial scientific conference on the greater Yellowstone ecosystem. Tech. Rep. NPS/NRYELL/NRTR-93/XX. Denver, CO: U.S. Department of the Interior, Park Service, Natural Resources Publication Office: 171–184.

Lefevre, R. 2004. [Personal communication]. Tucson, AZ: U.S. Department of Agriculture, Forest Service, Coronado National Forest.

Levno, A.; Rothacher, J. 1969. Increases in maximum stream temperatures after slash burning in a small experimental watershed. Res. Note PNW-110. Portland, OR: U.S. Department of Agriculture, Forest Service, Pacific Northwest Forest and Range Experiment Station. 7 p.

Lewis, W.M. 1974. Effects of fire on nutrient movement in a South Carolina pine forest. Ecology. 55: 1120–1127.

Lide, D.R. (ed.). 2001. CRC handbook of chemistry and physics. 82nd Edition. New York: CRC Press: 4–81.

Little, E.E.; Calfee, R.D. 2002. Effects of fire-retardant chemical products on fathead minnows in experimental streams. Final report to USDA Forest Service, Wildland Fire Chemical Systems, Missoula Technology and Development Center, Missoula, MT. Columbia, MO: U.S. Department of the Interior, U.S. Geological Survey, Columbia Environmental Research Center.

Little, S. 1946. The effects of forest fires on the stand history of New Jersey's pine region. Forest Management Paper No. 2. Upper Darby, PA: U.S. Department of Agriculture, Forest Service, Northeastern Expirement Station. 43 p.

Loftin, S.; Fletcher, R.; Luehring, P. 1998. Disturbed area rehabilitation review report. Unpublished report on file at: U.S. Department of Agriculture, Forest Service, Southwestern Region, Albuquerque, NM. 17 p.

Loftin, S.R.; White, C.S. 1996. Potential nitrogen contribution of soil cryptograms to post-disturbance forest ecosystems in Bandelier National Monument, NM. In: Allen, C.D. (tech. ed.). Fire effects in southwestern forests: Proceedings of the 2nd La Mesa fire symposium. Gen. Tech. Rep. RM-GTR-286. Fort Collins, CO: U.S. Department of Agriculture, Forest Service, Rocky Mountain Forest and Range Experiment Station: 140–148.

Longstreth, D.J.; Patten, D.T. 1975. Conversion of chaparral to grass in central Arizona: Effects on selected ions in watershed runoff. The American Midland Naturalist. 93: 25–34.

Lotspeich, F.B.; Mueller, E.W.; Frey, P.J. 1970. Effects of large scale forest fires on water quality in interior Alaska. Alaska Water Lab, Collage, AK: U.S. Department of the Interior, Federal Water Pollution Control Administration. 115 p.

Lucarotti, C.J.; Kelsey, C.T.; Auclair, A.N.D. 1978. Microfungal variations relative to post-fire changes in soil environment. Oecologia. 37: 1–12.

Luce, C.H. 1995. Chapter 8: Forests and wetlands. In: Ward, A.D.; Elliot, W.J. (eds.). Environmental hydrology. Boaca Raton, FL: Lewis Publishers: 263–284.

Lugo, A.E. 1995. Fire and wetland management. In: Cerulean, S.I.; Engstrom, R.T. (eds.). Fire in wetlands: a management perspective. Proceedings of the Tall Timbers fire ecology conference. Tallahassee, FL: Tall Timbers Research Station. 19: 1–9.

Lynch, J.A.; Corbett, E.S. 1990. Evaluation of best management practices for controlling nonpoint pollution from silvicultural operations. Water Resources Bulletin. 26(1): 41–52.

Lynch, J.M.; Bragg, F. 1985. Microorganisms and soil aggregation stability. Advances in Soil Science. 2: 133–71.

Lynham, J.T.; Wickware, G.M.; Mason, J.A. 1998. Soil chemical changes and plant succession following experimental burning in immature jack pine. Canadian Journal of Soil Science. 78: 93–104.

Lyon, L.J.; Crawford, H.S.; Czuhai, E.; Fredriksen, R.L.; Harlow, R.F.; Metz, L.J.; Pearson, H.A. 1978. Effects of fire on fauna. Gen. Tech. Rep. WO-6. Washington, DC: U.S. Department of Agriculture, Forest Service. 22 p.

Lyon, L.J.; Huff, M.H.; Telfer, E.S.; Schreinder, D.S.; Smith, J.K. 2000b. Fire effects on animal populations: Chapter 4. In: Smith, J.L. (ed.). Wildland fire in ecosystems: effects of fire on fauna. Gen. Tech. Rep. RMRS-GTR-42-Volume 1. Fort Collins, CO: U.S. Department of Agriculture, Forest Service, Rocky Mountain Forest and Range Experiment Station: 25–34.

Lyon, L.J.; Telfer, E.S.; Schreiner, D.S. 2000a. Direct effects of fire and animals responses: Chapter 3. In: Smith, J.K. (ed.) Wildland fire in ecosystems: effects of fire on fauna. Gen. Tech. Rep. RMRS-GTR-42-Volume 1. Fort Collins, CO: U.S. Department of Agriculture, Forest Service, Rocky Mountain Forest and Range Experiment Station: 17–24.

Maars, R.H.; Roberts, R.D.; Skeffinton, R.A.; Bradshaw, A.D. 1983. Nitrogen in the development of ecosystems. In: Lee, J.A.; McNeill, S.; Rorison, I.H., (eds.). Nitrogen as an ecological factor. Oxford, England: Blackwell Science Publishing: 131–137.

Macadam, A.M. 1987. Effects of broadcast slash burning on fuels and soil chemical properties in sub-boreal spruce zone of central British Columbia. Canadian Journal of Forest Research. 17: 1577–1584.

Marion, G.M.; Moreno, J.M.; Oechel, W.C. 1991. Fire severity, ash deposition, and clipping effects on soil nutrients in chaparral. Soil Science Society of America Journal. 55: 235–240.

Martin, D.A.; Moody, J.A. 2001. The flux and particle-size distribution of sediment collected in the hillslope traps after a Colorado wildfire. In: Procceedings of the 7th Federal interagency sedimentation conference; 2001 March 25–29; Reno, NV. Washington, DC: Federal Energy Regulatory Commission: III: 3–47.

Martin, R.E.; Miller, R.L.; Cushwa, C.T. 1975. Germination response of legume seeds subjected to moist and dry heat. Ecology. 56: 1441–1445.

Matson, P.A.; Vitousek, P.M.; Ewel, J.J.; Mazzarino, M.J.; Robertson, G.P. 1987. Nitrogen transformation following tropical forest felling and burning on a volcanic soil. Ecology. 68: 491–502.

Mausbach, M.J.; Parker, W.B. 2001. Background and history of the concept of hydric soils. In: Richardson, J.L.; Vepraskas, M.J.

(eds.). Wetland soils—genesis, hydrology, landscapes, and classification. Lewis Publishers: 19–33.

Maxwell, J.R.; Neary, D.G. 1991. Vegetation management effects on sediment yields. In: Shou-Shou, T.; Yung-Huang, K. (eds.). Proceedings of the 5th interagency sediment conference. Washington, DC: Federal Energy Regulatory Commission. 2: 12–63.

McArthur, A.G.; Cheney, N.P. 1966. The characterization of fires in relation to ecological studies. Australian Forest Research. 2(3): 36–45.

McCammon, B.P.; Hughes, D. 1980. Fire rehab in the Bend Municipal Watershed. In: Proceedings of the 1980 watershed management symposium; 1980 July 21–23; Boise, ID. New York: American Society of Civil Engineers: 252–259.

McClure, N.R. 1956. Grass and legume seedings on burned-over forest lands in northern Idaho and adjacent Washington. Moscow, ID: University of Idaho. 25 p. Thesis.

McColl, J.G.; Grigal, D.F. 1975. Forest fire: effects on phosphorus movement to lakes. Science. 185: 1109–1111.

McColl, J.G.; Cole, D.W. 1968. A mechanism of cation transport in a forest soil. North Science. 42: 135–140.

McCool, D.K.; Brown, L.C.; Foster, G.R.; Mutchler, C.K.; Meyer, L.D. 1987. Revised slope steepness factor for the universal soil loss equation. Transactions of the American Society of Agricultural Engineers 30(5): 1387–1395.

McCool, D.K.; Foster, G.R.; Mutchler, C.K.; Meyer, L.D. 1989. Revised slope length factor for the Universal Soil Loss Equation. Transactions of the American Society of Agricultural Engineers 32(5): 1571–1576.

McDonald, S.F.; Hamilton, S.J.; Buhl, K.J.; Heisinger, J.F. 1996. Acute toxicity of fire control chemicals to Daphnia magna (Straus) and Selenastrul capricornutum (Pintz). Ecotoxicology and Environmental Safety. 33: 62–72.

McKee, W.H., Jr. 1982. Changes in soil fertility following prescribed burning on coastal plain sites. Res. Pap. SE-234. Asheville, NC: U.S. Department of Agriculture, Forest Service, Southeastern Forest Experiment Station. 28 p.

McMahon, T.E.; de Calesta, D.S. 1990. Effects of fire on fish and wildlife: In: Walstad, J.D.; Radosevich, S.R.; Sandberg, D.V. (eds.) Natural and prescribed fire in Pacific Northwest forests. Corvallis: Oregon State University Press: 233–250.

McNabb, D.H.; Cromack, K. 1990. Effects of prescribed fire on nutrients and soil productivity. In: Walstad, J.D.; Radosevich, S.R.; Sandberg, D.V. (eds.). Natural and prescribed fire in pacific northwest forests. Corvallis: Oregon State University Press: 125–142.

McNabb, D.H., Gaweda, F.; Froehlich, H.A. 1989. Infiltration, water repellency, and soil moisture content after broadcast burning a forest site Southwest Oregon. Journal of Soil and Water Conservation. 44: 87–90.

Meeuwig, R.O. 1971. Infiltration and water repellency in granitic soils. Res. Pap. PSW-33. Berkeley, CA: U.S. Department of Agriculture, Forest Service, Pacific Southwest Forest and Range Experiment Station. 20 p.

Megahan, W.F. 1984. Road effects and impacts—watershed. In: Proceedings, forest transportation symposium; 1984 December 11–13; Casper, WY. Denver, CO: U.S. Department of Agriculture, Forest Service, Rocky Mountain Region, Engineering Staff Unit: 57–97.

Megahan, W.F.; Molitor, D.C. 1975. Erosion effects of wildfire and logging in Idaho. In: Watershed management symposium; 1975 August; Logan, UT. New York: American Society of Civil Engineers Irrigation and Drainage Division: 423–444.

Meginnis, H.G. 1935. Effect of cover on surface runoff and erosion in the loessial uplands of Mississippi. Circular 347. Washington, DC: U.S. Department of Agriculture, Soil Conservation Service. 15 p.

Mihuc, T.B.; Minshall, G.W.; Robinson, C.T. 1996. Response of benthicmacroinvertebrate populations in Cache Creek, Yellowstone National Park, to the 1988 wildfires. In: Greenlee, J. (ed.). Proceedings of the 2nd biennial conference on the greater Yellowstone ecosystem: The ecological implications of fire in greater Yellowstone. Fairfield, WA: International Association of Wildland Fire: 83–94.

Miles, S.R.; Haskins, D.M.; Ranken, D.W. 1989. Emergency burn rehabilitation: cost, risk, and effectiveness. In: Berg, N.H.

(tech.coord.). Proceedings of the symposium on fire and water-shed management; 1988 October 26-28; Sacramento, CA. Gen. Tech. Rep. PSW-109. Berkeley, CA: U.S.Department of Agriculture, Forest Service, Pacific Southwest Forest and Range Experiment Station: 97–102.

Miller, D.H. 1966. Transport of intercepted snow from trees during snow storms. Res. Pap. PSW-33. Berkeley, CA: U.S. Department of Agriculture, Forest Service, Pacific Southwest Forest and Range Experiment Station. 30 p.

Miller, E.L.; Beasley, R.S.; Lawson, E.R. 1988. Forest harvest and site preparation effects on erosion and sedimentation in the Ouachita Mountains. Journal of Environmental Quality. 17: 219–225.

Miller, M. 1977. Response of blue huckleberry to prescribed fires in a western Montana larch-fir forest. Res. Pap. INT-188. Ogden, UT: U.S. Department of Agriculture, Forest Service, Intermountain Forest and Range Experiment Station. 33 p.

Miller, M. 2000. Fire autecology. In: Brown, J.K.; Smith, J.K. (eds.). Wildland fire in ecosystems: effects of fire on flora. Gen. Tech. Rep. RMRS-GTR-42-Volume 2. Ogden, UT: U.S. Department of Agriculture, Forest Service, Rocky Mountain Forest and Range Experiment Station: 9–34.

Miller, S.L.; McClean, T.M.; Stanton, N.L.; Williams, S.E. 1998. Mycorrhization, physiognomy, and first-year survivability of conifer seedlings following natural fire in Grand Teton National Park. Canadian Journal of Forest Research. 28: 115–122.

Minshall, G.W.; Brock, J.T. 1991. Observed and anticipated effects of forest fire on Yellowstone stream ecosystems. In: Leiter, R.B.; Boyce, M.S. (eds.). The greater Yellowstone ecosystem: redefining American's wilderness heritage. New Haven, CT: Yale University Press: 123–135.

Minshall, G.W; Brock, J.T.; Varley, J.D. 1989a. Wildfires and Yellowstone's stream ecosystems: A temporal perspective shows that aquatic recovery parallel forest succession. BioScience 39(10):707-715.

Minshall, G.W.; Jensen, S.E.; Platt, W.S. 1989b The ecology of stream and riparian habits of the Great Basin Region: a community profile. Biol. Rep. 85(7.24). Denver, CO: U.S. Department of the Interior, Fish and Wildlife Service. 142 p.

Minshall, G.W.; Robinson, C.T.; Lawrence, D.E. 1997. Postfire responses of lotic ecosystems in Yellowstone National Park, USA. Canadian Journal of Fisheries and Aquatic Sciences. 54: 2509–2525.

Minshall, G.W.; Robinson, C.T.; Royer, T.V.; Rushforth, S.R. 1995. Benthiccommunity structure in two adjacent streams in Yellowstone National Park five years after the 1988 wildfires. Great Basin Naturalist. 55: 193–200.

Mitsch, W.J.; Gosselink, J.G. 1993. Wetlands. New York: Van Nostrand Reinhold. 722 p.

Molina, R.; O'Dell, T.; Dunham, S.; Pilz, D. 1999. Biological diversity and ecosystem functions of forest soil fungi: management implications. In: Meurisse, R. T.;Ypsilantis, W. G.; Seybold, C. (tech. eds.). Proceedings, Pacific Northwest forest and rangeland soil organism symposium. Gen. Tech. Rep. PNW-GTR-461. Portland, OR: U.S. Department of Agriculture, Forest Service, Pacific Northwest Forest and Range Experiment Station: 45–58.

Monleon, V.J.; Cromack, K., Jr. 1996. Long-term effects of prescribed underburning on litter decomposition and nutrient release in ponderosa pine stands in central Oregon. Forest Ecology and Management. 81: 143–152.

Monleon, V.J.; Cromack, K., Jr.; Landsberg, J.D. 1997. Short and long-term effects of prescribed underburning on nitrogen availability in ponderosa pine stands in central Oregon. Canadian Journal of Forest Research. 27: 369–378.

Moreno, J.M.; Oechel, W.C. 1989. A simple method for estimating fire intensity after a burn in California chaparral. Ecologia Plantarum. 10(1): 57–68.

Morgan, P.; Neuenschwander, L.F. 1988. Shrub response to high and low severity burns following clear-cutting in northern Idaho. Western Journal of Applied Forestry. 3: 5–9.

Moring, J.R.; Lantz, R.L. 1975. Alsea watershed study: Effects of logging on the aquatic resources of three headwater streams of the Alsea River, Oregon. In: Federal aid to fish restoration, Project AFS-58, final report. Fishery Report No. 9. Corvallis, OR: Oregon Department of Fish and Wildlife, Research Section. 56 p.

Morrison, P.H.; Swanson, F.J. 1990. Fire history and pattern in a Cascade Range landscape. Gen. Tech. Rep. PNW-254. Portland, OR: U.S. Department of Agriculture, Forest Service, Pacific Northwest Research Station. 77 p.

Morton, F.I. 1990. Studies in evaporation and their lessons for the environmental sciences. Canadian Water Resources Journal. 15: 261–286.

Mroz, G.D.; Jurgensen, M.F.; Harvey, A.E.; Larsen, M.J. 1980. Effects of fire on nitrogen in forest floor horizons. Soil Science Society of America Journal. 44: 395–400.

Musil, C.F.; Midgley, G.F. 1990. The relative impact of invasive Australian acacias, fire and season on the soil chemical status of a sand plain lowland fynbos community. South African Journal of Botany. 56: 419–427.

Myers. R. L. 2000. Fire in tropical and subtropical ecosystems. In: Brown, J. K.; Smith, J. K. (eds.). Wildland fire in ecosystems: Effects of fire on flora. Gen. Tech. Rep. RMRS-GTR-42-vol 2. Ogden, UT: U.S. Department of Agriculture, Forest Service, Rocky Mountain Research Station: 161–174.

National Academy of Science. 2002.Riparian areas: Functions and strategies for management. Washington, DC: National Academy Press. 428 p.

Neary, D.G. 1995. Effects of fire on watershed resource responses in the Southwest. Hydrology and Water Resources in Arizona and the Southwest. 26: 39–44.

Neary, D.G. 2002. Chapter 6: Environmental sustainability of forest energy production, 6.3 hydrologic values. In: Richardson, J.,; Smith, T.; Hakkila, P. Bioenergy from sustainable forestry: guiding principles and practices. Amsterdam: Elsevier:. 36–67.

Neary, D.G.; Gottfried, G.J. 2002. Fires and floods: post-fire watershed responses. In: Viegas, D.X. (ed.). Forest fire research and wildland fire safety. Proceedings of the 4th international forest fire research conference; 2002 November 18–22; Luso, Portugal. Rotterdam, The Netherlands: Mill Press: 203–208.

Neary, D.G.; Hornbeck, J.W. 1994. Chapter 4: Impacts of harvesting practices on off-site environmental quality. In: Dyck, W.J.; Cole, D.W.; Comerford, N.B., (eds.). Impacts of harvesting on long-term site productivity. London: Chapman and Hall: 81–118.

Neary, D.G.; Michael, J.L. 1996. Herbicides—protecting long-term sustainability and water quality in forest ecosystems. New Zealand Journal of Forestry Science. 26: 241–264.

Neary, D.G.; Klopatek, C.C.; DeBano, L.F.; Ffolliott, P.F. 1999. Fire effects on belowground sustainability: a review and synthesis. Forest Ecology and Management. 122: 51–71.

Neary, D.G.; Overby, S.T.; Haase, S.M. 2003. Effects of fire interval restoration on carbon and nitrogen in sedimentary- and volcanic-derived soils of the Mogollon Rim, Arizona. In: Omi, P.N.; Joyce, L.A. (tech. eds.). Fire, fuel treatments, and ecological restoration: conference proceedings; 2002 April 16–18; Fort Collins, CO. Proc. RMRS-P-29. Fort Collins, CO: U.S. Department of Agriculture, Forest Service, Rocky Mountain Research Station: 105–115.

Neary, D.G.; Ryan, K.C.; DeBano, L.F. (eds.). [In press]. Wildland fire in ecosystems: Effects of fire on soil and water. Gen. Tech. Rep. RMRS-GTR-42, Volume 4. Fort Collins, CO: U.S. Department of Agriculture, Forest Service, Rocky Mountain Research Station.

Newland, J.A.; DeLuca, T.H. 2000. Influence of fire on native nitrogen-fixed plants and soil nitrogen status in ponderosa pine—Douglas-fir forests in western Montana. Canadian Journal of Forest Research. 30: 274–282.

Niehoff, G.J. 1985. Effects of clearcutting and varying fire severity of prescribed burning on levels of organic matter and the mineralization of ammonium nitrogen in the surface layer of forest soils. Moscow: University of Idaho. 43 p. Thesis.

Noble, E.L. 1965. Sediment reduction through watershed rehabilitation. In: Proceedings of the federal inter-agency sedimentation conference, 1963. Washington, DC: U.S. Department of Agriculture, Misc. Pub. 1970: 114-123.

Noble, E.L.; Lundeen, L. 1971. Analysis of rehabilitation treatment alternatives for sediment environment. In: Symposium on forest land uses and stream environment: proceedings. 1971 October 19–21. Corvallis: Oregon State University, School of Forestry and Departement of Fisheries and Wildlife: 86–96.

Norris, L.A. 1990. An overview and synthesis of knowledge concerning natural and prescribed fire in the Pacific Northwest forest. In:

USDA Forest Service Gen. Tech. Rep. RMRS-GTR-42-vol. 4. 2005

225

Walstad, J.D.; Radosevich, S.R.; Sandberg, D.V. (eds.). Natural and prescribed fire in Pacific Northwest forests. Corvallis: Oregon State University Press: 7–22.

Norris, L.A.; Webb, W.L. 1989. Effects of fire retardant on water quality. In: Berg, N.H. (tech. coord.). Proceedings of a symposium on fire and watershed management; 1988 October 26–28. Gen. Tech. Rep. PSW-109, Berkeley, CA: U.S. Department of Agriculture, Forest Service, Pacific Southwest Forest and Range Experiment Station: 79–86.

Novak, M.A.; White, R.G. 1989. Impact of fire and floods on the trout population of Beaver Creek, Upper Missouri Basin, Montana. In: Richardson, F.; Hamre, R.H. (eds.). Wild trout IV, Proceedings of the symposium; 1989 September 18–19;. Mammoth, WY: 120–126.

Ojima, D.S.; Schimel, D.S.; Parton, W.J.; Owensby, C.E. 1994. Long- and short-term effects of fire on nitrogen cycling in tallgrass prairie. Biogeochemistry. 24: 67–84.

O'Loughlin, C.L.; Rowe, L.K.; Pearce, A.J. 1980. Sediment yield and water quality reponses to clearfelling of evergreen mixed forests in western New Zealand. In: The influence of man on the hydrological regime with a special reference to representative basins. Publication 130. Gentbrugge, Belgium: International Association of Hydrological Science: 285–292.

Opitz, W. 2003. Spermatophores and spermatophore producing internal organs of Clerida (Coleoptera: Clerinae): their biological and phylogenetic implications. Coleopterists Bulletin. 57: 167–190.

Packer, P.E.; Christensen, G.F. 1977. Guides for controlling sediment from secondary logging roads. Ogden, UT: U.S. Department of Agriculture, Forest Service, Intermountain Forest and Range Experiment Station; and Missoula, MT: U.S. Department of Agriculture, Forest Service, Northern Region. 42 p.

Packham, D.; Pompe, A. 1971. The radiation temperatures of forest fires. Australian Forest Research. 5(3): 1–8.

Page-Dumroese, D.S.; Harvey, A.E.; Jurgensen, M.F.; Graham, R.T. 1991. Organic matter function in the inland northwest soil system. In: Harvey, A. E.; Neuenschwander, L.F. (comps.). Proceedings: management and productivity of western montane forest soils. Gen. Tech. Rep. INT-280. Ogden, UT: U.S. Department of Agriculture, Forest Service, Intermountain Forest and Range Experiment Station: 95–100.

Palmborg, C.; Nordgren, A. 1993. Modeling microbial activity and biomass in forest soil with substrate quality measured using near infrared reflectance spectroscopy. Soil Biology and Biochemistry. 25: 1713–1718.

Pannkuk, C.D.; Robichaud, P.R. 2003. Effectiveness of needle cast at reducing erosion after forest fires. Water Resources Research. 39(12): 1333–1341.

Pannkuk, C.D.; Robichaud, P.R.; Brown, R.S. 2000. Effectiveness of needle cast from burnt conifer trees on reducing erosion. ASAE Paper 00-5018. Milwaukee, WI: American Society of Agricultural Engineers Annual Meeting. 15 p.

Parke, J.L.; Linderman, R.G.; Trappe, J.M. 1984. Inoculum potential of ectomycorrhizal fungi in forest soils of southwest Oregon and northern California. Forest Science. 30: 300–304.

Pase, P.C.; Lindenmuth, A.W., Jr. 1971. Effects of prescribed fire on vegetation and sediment in oak-mountain mahogany chaparral. Journal of Forestry. 69: 800–805.

Pase, P.C.; Ingebo, P.A. 1965. Burned chaparral to grass: early effects on water and sediment yields from two granitic soil watersheds in Arizona. In: Proceedings of the 9th annual Arizona watershed symposium; Tempe, AZ: 8–11.

Patric, J.H. 1976. Soil erosion in eastern forests. Journal of Forestry. 74: 671–676.

Patrick, W.H., Jr. 1982. Nitrogen transformations in submerged soils. In: Nitrogen in agricultural soils. Agronomy Monograph 22, Madison, WI: American Society of Agronomy, Inc.: 449–465.

Paul, E.A.; Clark, F.E. 1989. Soil microbiology and biochemistry. San Diego, CA: Academic Press. 177 p.

Paysen, T.E.; Ansley, R.J.; Brown, J.K.; Gottfried, G.J.; Haase, S.M.; Harrington, M.G.; Narog, M.G.; Sackett, S.S.; Wilson, R.C. 2000. Chapter 6: Fire in Western shrubland, woodland, and grassland ecosystems. In: Brown, J.K.; Smith, J.K. (eds.). Wildfire in ecosystems: Effects of fire on flora. Gen. Tech. Rep. RMRS-GTR-42-vol. 2. Ogden, UT: U.S. Department of Agriculture, Forest Service, Rocky Mountain Research Station: 121–157.

Perala, D.A.; Alban, D.H. 1982. Rates of forest floor decomposition and nutrient turnover in aspen, pine, and spruce stands on two soils. Gen. Tech. Rep. GTR-NC-227. St. Paul, MN: U.S. Department of Agriculture, Forest Service, North Central Forest Experiment Station. 5p.

Pérez B.; Moreno, J.M. 1998. Methods for quantifying shrubland-fires from an ecological perspective. Plant Ecology. 139(1998b): 91–101.

Peter, S. 1992. Heat transfer in soils beneath a spreading fire. Fredericton, New Brunswick, Canada: University of New Brunswick. 479 p. Dissertation.

Pierson, F.B.; Robichaud, PR; Spaeth, K. 2001a. Spatial and temporal effects of wildfire on the hydrology of a steep rangeland watershed. Hydrological Processes. 15: 2905–2916.

Pierson, F.B.; Spaeth, K.E.; Carlson, D.H. 2001b. Fire effects on sediment and runoff in steep rangeland watersheds. In: Proceedings of the 7th federal interagency sedimentation conference; 2001 March 25–29; Reno, NV. Washington, DC: Federal Energy Regulatory Commission. p: X-10 to X-40.

Pietikainen, J.; Fritze, H. 1993. Microbial biomass and activity in the humus layer following burning: short-term effects of two different fires. Canadian Journal of Forest Research. 23: 1275–1285.

Pietikainen, J.; Hiukka, R.; Fritze, H. 2000. Does short-term heating of forest humus change its properties as a substrate for microbes? Soil Biology and Biochemistry. 32: 277–288.

Prietro-Fernandez, A.; Acea, M.J.; Carballas, T. 1998. Soil microbial and extractable C and N after wildfire. Biology and Fertility of Soils. 27: 132–142.

Pillsbury, A.F.; Osborn, J.O.; Naud, P.E. 1963. Residual soil moisture below the root zone in southern California watersheds. Journal of Geophysical Research. 68: 1089–1091.

Pilz, D.P.; Perry, D.A. 1984. Impact of clearcutting and slash burning on ectomycorrhizal associations of Douglas-fir seedlings. Canadian Journal of Forest Research. 14: 94–100.

Ping, C.L.; Moore, J.P.; Clark, M.H. 1992. Wetland properties of permafrost soils in Alaska. In: Kimble, J.M. (ed.). Characterization, classification, and utilization of wet soils. Proceedings, 8th international soil correlation meeting (VIII ISCOM). Lincoln, NE: U.S. Department of Agriculture, Soil Conservation Service, National Soil Survey Center: 198–205.

Pope, J.B.; Archer, J.C.; Johnson, P.R. 1946. Investigations in erosion control and reclamation of eroded sandy clay lands of Texas, Arkansas, and Louisiana at the Conservation Experiment Station, Tyler, Texas, 1931–1940. Tech. Bull. 916. Washington, DC: U.S. Department of Agriculture, Soil Conservation Service. 76 p.

Potter, L.; Kidd, D.; Standiford, D. 1975. Mercury levels in Lake Powell: bioamplification of mercury in man-made desert reservoir. Environmental Science and Technology. 9: 41–46.

Precht, J.; Chrisphersen, J.; Hensel, H.; Larcher, W. 1973. Temperature and life. New York: Springer-Verlag. 779 p.

Prevost, M. 1994. Scalping and burning of Kalmia Angustifolia (Ericaceae) litter: Effects on Picea mariana establishment and ion leaching in a greenhouse experiment. Forest Ecology and Management. 63: 199–218.

Propst, D.L.; Stefferud, J.A.; Turner, P.R. 1992. Conservation and status of Gila trout, Oncorhynchus gilae. Southwestern Naturalist. 37(2): 117–125.

Pryor, L.D. 1963. Ash-bed growth response as a key to plantation establishment on poor sites. Australian Forestry. 27: 48–51.

Puppi, G.; Tartaglini, N. 1991. Mycorrhizal types in three Mediterranean communities affected by fire to different extents. Acta Oecologica. 12: 295–304.

Pyne, S.J. 1982. Fire in America: a cultural history of wildland and rural fire. Seattle: University of Washington Press. 654 p.

Pyne, S.J.; Andrews, P.L.; Laven, R.D. 1996. Introduction to wildland fire. New York: John Wiley & Sons, Inc. 769 p.

Rab, M.A. 1996. Soil physical and hydrological properties following logging and slash burning in the Eucalyptus regnans forest in southeastern Australia. Forest Ecology Management. 70: 215–229.

226

USDA Forest Service Gen. Tech. Rep. RMRS-GTR-42-vol. 4. 2005

Radek, K.J. 1996. Soil erosion following wildfires on the Okanogan National Forest—initial monitoring results. In: Erosion control technology—bringing it home: proceedings of conference XXVII; 1996 February 27–March 1; Seattle, WA. Steamboat Springs, CO: International Erosion Control Association: 499–504.

Raison, R.J. 1979. Modification of the soil environment by vegetation fires, with particular reference to nitrogen transformations: a review. Plant and Soil. 51: 73–108.

Raison, R.J.; McGarity, J.W. 1980. Effects of ash, heat, and the ash-heating interaction on biological activities in two contrasting soils. Plant and Soil. 55: 363–376.

Raison, R.J.; Keith, H.; Khanna, P.K. 1990. Effects of fire on the nutrient supplying capacity of forest soils. In: Dyck, W.J.; Meeg, C.A. (eds.). Impact of intensive harvesting on forest site productivity. Bull. No. 159. Rotorua, New Zealand: Forest Research Institute: 39–54.

Raison, R.J.; Khanna, P.K.; Woods, P.V. 1985a. Mechanisms of element transfer to the atmosphere during vegetation fires. Canadian Journal of Forest Research. 15: 132–140.

Raison, R.J.; Khanna, P.K.; Woods, P.V. 1985b. Transfer of elements to the atmosphere during low-intensity prescribed fires in three Australian sub-alpine eucalypt forests. Canadian Journal of Forest Research. 15: 657–664.

Raison, R.J.; O'Connell, A.M.; Khanna, P.K.; Keith, H. 1993. Effects of repeated fires on nitrogen and phosphorus budgets and cycling processes in forest ecosystems. In: Trabaud, L.; Prodon, P. (eds.). Fire in Mediterranean ecosystems. Ecosystem Res. Rep. 5. Brussels, Belgium: Commission of the European Countries: 347–363.

Raison, R.J.; Woods, P.V.; Jakobsen, B.F.; Bary, G.A.V. 1986a. Soil temperatures during and following low-intensity prescribed burning in a eucalyptus pauciflora forest. Australian Journal of Soil Research. 24: 33–47.

Raison, R.J.; Woods, P.V.; Khanna, P.K. 1986b. Decomposition and accumulation of litter after fire in sub-alpine eucalypt forests. Australian Journal of Ecology. 11:9–19.

Rashid, G.H. 1987. Effects of fire on soil carbon and nitrogen in a Mediterranean oak forest of Algeria. Plant Soil. 103: 89–93.

Ratliff, R.D.; McDonald, P.M. 1987. Postfire grass and legume seeding: what to seed and potential impacts on reforestation. In: Proceedings, 9th annual forest vegetation management conference; 1987 November 3–5. Redding, CA: Forest Vegetation Management Conference: 111–123.

Read, D.J. 1991. Mycorrhizae in ecosystems. Experiential. 47: 376–391.

Reddell, P.; Malajczuk, N. 1984. Formation of mycorrhizae by jarrah (Eucalyptus marginata Dom ex Smith) in litter and soil. Australian Journal of Botany. 32: 511–520.

Reed, C.C. 1997. Responses of prairie insects and other arthropods to prescription burns. Natural Areas Journal. 17: 380–385.

Reid, L.M. 1993. Research and cumulative watershed effects. Gen. Tech. Rep. PSW-GTR-141. Berkeley, CA: U.S. Department of Agriculture, Forest Service, Pacific Southwest Research Station. 118 p.

Reiman, B. E.; Clayton, J. 1997. Wildfire and native fish: issues of forest health and conservation of sensitive species. Fisheries. 22(11): 6–15.

Reiman, B.E.; Lee, D.; Chandler, G.; Myers, D. 1997. Does wildlife threaten extinction for salmonids: responses of redband trout and bull trout following recent large fires on the Boise National Forest. In: Greenlee, J. (ed.). Proceedings of the symposium on fire effects on threatened and endangered species and habitats. Fairfield, WA: International Association of Wildland Fire: 47–57.

Reinhardt, E.D. 2003. Using FOFEM 5.0 to estimate tree mortality, fuel consumption, smoke production, and soil heating from wildland fire. 2nd imternational wildland fire ecology and fire management congress; 2003 November 16-20; Orlando, FL. Boston, MA: American Meteorological Society. 6 p.

Reinhardt, E.D.; Keane, R.E.; Brown, J.K. 1997. First Order Fire Effects Model: FOFEM 4.0, User's Guide. Gen. Tech. Rep. INT-344. Ogden, UT: U.S. Department of Agriculture, Forest Service, Intermountain Research Station. 65 p.

Reinhardt, E.D.; Keane, R.E.; Brown, J.K. 2001. Modeling fire effects. International Journal of Wildland Fire. 10: 373–380.

Reinhart, K.G.; Eschner, A.R.; Trimble, G.R. 1963. Effects on streamflow of four forest practices in the mountains of West Virginia. Res. Pap. NE-1. U.S. Department of Agriculture, Forest Service, Northeastern Forest Experiment Station. 79 p.

Renard, K.G.; Simanton, J.R. 1990. Application of RUSLE to rangelands. Proceedings of the symposium on watershed planning and analysis in action; 1990 July 9–11; Durango, CO. New York: American Society of Civil Engineers: 164–173.

Renard, K.G.; Foster, G.R.; Weesies, G.A.; McCool, D.K.; Yoder, D.C. (coords). 1997. Predicting soil erosion by water: a guide to conservation planning with the revised universal soil loss equation (RUSLE). Agric. Handb. No. 703. Washington, DC: U.S. Department of Agriculture, Natural Resources Conservation Service. 404 p.

Renbuss, M.A.; Chilvers, G.A.; Pryor, L.D. 1973. Microbiology of an ashbed. Proceedings of the Linnaean Society of New South Wales 97:302-310.

Rice, R.M. 1974. The hydrology if chaparral watersheds. In: Rosenthal, M. (ed.). Proceedings of symposium on living with the chaparral. Riverside, CA: University of California: 27–34.

Rice, R.M.; Crouse, R.P.; Corbett, E.S. 1965. Emergency measures to control erosion after a fire on the San Dimas Experimental Forest. In: Proceedings of the Federal inter-agency sedimentation conference; 1963. Misc. Pub. 970. Washington, DC: U.S. Department of Agriculture: 123–130.

Rich, L.R. 1962. Erosion and sediment movement following a wildfire in a ponderosa pine forest of central Arizona. Res. Note 76. Fort Collins, CO: U.S. Department of Agriculture, Forest Service, Rocky Mountain Forest and Range Station. 12 p.

Richards, C.; Minshall, G.W. 1992. Spatial and temporal trends in stream macroinvertebrate communities: the influence of catchment disturbance. Hydrobiologia. 241: 173–184.

Richardson, J.L.; Vepraskas, M.J. 2001. Wetland soils: genesis, hydrology, landscapes, and classification. Boca Raton, FL: Lewis Publishers. 417 p.

Richardson, J.L.; Arndt, J.L.; Montgomery, J.A. 2001. Hydrology of wetland and related soils. In: Richardson, J.L.; Vepraskas, M.J. (eds.). Wetland soils: genesis, hydrology, landscapes, and classification. Boca Raton, FL: Lewis Publishers: 35–84.

Richter, D.D.; Ralson, C.W.; Harms, W.R. 1982. Prescibed fire: effects on water quality and forest nutrient recycling. Science. 215: 661–663.

Riekerk, H. 1983. Impacts of silviculture on flatwoods runoff, water quality, and nutrient budgets. Water Resources Bulletin. 19: 73–79.

Riggan, P.J.; Lockwood, R.N.; Jacks, P.M. 1994. Effects of fire severity on nitrate mobilization in watersheds subject to chronic atmospheric deposition. Environmental Science and Technology. 28: 369–375.

Rinne, J.N. 1988. Grazing effects on stream habitat and fishes: research design considerations. North American Journal Fish Management. 8: 240–247.

Rinne, J.N. 1994. Declining Southwestern aquatic habitats and fishes: are they sustainable? Gen. Tech. Rep. RM-247. Fort Collins, CO: U.S. Department of Agriculture, Forest Service, Rocky Mountain Forest and Range Experiment Station: 256–265.

Rinne, J.N. 1996. Shorterm effects of wildfire on fishes and aquatic macroinvertebrates in the Southwestern United States. North American Journal of Fish Management. 16: 653–658.

Rinne, J. N. 2003a. Flows, fishes, foreigners, and fires: Relative impacts on southwestern native fishes. Hydrology and Water Resources in the Southwest. 33: 79-84.

Rinne, J. N. 2003b. Wildfire in the southwestern United States: Effects on fishes and their habitats. 2nd International Fire Ecology Conference Proceedings, American Meteorological Society, Orlando, FL, 17-20 November 2003. CD ROM publication Paper 66000, 5 p. http://ams.confex.com/ams/FIRE2003/techprogram/paper_66000.htm

Rinne J. N. 2004. Forest, Fires and fishes: Lessons and management implications from the southwestern USA. Pp 151-156. In, G. J. Scrimgeour, G. Eisler, B. McCullock, U. Silins, and M. Morita (editors). Forest Lands – Fish II, Ecosystem stewardship through collaboration. Conference. Edmonton, Alberta, April 26-28, 2004

USDA Forest Service Gen. Tech. Rep. RMRS-GTR-42-vol. 4. 2005

227

Rinne, J.N. and B. Calamusso. In Press. Southwestern trouts. Distribution with reference to physiography, hydrology, distribution and threats. In: Symposium on Inland trouts. Western Division of American Fisheries Society, Feb 28-Mar 2, 2004, Salt Lake City Utah.

Rinne, J. N. and C. D. Carter. In Press. Short-Term Effects of Wildfires on Fishes in streams in the Southwestern United States, 2002: Management Implications. In: Narog, M.G., technical coordinator. Proceedings of the 2002 Fire Conference on Managing fire and fuels in the remaining wildlands and open spaces of the southwestern United States. December 2-5, 2002, San Diego, CA. Gen. Tech. Rep. PSW-189, Albany, CA: Pacific Southwest Research Station, Forest Service, U.S. Department of Agriculture:

Rinne, J.N.; Neary, D.G. 1996. Effects of fire on aquatic habitats and biota in Madrean-type ecosystems—Southwestern USA. Gen. Tech. Rep. RM-289. Fort Collins, CO: U.S. Department of Agriculture, Forest Service, Rocky Mountain Forest and Range Experiment Station: 135–145.

Roberts, W.B. 1965. Soil temperatures under a pile of burning logs. Australian Forest Research. 1(3): 21–25.

Robichaud, PR. 2000. Fire and erosion: evaluating the effectiveness of a post-fire rehabilitation treatment, contour-felled logs. In: Proceedings, watershed management and operations management conference; 2000 June 20–24; Fort Collins, CO. Reston, VA: American Society of Civil Engineers. 11 p.

Robichaud, P.R. 2002. Wildfire and erosion: when to expect the unexpected. Symposium on the geomorphic impacts of wildfire. Paper 143-10. Boulder, CO: Geological Society of America. 1 p.

Robichaud, P.R.; Brown, R.E. 1999. What happened after the smoke cleared: onsite erosion rates after a wildfire in eastern Oregon. In: Olsen, D.S.; Potyondy, J.P. (eds.). Proceedings, wildland hydrology conference; 1999 June; Bozeman, MT. Hernon, VA: American Water Resource Association: 419–426.

Robichaud, P.R.; Brown, R.E. 2002. Silt fences: an economical technique for measuring hillslope soil erosion. Gen. Tech. Rep. RMRS-GTR-94. Fort Collins, CO, U.S. Department of Agriculture, Forest Service, Rocky Mountain Research Station. 24 p.

Robichaud, P.R.; Hungerford, R. 2000. Water repellency by laboratory burning of four northern Rocky Mountain forest soils. Journal of Hydrology. 231–232: 207–219.

Robichaud, P.R.; Beyers, J.L.; Neary. D.G. 2000. Evaluating the effectiveness of postfire rehabilitation treatments. Gen. Tech. Rep. RMRS-GTR-63. Fort Collins, CO: U.S. Department of Agriculture, Forest Service, Rocky Mountain Research Station. 85 p.

Robichaud, P.R.; MacDonald, L.; Freeouf, J.; Neary, D.; Martin, D.; Ashmun, L. 2003. Postfire rehabilitation. In: Graham, R.T., (tech. ed.). Hayman Fire case study. Gen. Tech. Rep. RMRS-GTR-114. Ogden, UT: U.S. Department of Agriculture, Forest Service, Rocky Mountain Research Station: 293–313.

Robinson, C.T.; Minshall, G.W.; Rushforth, S.R. 1994. The effects of the 1988 wildfires on diatom assemblages in streams of Yellowstone National Park. In: Despain, D.G. (ed.). Plants and their environments: proceedings of the first biennial scientific conference on the Greater Yellowstone Ecosystem. Tech. Rep. NPS/NRYELL/NRTR-93/XX. Denver, CO: U.S. Department of the Interior, National Park Service, Natural Resources Publication Office: 247–257.

Roby, K.B. 1989. Watershed response and recovery from the Will Fire: ten years of observation. Gen. Tech. Rep. PSW-109. Berkeley, CA: U.S. Department of Agriculture, Forest Service, Pacific Southwest Forest and Range Experiment Station: 131–136.

Roby, K.B.; Azuma, D.L. 1995. Changes in a reach of a northern California stream following wildfire. Environmental Management. 19: 591–600.

Romanya, J.; Khanna, P.; Raison, R.J. 1994. Effects of slash burning of soil phosphorus fractions and sorption and desorption of phosphorus. Forest Ecology and Management. 65: 89–103.

Roscoe, R.; Buurman, P.; Velthorst, E.J.; Pereira, J.A.A. 2000. Effects of fire on soil organic matter in a "cerrado sensu-stiricto" from southeast Brazil as revealed by changes in ^{13}C. Geoderma 95: 141–161.

Rosen, K. 1996. Effect of clear-cutting on streamwater quality in forest catchments in central Sweden. Forest Ecology and Management. 83: 237–244.

Rosentreter, R. 1986. Compositional patterns within a rabbitbrush (Chrysothamnus) community of the Idaho Snake River Plain. In: McArthur, E.D.; Welch, B.L. (comps.). Symposium proceedings on the biology of Artemisia and Chrysothamnus. Gen. Tech. Rep. INT-200. Ogden, UT: U.S. Department of Agriculture, Forest Service, Intermountain Forest and Range Experiment Station. 273–277.

Rosgen, D.L. 1994. A classification of natural rivers. Catena. 22: 169–199.

Rosgen, D.L. 1996. Applied river morphology. Pagosa Springs, CO: Wildland Hydrology. 364 p.

Ross, D.J.; Speir, T.W.; Tate, K.; Feltham, C.W. 1997. Burning in a New Zealand snow-tussock grassland: effects on soil microbial biomass and nitrogen and phosphorus availability. New Zealand Journal of Ecology. 21: 63–71.

Roth, F.A., Chang, M. 1981. Throughfall in planted stands of four southern pine species in east Texas. Water Resources Bulletin. 17: 880–885.

Rothacher, J. 1963. Net precipitation under a Douglas-fir forest. Forest Science. 9: 423–429.

Rothermel, R.C. 1972. A mathematical model prediction fire spread in wildland fuels. Res. Pap. INT-115. Ogden, UT: U.S. Department of Agriculture, Forest Service, Intermountain Forest and Range Experiment Station. 40 p.

Rothermel, R.C. 1991. Predicting behavior and size of crown fires in the Northern Rocky Mountains. Res. Pap. INT-438. Ogden, UT: U.S. Department of Agriculture, Forest Service, Intermountain Forest and Range Experiment Station. 46 p.

Rothermel, R.C.; Deeming, J.E. 1980. Measuring and interpreting fire behavior for correlation with fire effects. Gen. Tech. Rep. INT-93. Ogden, UT: U.S. Department of Agriculture, Forest Service, Intermountain Forest and Range Experiment Station. 4 p.

Rowe, J.S. 1983. Concepts of fire effects on plant individuals and species. In: Wein, R.W.; MacLean, D.A. (eds.). The role of fire in northern circumpolar ecosystems. Scope 18. New York: John Wiley & Sons: 135–154.

Rowe, J.S.; Bergstiensson, J.L.; Padbury, G.A.; Hermesh, R. 1974. Fire studies in the MacKenzie Valley. ALUR Rep 73. Canadian Department of Indian and Northern Development: 71–84.

Ruby, E. 1997. Observations on the Dome Fire emergency rehabilitation seeding. Unpublished report on file at: U.S. Department of Agriculture, Forest Service. Tonto National Forest, Phoenix, AZ. 3 p.

Rundel, P.W. 1977. Water balance in Mediterranean sclerophyll ecosystems. In: Mooney, H.A.; Conrad, C.E. (tech. coords.). Proceedings of the symposium on the environmental consequences of fire and fuel management in Mediterranean ecosystems. Gen. Tch. Rep. WO-3. Washington, DC: U.S. Department of Agriculture, Forest Service: 95–106.

Russell, K.R.; Van Lear, D.H.; Guynn, D.C., Jr. 1999. Prescribed fire effects on herpetofauna: review and management implications. Wildlife Society Bulletin. 27(2): 374–384.

Russell, J.D.; Fraser, A.R; Watson, J.R.; Parsons, J.W. 1974. Thermal decomposition of protein in soil organic matter. Geoderma. 11: 63–66.

Ryan, K.C. 2002. Dynamic interactions between forest structure and fire behavior in boreal ecosystems. Silva Fennica. (36(1): 13–39.

Ryan, K.C.; Noste, N.V. 1985. Evaluating prescribed fires. In: Lotan, J.E.; Kilgore, B.M.; Fischer, W.C.; Mutch, R.W. (eds.). Proceedings—symposium and workshop on wilderness fire. Gen. Tech. Rep. INT-182. Ogden, UT: U.S. Department of Agriculture, Forest Service, Intermountain Forest and Range Experiment Station: 230–238.

Ryan, K.C.; Frandsen, W.H. 1991. Basal injury from smoldering fires in mature *Pinus ponderosa* laws. International Journal of Wildland Fire. 1: 107–118.

Ryan, P.W.; McMahon, C.K. 1976. Some chemical and physical characteristics of emissions from forest fires. In: Proceedings of the 69th annual meeting of the Air Pollution Control Association. Paper Number 76-2.3. Portland, OR. 15 p.

Saa, A.; Trasar-Cepeda, M.C.; Carballas, T. 1998. Soil phosphorus status and phosphomonoesterase activity of recently burnt and unburnt soil following laboratory incubation. Soil Biology and Biochemistry. 30: 419–428.

Saa, A.; Trasar-Cepeda, M.C.; Gil-Sotres, F.; Carballas, T. 1993. Changes in soil phosphorus and acid phosphatase activity immediately following forest fires. Soil Biology Biochemistry. 25: 1223–1230.

Sackett, S.S. 1980: Reducing natural ponderosa pine fuels using prescribed fire: two case studies. Rese. Pap. RM-392. Fort Collins, CO: U.S. Department of Agriculture, Forest Service, Rocky Mountain Forest and Range Experiment Station. 6 p.

Sackett, S.S.; Haase, S.M. 1992. Measuring soil and tree temperatures during prescribed fires with thermocouple probes. Gen. Tech. Rep. PSW-131. U.S. Department of Agriculture, Forest Service, Pacific Southwest Forest and Range Experiment Station. 15 p.

Sackett, S.S.; Haase, S.M.; Harrington, M.G. 1996. Lessons learned from fire use restoring Southwestern ponderosa pine ecosystems. Gen. Tech. Rep. RM-278. Fort Collins, CO: U.S. Department of Agriculture, Forest Service, Rocky Mountain Forest and Range Experiment Station: 54–61.

San Dimas Technology Development Center (SDTDC). 2003. Helicopter straw mulching: planning and implementation. [Online] Available: http://fsweb.sdtdc.wo.fs.fed.us/programs/wsa/helimulch_etip/

Sandberg, D.V.; Ottmar, R.D.; Peterson, J.L.; Core, J. 2002. Wildland fire on ecosystems: effects of fire on air. Gen. Tech. Rep. RMRS-GTR-42-vol. 5. Ogden, UT: U.S. Department of Agriculture, Forest Service, Rocky Mountain Research Station. 79 p.

Sandberg, D.V.; Pierovich, J.M.; Fox, D.G.; Ross, E.W. 1979. Effects of fire on air: a State-of-knowledge review. Gen. Tech. Rep. WO-9. Washington, D.C.: U.S. Department of Agriculture, Forest Service. 40 p.

Sands, R. 1983. Physical changes to sandy soils planted to radiata pine. In: Ballard, R; Gessel, S.P. (eds.). IUFRO symposium on forest site and continuous productivity. Gen. Tech. Rep. PNW-163. Berkeley, CA: U.S. Department of Agriculture, Forest Service, Pacific Southwest Forest and Range Experiment Station: 146–152.

Sartz, R.S. 1953. Soil erosion on a fire-denuded forest area in the Douglas-fir region. Journal of Soil and Water Conservation. 8(6): 279–281.

Satterlund, D.R.; Adams, P.W. 1992. Wildland watershed management. New York: John Wiley & Sons, Inc. 436 p.

Satterlund, D.R.; Haupt, H.F. 1970. The disposition of snow caught by conifer crowns. Water Resources Research. 6(2): 649–652.

Scatena, F.N. 2000. Chapter 2: Drinking water quality. In: Dissmeyer, G.E. (ed.). Drinking water from forests and grasslands: a synthesis of the scientific literature. Gen. Tech. Rep. SRS-39. Asheville, NC: U.S. Department of Agriculture, Forest Service, Southern Research Station: 7–25.

Schimmel, J.; Granström, A. 1996. Fire severity and vegetation response in the boreal Swedish forest. Ecology. 77(5): 1436–1450.

Schlesinger, W.H. 1991. Biogeochemistry—an analysis of global change. San Diego: Academic Press. 443 p.

Schmalzer, P.A.; Hinkle, C.R. 1992. Soil dynamics following fire in Juncus and Spartina marshes. Society of Wetlands Scientists. Wetlands. 12(1): 8–21.

Schmidt, L. 2004. [Personal communication]. Fort Collins, CO: U.S. Department of Agriculture, Forest Service (retired). Stream Systems Technology Center.

Schmidt, K.M.; Menakis, J.P.; Hardy, C.C.; Hann, W.J.; Bunnell, D.L. 2002. Development of coarse-scale spatial data for wildland fire and fuel management. Gen. Tech. Rep. RMRS-87. Fort Collins, CO: U.S. Department of Agriculture, Forest Service, Rocky Mountain Research Station. 41 p.

Schnitzer, M.; Hoffman, I. 1964. Pyrolysis of soil organic matter. Soil Science Society of America Proceedings. 28: 520–525.

Schoch, P.; Binkley, D. 1986. Prescribed burning increased nitrogen availability in a mature loblolly pine stand. Forest Ecology and Management. 14: 13–22.

Schoeneberger, M.M.; Perry, D.A. 1982. The effect of soil disturbance on growth and ectomycorrhizae of Douglas-fir and western hemlock seedlings: a greenhouse bioassay. Canadian Journal of Forest Research. 12: 343–353.

Schumm, S.A.; Harvey, M.D. 1982. Natural erosion in the USA. In: Schmidt, B.L. (ed.). Determinants of soil loss tolerance. Special Publ. 45. Madison, WI: American Society of Agronomy: 23–39.

Scott, D.F. 1993. The hydrological effects of fire in South African mountain catchments. Journal Hydrology. 150: 409–432.

Scott, D.F.; Van Wyk, D.B. 1990. The effects of wildfire on soil wettability and hydrologic behavior of an afforested catchment. Journal of Hydrology. 121: 239–256.

Scott, J.H. 1998. Sensitivity analysis of a method for assessing crown fire hazard in the northern Rocky Mountains, USA. International conference on forest fire research, 14th conference on fire and forest meteorology. Vol. II: 2517–2532.

Scott, J.H. 2001. Quantifying surface fuel characteristics in pocosin plant communities at the Dare County Bombing Range. Final Report RJVA INT-96093. Unpublished report on file at: U.S. Department of Agriculture, Forest Service, Rocky Mountain Research Station, Fire Sciences Laboratory, Missoula, MT. 18 p.

Scott, J.H.; Reinhardt, E.D. 2001. Assessing crown fire potential by linking models of surface and crown fire behavior. Res. Pap. RMRS-29. Fort Collins, CO: U.S. Department of Agriculture, Forest Service, Rocky Mountain Research Station. 59 p.

Service, R.F. 2004. As the West goes dry. Science. 303: 1124–1127.

Severson, K.E.; Rinne, J.N. 1988. Increasing habitat diversity in Southwestern forests and woodlands via prescribed fire. In: Krammes, J.S. (tech. coord.). Proceedings of a symposium: effects of fire management on Southwestern natural resources; 1988 November 15–17. Tucson, AZ. Gen. Tech. Rep. RM-191. Fort Collins, CO: U.S. Department of Agriculture, Forest Service, Rocky Mountain Forest and Range Experiment Station: 95–104.

Sgardelis, S.P.; Pantis, J.D.; Argyropoulou, M.D.; Stamou, G.P. 1995. Effects of fire on soil macroinvertebrates in a Mediterranean Phryganic ecosystem. International Journal of Wildland Fire. 5: 113–121.

Sharitz, R.R.; Gibbons, J.W. 1982. The ecology of southeastern shrub bogs (pocosins) and Carolina bays: a community profile. Tech. Rep.FWS/OBS-82/04. Washington, DC: U.S. Department of the Interior, Fish and Wildlife Service. 93 p.

Shea, R.W. 1993. Effects of prescribed fire and silvicultural activities on fuel mass and nitrogen redistribution in *Pinus ponderosa* ecosystems of central Oregon. Corvallis: Oregon State University, 163 p. Thesis.

Sidle, R.C. 1985. Factors influencing the stability of slopes. In: Swanston, D. (ed.). Proceedings of the workshop on slope stability: problems and solutions in forest management. Gen. Tech. Rep. PNW-180. Portland, OR: U. S. Department of Agriculture, Forest Service, Pacific Northwest Forest and Range Experiment Station: 17–25.

Sidle, R.C.; Drlica, D.M. 1981. Soil compaction from logging with a low-ground pressure skidder on the Oregon coast ranges. Soil Science Society of America Journal. 45: 1219–1224.

Simard, A.J. 1991. Fire severity, changing scales, and how things hang together. International Journal of Wildland Fire. 1: 23–34.

Sims, B.D.; Lehman, G.S.; Ffolliott, P.F. 1981. Some effects of controlled burning on surface water quality. Hydrology and Water Resources in Arizona and the Southwest. 11: 87–89.

Sinclair, J.D.; Hamilton, E.L. 1955. Streamflow reactions of a fire-damaged watershed. In: Proceedings, American Society of Civil Engineers, Hydaulics Division. 81(629): 1–17.

Singer, M.J.; Munns, D.N. 1996. Soils: an introduction. 3rd edition. Upper Saddle River, NJ: Prentice Hall. 480 p.

Skau, C.M. 1964a. Interception, throughfall, and stemflow in Utah and alligator juniper cover types of northern Arizona. Forest Science. 10: 283–287.

Skau, C.M. 1964b. Soil water storage under natural and cleared stands of alligator and Utah juniper in northern Arizona. Res. Note RM-24. Fort Collins, CO: U.S. Department of Agriculture, Forest Service, Rocky Mountain Forest and Range Experiment Station. 3 p.

Smith, J. K. (ed.). 2000. Wildland fire in ecosystems: effects of fire on fauna. Gen. Tech. Rep. RMRS-GTR-42-Vol. 1, Ogden, UT: U.S. Department of Agriculture, Forest Service, Rocky Mountain Forest and Range Experiment Station 83 p.

Smith, J.K.; Fischer, W.C. 1997. Fire ecology of the forest habitat types of northern Idaho. Gen. Tech. Rep. INT-363. Ogden, UT: U.S. Department of Agriculture, Forest Service, Intermountain Research Station. 142 p.

USDA Forest Service Gen. Tech. Rep. RMRS-GTR-42-vol. 4. 2005

229

Smith, R. D.; Sidle, R.C.; Porter, P.E.; Noel, J.R. 1993. Effects of experimental removal of woody debris on the channel morphology of a forest, gravel-bed stream. Journal of Hydrology. 152: 153–178.

Snyder, G.G.; Haupt, H.F.; Belt, G.H., Jr. 1975. Clearcutting and burning slash alter quality of stream water in northern Idaho. Res. Pap. INT-168. Ogden, UT: U.S. Department of Agriculture, Forest Service, Intermountain Forest and Range Experiment Station. 34 p.

Soil Science Society of America. 1997. Glossary of soil science terms. Madison, WI: Soil Science Society of America. 134 p.

Soil Science Society of America. 2001. Glossary of soil science terms. Madison, WI: Soil Science Society of America. 140 p.

Soto, B.; Diaz-Fierros, F. 1993. Interactions between plant ash leachates and soil. International Journal of Wildland Fire. 3(4): 207–216.

Springett, J.A. 1976. The effect of prescribed burning on the soil fauna and on litter decomposition in western Australia forests. Australian Journal of Ecology. 1: 77–82.

St. Clair, L.L.; Webb, B.L.; Johansen, J.R.; Nebecker, G.T. 1984. Cryptogamic soil crusts: enhancement of seeding establishment in disturbed and undisturbed areas. Reclamation Revegetation Research. 3: 129–136.

St. John, T.V.; Rundel, P.W. 1976. The role of fire as a mineralizing agent in the Sierran coniferous forest. Oecologia. 25: 35–45.

Staddon, W.J.; Duchesne, L.C.; Trevors, J.T. 1998. Acid phosphatase, alkaline phosphatase and arylsulfatase activities in soils from a jack pine (Pinus banksiana Lamb.) ecosystem after clear-cutting, prescribed burning, and scarification. Biology and Fertility of Soils. 27: 1–4.

Stefan, D.C. 1977. Effects of a forest fire upon the benthic community of a mountain stream in northeast Idaho. Missoula: University of Montana, 205 p. Thesis.

Stevenson, F.J. 1986. Cycles of soil—carbon, nitrogen, phosphorus, sulfur, and micronutrients. New York, NY: John Wiley & Sons, Inc. 427 p.

Stevenson, F. J.; Cole, M. A. 1999. Cycles of soil: carbon, nitrogen, phosphorus, sulfur, micronutirents. 2d ed. New York: John Wiley & Sons. 427 p.

Stocks, B.D.; Lawson, M.E.; Alexander, C.E.; Van Wagner, R.S.; McAlpine, T.J.; Lynham,; Dube, D.E. 1989. The Canadian forest fire danger rating system: an overview. The Forestry Chronicle. 65: 258–265.

Stocks, B.J. 1991. The extent and impact of forest fires in northern circumpolar countries. In: Levine, J.S. (ed.). Global biomass burning: atmospheric climate and biospheric implications Cambridge: Massachusetts Institute of Technology Press: 197–202.

Sutherland, D.R.; Haupt, D.R. 1970. The deposition of snow caught by conifer crowns. Water Resources Research. 6: 649–652.

Swank, W.T.; Crossley, D.A., Jr. 1988. Forest hydrology and ecology at Coweeta. New York: Springer-Verlag. 469 p.

Swank, W.T.; Miner, N.H. 1968. Conversion of hardwood-covered watersheds to white pine reduces water yield. Water Resources Research. 4: 947–954.

Swanson, F.J. 1981. Fire and geomorphic processes. In: Mooney, H.A.; Bonnicksen, T.M.; Christensen, N.L.; Lotan, J.E.; Reiners, W.A., (tech.coords.). Fire regimes and ecosystem properties; proceedings; 1979 December 11–5; Honolulu, HI. Gen. Tech. Rep. WO-26.Washington, DC: U.S.Department of Agriculture, Forest Service: 410–420.

Swift, L.W., Jr. 1984. Gravel and grass surfacing reduces soil loss from mountain roads. Forest Science. 30: 657–670. [Should this be Swift 1984a?]

Swift, L.W., Jr. 1984. Soil losses from roadbeds and cut and fill slopes in the southern Appalachian Mountains. Southern Journal of Applied Forestry. 8: 209–213. [Should this be Swift 1984b?]

Swift, L.W.; Messner, J.B. 1971. Forest cuttings raise temperature of small streams in the southern Appalachians. Journal of Soil and Water Conservation. 26: 111–116.

Tarapchak, S.J.; Wright, R.F. 1977. Three oligotrophic lakes in northern Minnesota. In: Seyb, L.; Randolph, K. (eds.). North American project—a study of U.S. water bodies. Corvallis, OR: Environmental Protection Agency: 64–90.

Tate, R.L., III. 1987. Soil organic matter: Biological and ecological effects. New York, NY: John Wiley & Sons, Inc. 304 p.

Taylor, M.J.; Shay, J.M.; Hamlin, S.N. 1993. Changes in water quality conditions in Lexington Reservoir, Santa Clara County, California, following a large fire in 1985 and flood in 1986. Water Resources Investigations Report, Geological Survey, U.S. Department of the Interior, Denver, Colorado. 23 p.

Tecle, A. 2004. [Personal communication]. Flagstaff: Northern Arizona University.

Tennyson, L.C.; Ffolliott, P.F.; Thorud, D.B. 1974. Use of time-lapse photography to assess potential interception in Arizona ponderosa pine. Water Resources Bulletin. 10: 1246–1254.

Terry, J.P.; Shakesby, R.A. 1993. Soil hydrophobicity effects on rain splash: simulated rainfall and photographic evidence. Earth Surface Processes and Landforms. 18: 519–525.

Theodorou, C.; Bowen, G.D. 1982. Effects of a bushfire on the microbiology of a South Australian low open (dry sclerophyll) forest soil. Australian Forestry Research. 12: 317–327.

Tiedemann, A.R. 1973. Stream chemistry following a forest fire and urea fertilization in north-central Washington. Res. Note PNW-203. Portland, OR: U.S. Department of Agriculture, Forest Service, Pacific Northwest Forest and Range Experiment Station. 20 p.

Tiedemann, A.R. 1987. Combustion losses of sulfur and forest foliage and litter. Forest Science. 33: 216–223.

Tiedemann, A.R.; Klock, G.O. 1976. Development of vegetation after fire, reseeding, and fertilization on the Entiat Experimental Forest. In: Proceedings, Tall Timbers fire ecology conference, Pacific Northwest; 1974 October 16-17; Portland, OR. Tallahassee, FL: Tall Timbers Research Station. 15: 171–192.

Tiedemann, A.R.; Conrad, C.E.; Dieterich, J.H.; Hornbeck, J.W.; Megahan, W.F.; Viereck, L.A.; Wade, D.D. 1979. Effects of fire on water: a state-of-knowledge review. National fire effects workshop. Gen. Tech. Rep. WO-10. Washington, DC: U.S. Department of Agriculture, Forest Service. 28 p.

Tiedemann, A.R.; Helvey, J.D.; Anderson, T.D. 1978. Stream chemistry and watershed nutrient economy following wildfire and fertilization in eastern Washington. Journal of Environmental Quality. 7: 580–588.

Tiedemann, A.R.; Klemmedson, J.O.; Bull, E.L. 2000. Solution of forest health problems with prescribed fire: are forest productivity and wildfire at risk? Forest Ecology and Management. 127: 1–18.

Tiedemann, A.R.; Quigley, T.M.; Anderson, T.D. 1988. Effects of timber on stream chemistry and dissolved nutrient losses in northeast Oregon. Forest Science. 34: 344–358.

Tomkins, I.B.; Kellas, J.D.; Tolhurst, K.G.; Oswin, D.A. 1991. Effects of fire intensity on soil chemistry in a eucalypt forest. Australian Journal Soil Research. 29: 25–47.

Tongway, D.J.; Hodgkinson, K.C. 1992. The effects of fire on the soil in a degraded semi-arid woodland. III. Nutrient pool sizes, biological activity and herbage response. Australian Journal of Soil Research. 30: 17–26.

Torsvik, V.; Goksoyr, J.; Daae, F.L. 1990. High diversity of DNA of soil bacteria. Applied Environmental Microbiology. 56: 782–787.

Trappe, J.M.; Bollen, W.B. 1979. Forest soil biology. In: Heilman, P.E.; Anderson, H.A; Baumgartner, D.M., (comps). Forest soils of the Douglas-fir region. Pullman: Washington State University, Cooperative Extension Service: 145–151.

Trettin, C.C.; Song, B.; Jurgensen, M.F.; Li, C. 2001. Existing soil carbon models do not apply to forested wetlands. Gen. Tech. Rep. GTR SRS-46. Asheville, NC: U.S. Department of Agriculture, Forest Service, Southern Forest Experiment Station. 16 p.

Trettin, C.C.; Davidian, M.; Jurgensen M.F.; Lea, R. 1996. Organic matter decomposition following harvesting and site preparation of a forested wetland. Soil Science Society of America Journal. 60: 1994–2003.

Troendle, C.A. 1983. The potential for water yield augmentation from forest management in the Rocky Mountain region. Water Resources Bulletin. 19: 359–373.

Troendle, C.A.; Meiman, J.R. 1984. Options for harvesting timber to control snowpack accumulations. Proceedings of the annual Western Snow Conference. 52: 86–97.

Troendle, C.A.; King, R.M. 1985. The effect of timber harvest on the Fool Creek watershed: 30 years later. Water Resources Research. 21: 1915–1922.

230

USDA Forest Service Gen. Tech. Rep. RMRS-GTR-42-vol. 4. 2005

Tromble, J.M. 1983. Interception of rainfall by creosotebush (*Larrea tridentata*). In: Proceedings of XIV international grassland congress; Lexington, KY: 373–375.

Tunstall, B.R.; Walker, J.; Gill, A.M. 1976. Temperature distribution around synthetic trees during grass fires. Forest Science. 22(3): 269–276.

Turetsky, M. R.; Wieder, R.K. 2001. A direct approach to quantifying organic matter lost as a result of peatland wildfire. Canadian Journal of Forest Research. 31: 363–366.

Turner, M.G.; Romme, W.H.; Gardner, R.H.; Hargrove, W.W. 1994. Influence of patch size and shape on post-fire succession on the Yellowstone Plateau. Bulletin of the Ecological Society of America. 75(Part 2): 255.

Tyrrel, R.R. 1981. Memo, Panorama burn rehabilitation. Unpublished report on file at: U.S. Department of Agriculture, Forest Service, Pacific Southwest Region, San Bernardino National Forest, San Bernardino, CA. 16 p.

Ursic, S.J. 1970. Hydrologic effects of prescribed burning, and deadening upland hardwoods in northern Mississippi. Res. Pap. SO-54. New Orleans, LA: U.S. Department of Agriculture, Forest Service, Southern Forest Experiment Station. 15 p.

USDA Forest Service. 1989a. Final environmental impact statement, vegetation management in the Coastal Plain and Piedmont. Mgmt. Bull. R8MB23. Atlanta, GA: U.S. Department of Agriculture, Forest Service, Southern Region. 1248 p.

USDA Forest Service. 1989b. Final environmental impact statement, vegetation management in the Appalachian Mountains. Mgmt. Bull. R8MB38. Atlanta, GA: U.S. Department of Agriculture, Forest Service, Southern Region. 1638 p.

USDA Forest Service. 1990a. Final environmental impact statement, vegetation management in the Ozark-Ouachita Mountains. Mgmt. Bull. R8MB45. Atlanta, GA: U.S. Department of Agriculture, Forest Service, Southern Region. 1787 p.

USDA Forest Service. 1990b. R1-WATSED: Region 1 water and sediment model. Missoula, MT: U.S. Department of Agriculture, Forest Service, Montana Region.

USDA Forest Service. 1995. Burned area emergency rehabilitation handbook. FSH 2509. Washington, DC: U.S. Department of Agriculture, Forest Service. 13 p.

USDA Forest Service. 1988. Vegetation management Final Environmental Impact Statement for Washington, Oregon, California. USDA Forest Service, Pacific Northwest Region, General Water Quality Best Management Practices, November 1988. 86 p.

USDA Natural Resources Conservation Service. 1998. Keys to soil taxonomy. Washington, DC: U.S. Department of Agriculture, Natural Resources Conservation Service, Soil Survey Staff. 326 p.

USDA 2000 Download for WEPP Windows. http://topsoil.nserl. perdue.edu/weppmain/wepp.html

U.S. Environmental Protection Agency. 1993. Chapter 3: Management measures for forestry. In: Guidance specifying management measures for sources of nonpoint pollution in coastal waters. Publ. EPA-840-B-92-002. Washington, DC: U.S. Environmental Protection Agency: 3-1 to 3-121.

U.S. Environmental Protection Agency. 1999. National primary drinking water regulations. Primary drinking water standards. [Available at http://www.epa.gov.OGWDW/wet /appa.html].

U.S. Geological Survey. 2002. NWIS webdata for the nation. Water data, surface water, water quality. http://waterdata.usgs.gov/ nwis/qwdata. U.S. Reston, VA: U.S. Deparment of the Interior, U.S. Geological Survey.

Van Cleve, K.; Viereck, L.A. 1983. A comparison of successional sequences following fire in permafrost-dominated and permafrost-free sites in interior Alaska. In: Permafrost: fourth international conference proceedings. Washington, DC: National Academy Press: 1286–1290.

Van Cleve, K.; Powers, R.F. 1995. Soil carbon, soil formation, and ecosystem development. In: McFee, W.W.; Kelly, J.M. (eds.). Carbon forms and functions in forest soils. Madison, WI: Soil Science Society of America: 155–200.

Van Cleve, K.; Viereck, L.A.; Schlentner, R.L. 1971. Accumulation of nitrogen in alder (Alnus) ecosystems near Fairbanks, Alaska. Arctic and Alpine Research. 3: 101–114.

Van Lear, D.H.; Douglass, J.E.; Fox, S.K.; Ausberger, M.K. 1985. Sediments and nutrient export in runoff from burned and harvested pine watersheds n the South Carolina Piedmont. Journal of Environmental Quality. 14: 169–174.

Van Meter, W.P.; Hardy, C.E. 1975. Predicting effects on fish of fire retardants in streams. Res. Pap. INT-166. Ogden, UT: U.S. Department of Agriculture, Forest Service, Intermountain Forest and Range Experiment Station. 16 p.

van Reenen, C.A.; Visser, G.J.; Loos, M.A. 1992. Soil microorganisms and activities in relation to season, soil factors and fire. In: Van Wilgen, B.W.; and others (eds.), Fire in South Africa mountain fynbos: Ecosystem, community and species response at Swartboskloof. Berlin: Springer-Verlag. 93: 258–272.

van Veen, J.A.; Kuikman, P.J. 1990. Soil structural aspects of decomposition of organic matter by mico-organisms. Biogeochemistry. 11: 213–233.

van Wagner, C.E. 1973. Height of crown scorch in forest fires. Canadian Journal of Forest Research. 3: 373–378.

van Wagner, C.E. 1977. Conditions for the start and spread of crown fire. Canadian Journal of Forest Research. 7(1): 23–34.

van Wagner, C.E. 1983. Fire behavior in northern conifer forests and shrublands. In: Wein, R.W.; MacLean, D.A. (eds.). The role of fire in northern circumpolar ecosystems. Scope 18. New York: John Wiley & Sons, Inc.: 65–80.

Varnes, D.J. 1978. Slope movement types and processes. In: Schuster, R.L.; Krizek, R.J. (eds.). Landslides analysis and control. Special Report 176. Washington, DC: National Academy of Science: 11–33.

Vasander, H.; Lindholm, T. 1985. Fire intensities and surface temperatures during prescribed burning. Silva Fennica. 19(1): 1–15.

Vazquez, F.J.; Acea, M.J.; Carballas, T. 1993. Soil microbial populations after wildfire. Microbial Ecology. 13: 93–104.

Vepraskas, M.J.; Faulkner, S.P. 2001. Redox chemistry of hydric soils. In: Richardson, J.L.; M.J. Vepraskas, M.J. (eds.). Wetland soils. Boca Raton, FL: CRC Press. 85–105.

Verhoef, H.A.; Bussard, L. 1990. Decomposition and nitrogen mineralization in natural and agroecosystems: The contribution of soil animals. Biogeochemistry 11: 175–211.

Viereck, L.A. 1973. Wildfire in the taiga of Alaska. Quaternary Research. 3: 465–495.

Viereck, L.A. 1982. Effects of fire and fireline construction on active layer thickness and soil temperature in interior Alaska. In: The Rodger Brown Memorial Volume. Proceedings of th 4th Canadian permafrost conference. Ottawa Canada: Natural Resource Council: 123–135.

Viereck, L.A. 1983. The effects of fire on black spruce ecosystems of Alaska and Northern Canada. In: Wein, R.W.; Maclean, D.A. (eds.). The role of fire in northern circumpolar ecosystems. Scientific committee on problems of the environment. Scope 18. New York: John Wiley & Sons, Inc.: 201–220.

Viereck, L.A.; Dyrness, C.T. 1979. Ecological effects of the Wickersham Dome fire near Fairbanks, Alaska. Gen. Tech. Rep. PNW-90. Portland, OR: U.S. Department of Agriculture, Forest Service, Pacific Northwest Forest and Range Experiment Station. 71 p.

Viereck, L.A.; Schandelmeier, L.A. 1980. Effects of fire in Alaska and adjacent Canada: literature review. Tech. Rep. No. 6. Alaska: U.S. Department of the Interior, Bureau of Land Management. 124 p.

Vilarino, A.; Arines, J. 1991. Numbers and viability of vesiculararbuscular fungal propagules in field soil samples after wildfire. Soil Biology and Biochemistry. 23: 1083–1087.

Visser, S. 1995. Ectomycorrhizal fungal succession in jack pine stands following wildfire. New Phytologist. 129: 389–401.

Vitousek, P.M.; Melillo, J.M. 1979. Nitrate losses from disturbed forests: Patterns and mechanisms. Forest Science. 25: 605–619.

Vlamis, J.; Biswell, H.H.; Schultz; A.M. 1955. Effects of prescribed burning on soil fertility in second growth ponderosa pine. Journal of Forestry. 53: 905–909.

Vose, J.M. 2000. Perspectives on using prescribed fire to achieve desired ecosystem conditions. In: Moser, W.K.; Moser, C.F. (eds.). Fire and forest ecology: innovative silviculture and vegetation management. Proceedings, Tall Timbers fire ecology conference. Tallahassee, Fl: Tall Timbers Research Station. 21: 12–17.

USDA Forest Service Gen. Tech. Rep. RMRS-GTR-42-vol. 4. 2005

231

Vose, J.M.; Swank, W.T. 1993. Site preparation burning to improve southern Appalachian pine-hardwood stands: aboveground biomass, forest floor mass, and nitrogen and carbon pools. Canadian Journal of Forest Research. 23: 2255–2262.

Vose, J.M.; Swank, W.T.; Clinton, B.D.; Knoepp, J.D.; Swift, L.W. 1999. Using stand replacement fires to restore southern Appalachian pine-hardwood ecosystems: effects on mass, carbon, and nutrient pools. Forest Ecology and Management. 114: 215–226.

Wade, D.D.; Ward, D.E. 1973. An analysis of the Air Force Bomb Range Fire. Special Report. Asheville, N.C: U.S. Department of Agriculture, Forest Service, Southeastern Forest Experiment Station. 38 p.

Wade, D.D.; Brock, B.L.; Brose, P.H.; Grace, J.B.; Hoch, G.A.; Patterson, W.A., III. 2000. In: Brown, J.K.; Smith, J.K. (eds.). Wildfire in ecosystems: effects of fire on flora. Gen. Tech. Rep. RMRS-GTR-42-vol. 2. Ogden, UT: U.S. Department of Agriculture, Forest Service, Rocky Mountain Research Station: 53–96.

Wade, D.D.; Ewel, J.; Hofstetter, R. 1980. Fire in south Florida ecosystems. Gen. Tech. Rep. SE-17. Asheville, NC: U.S. Department of Agriculture, Forest Service, Southeastern Forest Expirement Station. 125 p.

Wagenbrenner, J.W. 2003. Effectiveness of burned area emergency rehabilitation treatments in the Colorado Front Range. Fort Collins: Colorado State University. 193 p. Thesis.

Wallwork, J.A. 1970. Ecology of soil animals. Maidenhead, Berkshire, England: McGraw-Hill. 283 p.

Walsh, R.P.D.; Boakes, D.; Coelho, C.O.A.; Goncalves, A.J.B.; Shakesby, R.A.; Thomas, A.D. 1994. Impact of fire-induced hydrophobicity and post-fire forest litter on overland flow in northern and central Portugal. In: Viegas, D.X. (ed.).Volume II. 2nd international conference on forest fire research; 1994 November 21-24; Coimbra, Portugal: 1149–1190.

Walstad, J. D.; Radosevich, S.R.; Sandberg, D.V. (eds.). Natural and prescribed fire in Pacific Northwest forests. Corvallis: Oregon State University Press. 317 p.

Warcup, H.H. 1981. Effect of fire on the soil microflora and other non-vascular plants. In: Gill, A.M.; Groves, B.H.; Noble, I.R. (eds). Fire and the Australian biota. Canberra, Australia: Australian Academy of Science: 203–214.

Warners, D.P. 1997. Plant diversity in sedge meadows: effects of groundwater and fire. Ann Arbor: University of Michigan. 231 p. Dissertation.

Washington Forest Practices Board. 1997. Standard methodology for conducting watershed analysis. Olympia, WA. 95 p.

Wein, R.W. 1983. Fire behavior and ecological effects in organic terrain. In: Wein, R.W.; Maclean, D.A. (eds). The role of fire in northern circumpolar ecosystems. Scientific committee on problems of the environment. Scope 18. New York: John Wiley & Sons, Inc.: 81–96.

Wells, C.G., Campbell, R.E.; DeBano, L.F.; Lewis, C.E.; Fredrickson, R.L.; Franklin, E.C.; Froelich, R.C.; Dunn, P.H. 1979. Effects of fire on soil: a state-of-the-knowledge review. Gen. Tech. Rep. WO-7. Washington, DC: U.S. Department of Agriculture, Forest Service. 34 p.

Wells, W.G., II. 1981. Some effects of brushfires on erosion processes in coastal southern California. In: Davies, T.R.R. (ed.) Proceedings, erosion and sediment transport in Pacific rim steeplands; 1981 January 25–31; Christchurch, New Zealand. Publ. No.132. Washington, DC: International Association of Scientific Hydrology: 305–342.

Wells, W.G., II. 1987. The effects of fire on the generation of debris flows in southern California. Reviews in Engineering Geology. 7: 105–114.

West, N.E.; Skujins, J. 1977. The nitrogen cycle in North America cold-winter semidesert ecosystems. Oecologia Plant.12: 45–53.

Wetzel, R.G. 1983. Attached algae-substrate interactions: fact or myth, and when and how? In: R.G. Wetzel (ed.). Peripyton in freshwater ecosystems. Junk. The Hague, Netherlands: 207–215.

Whelan, R.J. 1995. The ecology of fire. Cambridge England: Cambridge University Press. 346 p.

Whisenant, S.G. 1990. Changing fire frequencies on Idaho's Snake River Plains: ecological and management implications. In: McArthur, E.D.; Romney, E.M.; Smith, S.D.; Tueller, P.T. (eds.). Symposium on cheatgrass invasion, shrub die-off, and other aspects of shrub biology and management. Gen. Tech. Rep. INT-

276. Ogden, UT: U.S. Department of Agriculture, Forest Service, Intermountain Forest and Range Experiment Station: 4–10.

White, C.S. 1986. Effects of prescribed fire on rates of decomposition and nitrogen mineralization in a ponderosa pine ecosystem. Biology and Fertility of Soil. 2: 87–95.

White, C.S. 1991. The role of monoterpenes in soil nitrogen cycling processes in ponderosa pine. Biogeochemistry. 12: 43–68.

White, C.S. 1996. The effects of fire on nitrogen cycling processes within Bandelier National Monument, NM. In: Allen, C.D. (tech. ed.) Fire effects in Southwestern forests: Proceedings of the 2nd La Mesa fire symposium, Los Alamos, New Mexico. Gen. Tech. Rep. RM-GTR-286: Fort Collins, CO: U.S. Department of Agriculture, Forest Service, Rocky Mountain Forest and Range Experiment Station: 123–139.

White, E.M.; Thompson, W.W.; Gartner, F.R. 1973. Heat effects on nutrient release from soils under ponderosa pine. Journal of Range Management. 26: 22–24.

White, J.D.; Ryan, K.C.; Key, C.C.; Running, S.W. 1996. Remote sensing of forest fire severity and vegetation recovery. International Journal of Wildland Fire. 6(3): 125–136.

White, P. S.; Pickett, S.T.A. 1985. Chapter 1. Natural disturbance and patch dynamics: an introduction. In: Pickett, S.T.A.; White, P.S. (eds.). The ecology of natural disturbance and patch dynamics. San Francisco, CA: Academic Press. 472 p.

Whitehead, P.G.; Robinson, M. 1993. Experimental watershed studies—an international and historical perspective for forest impacts. Journal of Hydrology. 145: 217–230.

Widden, P.; Parkinson, D. 1975. The effects of forest fire on soil microfungi. Soil Biology and Biochemistry. 7: 125–138.

Wikars, L.O.; Schimmel, J. 2001. Immediate effects of fire severity on soil invertebrates in cut and uncut pine forests. Forest Ecology and Management. 141(3): 189–200.

Wilbur, R.B. 1985. The effects of fire on nitrogen and phosphorus availability in a North Carolina coastal plain pocosin. Chapel Hill, NC: Duke University. 143 p. Dissertation.

Wilbur, R.B.; Christensen, N.L. 1983. Effects of fire on nutrient availability in a North Carolina coastal plain pocasin. American Midland Naturalist. 119: 54–61.

Willard, E.E.; Wakimoto, R.H.; Ryan, K.C. 1995. Vegetation recovery in sedge meadow communities within the Red Bench Fire, Glacier National Park. In: Cerulean, S.I.; Engstrom, R.T. (eds.). Fire in wetlands: a management perspective. Proceedings of the Tall Timbers Fire Ecology Conference. Tallahassee, FL: Tall Timbers Research Station: 19: 102–110.

Wischmeier, W.H.; Smith, D.D. 1978. Predicting rainfall erosion losses: a guide to conservation planning. Agric. Handb. No. 282. Washington, DC: U.S. Department of Agriculture, Soil Conservation Service. 58 p.

Wohlgemuth, P.M. 2001. Prescribed fire as a sediment management tool in southern California chaparral watersheds. In: Proceedings of the 7th Federal interagency sedimentation conference;2001 March 25–29; Reno, NV. Washington, DC: Federal Energy Regulatory Commission. 2: X-49 to X-56.

Wohlgemuth, P.M.; Beyers, J.L.; Wakeman, C.D.; Conard, S.G. 1998. Effects of fire and grass seeding on soil erosion in southern California chaparral. In: Proceedings, 19th annual forest vegetation management conference: wildfire rehabilitation; 1998 January 20-22. Redding, CA: Forest Vegetation Management Conference: 41–51.

Wolman, M.G. 1977. Changing needs and opportunities in the sediment field. Water Resources Research. 13: 59–54.

Wright, H.A.; Bailey, A.W. 1982. Fire ecology—United States and Southern Canada. New York: John Wiley & Sons, Inc. 501 p.

Wright, H.A.; Churchill, F.M.; Stevens, W.C. 1976. Effect of prescribed burning on sediment, water yield, and water quality from juniper lands in central Texas. Journal of Range Management. 29: 294–298.

Wright, H.A.; Churchill, F.M.; Stevens, W.C. 1982. Soil loss and runoff on seeded vs. non-seeded watersheds following prescribed burning. Journal of Range Management 35: 382-385.

Wright, H.E., Jr. 1981. The role of fire in land/water interactions. In: Mooney, H.A.; Bonnicksen, T.M.; Christensen, N.L.; Lotan, J.E.; Reiners, W.A. (tech. cords.). Proceedings of the conference: fire regimes and ecosystem properties. Gen. Tech. Rep. WO-26. Washington, DC: U.S. Department of Agriculture, Forest Service: 421–444.

Wright, H.E., Jr.; Bailey, A.W. 1982. Fire ecology and prescribed burning in the Great Plains: a research review. Gen. Tech. Rep. WO-26. Washington, DC: U.S. Department of Agriculture, Forest Service. 61 p.

Wright, R.J.; Hart, S.C. 1997. Nitrogen and phosphorus status in a ponderosa pine forest after 20 years of interval burning. Ecoscience. 4: 526–533.

Wu, X.B.; Redeker, E.J.; Thurow, T.L. 2001. Vegetation and water yield dynamics in an Edwards Plateau watershed. Journal of Range Management. 54: 98–105.

Yokelson, R.J.; Susott, R.; Ward, D.E.; Reardon, J.; Griffith, D.W.T. 1997. Emissions from smoldering combustion of biomass measured by open-path Fourier transform infrared spectroscopy. Journal of Geophysical Restoration. 102(D15): 18,865–18,877.

Young, M.K. 1994. Movement and characteristics of stream-borne coarse woody debris in adjacent burned and undisturbed watersheds in Wyoming. Canadian Journal of Forest Research. 24: 1933–1938.

Zasada, J.C.; Norum, R.A.; van Veldhuizen, R.M.; Teutsch, C.E. 1983. Artificial regeneration of trees and tall shrubs in experimentally burned upland black spruce/feather moss stands in Alaska. Canadian Journal of Forest Research. 13(5): 903–913.

Zhang, Q.L.; Hendrix, P.F. 1995. Earthworm (*Lumbricus rubellus* and *Aporrectodea caliginosa*) effects on carbon flux in soil. Soil Science Society of America Journal. 59: 816–823.

Zhang, Y.Q. 1997. Biogeochemical cycling of selenium in Benton Lake, MT. Missoula: University of Montana, 222 p. Thesis.

Zwolinski, M.J. 1990. Fire effects on vegetation and succession. In: Krammes, M.J. (tech. coord.). 1990. Proceedings of a symposium Effects of fire management of Southwestern Natural Resources. Gen. Tech. Rep. RM-191: Fort Collins, CO: U.S. Department of Agriculture, Forest Service, Rocky Mountain Forest and Range Experiment Station: 18–24.

Zwolinski, M.J. 1971. Effects of fire on water infiltration rates in a ponderosa pine stand. Hydrology and Water Resources in Arizona and the Southwest. 1: 107–113.

USDA Forest Service Gen. Tech. Rep. RMRS-GTR-42-vol. 4. 2005

233

Appendix A: Glossary _____

aerial fuels: Fuels more than 6.5 feet (2 m) above the mineral soil surface.

ammonification: Transformation of organic nitrogen (N) compounds, such as proteins and amino acids, into ammonia (NH_4). Ammonification is a process involved in the mineralization of N that is affected by fire.

ashbed effect: The accumulation of thick layers of ashy residue on the soil surface after fire resulting from combustion of concentrated fuels such as deep litter layers, piled slash, and windrows.

available fuel: Amount of fuel available for burning in a particular fire, a value varying widely in magnitude with the environmental conditions on a site.

average runoff efficiencies: The ratio of runoff to precipitation.

back fire: Fire set against an advancing fire to consume fuels and (as a consequence) prevent further fire. PAH formation increases since gaseous fuels have longer residence times in these types of combustion conditions.

backing fire: A fire that is burning against the slope or wind, i.e., backing down slope (this type of fire typically has the lowest fireline intensity but often longer flaming durations), or burning into the wind (the lee side of the fire, but again with the lowest fireline intensity on the perimeter).

baseflow: Streamflow sustained by subsurface flow and groundwater flow between precipitation events.

basidiomycetes: The phylum basidiomycota consists of fungi that produce spores that are formed outside a pedestal-like structure, the basidium. The members of this phylum, known as basidiomycetes, include all the fungi with gills or pores, including the familiar mushrooms and bracket fungi.

benefits: Favorable effects of fire-caused changes in the ecosystem.

biomass: All of the vegetative materials available for burning in natural ecosystems.

buffer strips: Vegetated bands along streams or around water.

burn area: The area over which a fire has spread.

burning: Refers to being set on fire.

Byram's intensity: The product of the available heat of combustion per unit area of ground surface and the rate of spread of the fire. It is also referred to as fireline intensity.

cellulose: Long-chain polysaccharides such as glucose, mannose, galactose, and xylose sugar derivatives with high oxidation levels that are found in both cell walls, but primarily in the secondary wall constituting the largest group of carbohydrates in wood (40 to 50 percent).

channel interception: Interception of precipitation that falls directly into a stream channel that contributes to streamflow.

char: A carbonaceous residue left on the surface of fuels by pyrolysis that is neither intact organic compound nor pure carbon.

char-height: The height of stem charring as a proportion of total tree height. Also is an indicator of postfire tree mortality in some species.

chemical energy: Solar energy fixed by plants in the synthesis of organic molecules and compounds.

chemical properties: Properties of fuels that affect the heat content and the types of pollutants emanating from a fire.

coarse woody debris: Made up of tree limbs, boles, and roots in various stages of decay, and that are greater than 3 inches (7.5) cm in diameter.

combustion: Is a rapid physical-chemical process, commonly called fire, that releases the solar energy stored in a chemical form in various fuels as heat and a variety of gaseous and particulate by-products.

combustion rate: The mass of fuel consumed by the combustion process (e.g. tons/minute, kg/sec,etc.) or the average speed (ha/min or acres/min) at which fuels are being burned.

condensation: Is the process whereby water changes from a gas to a liquid and releases heat.

USDA Forest Service Gen. Tech. Rep. RMRS-GTR-42-vol. 4. 2005

235

controlled burning: Also called prescribed burning, is the controlled application of fire to fuel buildups, in either their natural or modified state, in specified environmental conditions that allow the fire to be confined to a predetermined area and, at the same time, to produce the fireline intensity and rate of spread required to attain the planned management objectives.

convection: Process whereby heat is transferred from one point to another by the mixing of one portion of a fluid with another fluid.

conversion burning: One vegetative community on a site is replaced by another because of fire.

crown fire: Fire that advances from top-to-top of trees or shrubs more or less independently of the surface fire.

crown fuels: Tree and shrub crowns.

cryptogamic crusts: Communities of lichens, blue-green bacteria (cyanobacteria), fungi, mosses, and algae that are found on the surface of rocks and soil in dryland regions throughout the world.

damages: Unfavorable effects of fire-caused changes in a resource system.

dead fuels: Grouped by size class as 1, 10, 100, or 1,000 hour timelag classes; size classes are often separated into sound or rotten.

decomposition: The breakdown of organic matter, results in catabolism of organic matter into smaller organic materials.

denitrification: Process of reducing nitrate (NO_3–N) to nitrogen gas (N_2) and nitrous oxide (N_2O) by biological means.

dependent crown fire: Fire in the crowns that moves only in spurts and is dependent on intense heating from the surface fire. This type of fire, which generally causes the most severe impact on a natural ecosystem, occurs mostly in coniferous forests and woodlands, and in shrublands comprised of waxy-leaved species.

downstream fire effects: Occur when hydrologic processes are altered for a long enough time that the changes can accumulate through time, when responses from a number of sites are transported to the same site, or when a transported response interacts with an on-site change at another site.

duff: The F and H layers of organic matter. Therefore, the top of the duff is where leaves, needles, and other castoff vegetation have begun to decompose, while the bottom of the duff is where decomposed organic matter is mixed with mineral soil.

ecto-mycorrhizae: One of the two types of mycorrhizae found in soil.

ecotones: Abrupt edges between adjacent vegetative types.

efficient burns: Flaming combustion with high intensity consuming most, if not all, of the available fuels. With low-intensity smoldering fires fuel consumprion is only about 50%.

endo-mycorrhizae: (arbuscular) One of the two types of mycorrhizae found in soil.

endothermic: Heat-absorbing reactions that pyrolysis and combustion start with.

evapotranspiration (ET): Evaporation from soils, plant surfaces, and water bodies, together with the water losses from transpiring plants.

excess water: Portion of total precipitation that flows off the land surface plus that which drains from the soil and, therefore, is neither consumed by ET nor leaked into deep groundwater aquifers.

exothermic: Heat-producing reaction that pyrolysis and combustion progress to.

extinction: The fifth and last phase of a fire where combustion ceases.

F layer: Fermentation layer, the accumulation of dead organic plant matter above mineral soil consisting of partly decomposed matter.

favorable effects: Effects that contribute to the attainment of fire management objectives.

field capacity: The maximum amount of water that a soil mantle retains against the force of gravity.

fine fuels: Fast-drying dead fuels, characterized by a high surface area-to-volume ratio, less than 1 cm in diameter; these fuels (grasses, leaves, needles, and so forth) ignite readily and are consumed rapidly.

fire: A manifestation of a series of chemical reactions that result in the rapid release of the heat energy stored in (living and dead) plants by photosynthesis.

fire behavior: Manner in which a fire reacts to its environment—to the fuels available for burning, climate, local weather conditions, and topography. Fire behavior changes in time, space, or both in relation to changes in these environmental components. Common terms used to describe fire behavior include smoldering, creeping, running, spotting, torching.

fire climax: A plant community maintained by periodic fire.

fire cycle: Also called the fire-return interval, is the length of time necessary for an area equal to the entire area of interest to burn—the size of the area should be specified.

fire-dependent ecosystems: Those ecosystems where fire plays a vital role in determining the composition, structure, and landscape patterns.

fire ecology: The study of relationships among fire, the environment, and living organisms.

fire effects: The physical, chemical, and biological impacts of fire on the environment and ecosystem resources.

fire frequency: Also referred to as fire occurrence, is the number of fires in a specified time and area.

fire intensity: Describes the rate at which a fire produces thermal energy. When it is based on a line (of implied depth, D) it is Byram's fireline intensity, and when it is defined as a heat per unit area it is Rothermel's intensity.

fire interval: Also referred to as fire-free interval, is the time between two successive fires in a designated area—the size of the area should be clearly specified.

fireline intensity: The product of the available heat of combustion per unit area of ground surface and the rate of spread of the fire. It is also referred to as Byram's intensity.

fire occurrence: Also referred to as fire frequency, this is the number of fires in a specified time and area.

fire regime: Largely determined by the combinations of three factors: how often fire occurs (frequency), when it occurs (season), and how fiercely it burns (intensity).

fire resistance: The ability of vegetation to survive the passage of fire.

fire-return interval: Also called the fire cycle, is the length of time necessary for an area equal to the entire area of interest to burn—the size of the area should be specified.

fire severity: Describes ecosystems responses to fire and can be used to describe the effects of fire on the soil and water system, ecosystem flora and fauna, the atmosphere, and society. It reflects the amount of energy (heat) that is released by a fire which affects resource responses. Fire severity, loosely, is a product of fire intensity and residence time and is generally considered to be light, moderate, or high.

fire triangle: Fuels available for burning, along with heat and oxygen, represent the components needed for fire to occur.

fire type: Type of vegetation that commonly follows a fire, or otherwise is dependent upon the occurrence of fire.

flame: A gas-phase phenomenon of fire.

flaming: The second phase of a fire involving combustion in which pyrolysis continues.

flammability: The relative ease with which a substance ignites and sustains combustion.

forage: Grass and grasslike plants, forbs, and half-shrubs available to, and eaten by, livestock or other herbivores.

forced convection: Occurs when external mechanical forces alter the fluid flow from its natural "free" direction and velocity.

free convection: Occurs when the fluid motion of the gases is dependent upon the differences in densities resulting from temperature differences.

free radicals: Molecules that do not have a balanced charge due to excess electrons.

fuel: Term used interchangeably with *fuel available for burning* when referring to the biomass that decomposes as a result of ignition and combustion.

fuel available for burning: Term used interchangeably with *fuel* when referring to the biomass that decomposes as a result of ignition and combustion.

fuel loading: Total dry weight of fuel per unit of surface area, a measure of the potential energy that might be released by a fire.

fuel reduction burning: Removes fuel buildups to reduce the likelihood of ignition or lessen potential damage and the resistance to control of fire when it occurs on a site.

fuel state: The moisture condition of the fuel that largely determines the amount of fuel available for burning at any given time and on which fuel classifications can also be based.

glowing: The fourth phase of a fire also involving combustion in which pyrolysis virtually ceases.

good condition watershed: Precipitation infiltrates into the soil and does not contribute excessively to erosion, since the resultant overland flow (when a pathway of flow on the watershed) does not dislodge and move soil particles. Streamflow response to precipitation is relatively slow and baseflow (when a pathway of flow) is sustained between storms.

greenhouse gas: Gas that has potential to impact the global climate by warming Earth's atmosphere.

ground fire: Fire that burns the organic material in the upper soil layer and, at times, the surface litter and low-growing plants.

ground fuels: Fuels generally defined as lying below the litter (L or O_i) layer (i.e.,fermentation (For O_e) and humus (H or O_a) layers, logs with their center axis below the surface of the F-layer, peat and muck soils.

H layer: Humus layer (O_a layer), the accumulation of dead organic plant matter above the minerals soil consisting of well-decomposed organic matter.

healthy riparian ecosystem: Maintains a dynamic equilibrium between the streamflow forces acting to produce change and the resistance of vegetative, geomorphic, and structural features.

heat tolerance: The ability of plant tissue to withstand high temperatures.

heavy fuels: Snags, logs, large branches, peat of larger diameter (>3.1 inches or 8 cm) that ignite and burn more slowly than fine fuels.

heterotrophs: Organisms that are able to derive C and energy for growth and cell synthesis by utilizing organic compounds.

high fire severity: High soil heating, or deep ground char occurs, where the duff is completely consumed and the top of the mineral soil is visibly reddish or orange on severely burned sites. Less than 20 percent of the trees exhibit no visible damage, with the remainder fire-damaged, largely by root-kill; less than 40 percent of the fire-damaged trees survive.

high severity burn: All of the organic material is removed from the soil surface and organic material below the surface is consumed or charred. More than 10 percent of the area has spots that are burned at high severity, more than 80 percent moderately severe or severely burned, and the remainder is burned at a low severity.

human-caused fire: Fire caused directly or indirectly by a person or people.

hydrograph: A graphical relationship of streamflow discharge (ft^3/sec or m^3/sec) (m^3/sec) to time.

hydrologic function: Relates to the ability of a watershed to receive and process precipitation into streamflow without ecosystem deterioration.

ignition: The initiation of self-sustaining pyrolysis and flaming combustion, marks the transition point between the mainly endothermic preignition and exothermic flaming phases.

infiltration: The process of water entering the soil.

infiltration capacity: Maximum rate at which water can enter the soil.

instream flow: The streamflow regime required to satisfy the conjunctive demands being placed on water while the water remains in the stream channel.

interception: The process in which vegetative canopies, litter accumulations, and other decomposed organic matter on the soil surface interrupt the fall of precipitation (rain or snow) to the soil surface. It plays a hydrologic role of protecting the soil surface from the energy of falling raindrops.

interflow: Also called subsurface flow, it is that part of the precipitation input that infiltrates into the soil and then flows to a stream channel in a time short enough to be part of the stormflow.

intermittent stream: A stream that flows periodically, fed by channel interception, overland flow or surface runoff, and subsurface flow.

L layer: Litter layer (O_i layer), the accumulation of dead organic plant matter above the mineral soil consisting of unaltered leaves, needles, branches, and bark.

ladder fuels: Fuels continuous between ground fuels and crown fuels, forming a ladder by which a fire can spread into tree or shrub crowns.

latent heat of vaporization: The amount of heat and energy involved in the change in physical state of water. The latent heat vaporization of water is 560 cal/g, and this same amount is released during condensation.

light severity burn: A fire that leaves the soil covered with partially charred organic material.

live fuels: Living plants grouped by category as woody or herbaceous fuels.

low fire severity: Low soil heating, or light ground char, occurs where litter is scorched, charred, or consumed. The duff is left largely intact, although it can be charred on the surface. Woody debris accumulations are partially consumed or charred. Mineral soil is not changed. At least 50 percent of the trees exhibit no visible damage, with the remainder fire-damaged by scorched crowns, shoot-kill (top kill but sprouting), or root-kill (top kill and no sprouting); over 80 percent of the fire-damaged trees survive.

low severity burn: Less than 2 percent of the area is severely burned, less than 15 percent moderately burned, and the remainder of the area burned at a low severity or unburned.

macro-nutrients: Nutrients that are needed in the largest concentrations required for plant growth such as phosphorus, nitrogen, su;fur, iron, calcium, potassium, and magnesium.

mass transport of heat: Occurs during fires by air-borne spotting and downslope rolling.

methane: The third most abundant greenhouse gas contributing to global warming.

micro-nutrients: Nutrients needed in trace amounts for plant growth such as zinc, manganese, cobalt, molybdenum, and nickel.

mineralization: This is conversion of an element from an organic form to an inorganic state as the result of microbial activity. Mineralization includes the transformation of organic N compounds (such as proteins and amino acids) into ammonia (ammonification) and, subsequently, into nitrite and nitrate (nitrification); and the conversion of organic C into carbon dioxide.

moderate fire severity: Moderate soil heating, or moderate ground char, occurs where the litter on forest sites is consumed and the duff is deeply charred or consumed, but the underlying mineral soil surface is not visibly altered. Some 20 to 50 percent of the trees exhibit no visible damage, with the remainder fire-damaged; 40 to 80 percent of the fire-damaged trees survive.

moderate severity burn: Less than 10 percent of the area is severely burned, but over 15 percent is burned moderately severe, and the remainder is burned at low severity or unburned.

mycorrhizae: The plant root zone contains these fungi that enhance nutrient uptake by plants and contribute directly to the productivity of the terrestrial ecosystem.

natural fire: Fire of natural origin—lightning, spontaneous combustion, or volcanic activity.

natural fuels: Result from natural processes and, therefore, are not generated by management practices.

net precipitation: Precipitation that reaches the soil surface, moves into the soil, forms puddles of water on the soil surface, or flows over the surface of the soil.

nitrification: Transformation of organic N compounds into nitrite (NO_2) and nitrate (NO_3). Nitrification is a process involved in the mineralization of N that is affected by fire.

Nitrobacter: Responsible for the oxidation of nitrite to nitrate is almost exclusively in natural systems.

Nitrosolobus: Genera of autotrophic bacteria are able to oxidize ammonium nitrogen (NH_4-N) to nitrite.

Nitrosomonas: Genera of autotrophic bacteria are able to oxidize NH_4-N to nitrite.

Nitrospira: Genera of autotrophic bacteria are able to oxidize NH_4-N to nitrite.

O horizon: Organic matter overlying mineral soil made up of fresh litter (O_i horizon), partially decomposed litter (O_e horizon), and completely decomposed litter (O_a horizon).

old-field successions: Successional sequences on abandoned agricultural fields.

on-site fire effects: Include impacts on vegetation, soil, and nutrient cycling.

overland flow: Also called surface runoff, this is waterflow that has not infiltrated into the mineral soil and flows off the surface to a stream channel.

packing ratio: The proportion of a fuel bed volume actually occupied by fuel. The tighter fuels are packed together, the higher the packing ratio, and the lower the combustion efficiency because air supply to fire is restricted by the fuel density.

perennial stream: Stream that flows continuously throughout the year. It is fed by groundwater or baseflow, that sustains flow between precipitation events.

physical properties: Properties of fuels that affect the manner in which a fire burns and, ultimately, the generation of energy and production of air pollutants by the fire. Physical properties of interest to a fire manager generally include the quantity (fuel loading), size and shape, compactness, and arrangement.

phytobiomass: Aboveground vegetative material available for burning in natural ecosystems—often considered to be the total fuel available to burning.

polymer: Organic compound formed of repeating structural units such as simple sugars.

poor condition watershed: Precipitation flows on the soil surface and excessive erosion occurs during precipitation events. Streamflow response to precipitation is rapid and there is little or no baseflow between storms.

potential fuel: Material that might burn during an intense fire and is generally less than the total fuel.

preignition: The first phase of a fire involving fuel heating that results in dehydration and pyrolysis.

prescribed burning: Also called controlled burning, is the controlled application of fire to fuel buildups, in either their natural or modified state, in specified environmental conditions that allow the fire to be confined to a predetermined area and, at the same time, to produce the fireline intensity and rate of spread required to attain the planned management objectives.

prescribed fire: Fire burning with prescription, resulting from planned ignition that meets management objectives.

prescribed natural fire: Fire of natural origin that is allowed to burn as long as it is accomplishing one or more management objectives.

prescription: A statement specifying the management objectives to be attained, and the air temperature, humidity, wind direction and wind speed, fuel moisture conditions, and soil moisture conditions in which a fire will be allowed to burn.

primary succession: Progression to a climax plant community that is initiated on lava flow, sand dunes, alluvial deposits, and other newly exposed sites.

protective cover: Cover important to wildlife when escape from predators becomes necessary.

pyrolysis: A chemical decomposition process brought about by heating by which the fuel is converted to gases. An endothermic reaction set off by thermal radiation or convection from an advancing fire front that drives water from the surface of a fuel, elevates fuel temperatures, and then decomposes long-chain organic molecules in plant cells into shorter ones.

radiant intensity: Rate of thermal radiation emission that is intercepted at (or near) the ground surface, or at some specified distance ahead of the flame front.

radiation: Transfer of heat from one body to another not in contact with it by electromagnetic wave motion. All bodies at temperatures above $0^{\circ}K$ produce radiant energy.

reaction intensity: Total heat release per unit area of fuelbed divided by the burning time. It is the time-averaged rate of heat release of the active fire front that is calculated in the field by estimating the amounts of fuels burned per second and assuming heat yields for the fuels.

rhizosphere: The root environment, provides a favorable environment for soil microorganisms and is an important site of microbial activity.

riparian ecosystems: Areas that are situated in the interfaces between terrestrial and aquatic ecosystems that can be found along open bodies of water, such as the banks of rivers and ephemeral, intermittent, and perennial streams, and around lakes, ponds, springs, bogs, and meadows.

running crown fire: Fire in the crowns that races ahead of the fire on the surface in what is called a running crown fire.

secondary succession: Progression to a climax plant community that follows disturbance such as fire.

sediment: Eroded soil that is transported from watershed surfaces to stream channels by overland flow, and then through stream systems in streamflow and therefore, is the product of erosion.

sedimentation: The process of deposition of sediment in stream channels or downstream reservoirs.

sediment yield: The amount of sediment outflow from a watershed in a stream.

slash: Concentrations of downed fuels resulting from either natural events (wind, fire, snow breakage, and so forth) or management activities (logging, road construction, and so forth).

slash disposal: Treatment of slash (by burning or otherwise) to reduce the fire hazard or meet other purposes.

smoldering: The third phase of a fire also involving combustion in which pyrolysis beings to diminish.

snags: Cavities resulting from wood decay, or holes created by other species in deteriorating or dead trees used by numerous species of mammals, reptiles, amphibians, and invertebrates.

soil erosion: The dislodgement and transport of soil particles and small aggregates of soil by the actions of water and wind.

soil mass movement: The process where cohesive masses of soil are displaced by downslope movement driven by the force of gravity of soil, rock, and debris masses. This movement might be rapid (landslides) or relatively slow (creep).

soil productivity: Reflects the capabilities of a watershed for supporting sustained plant growth and plant communities, or the natural sequences of plant communities.

soil wood: Consists of buried or partially buried woody debris.

spotting: Involves the physical removal of burning material by thermal updrafts from flaming fuels and their subsequent deposition in unignited fuels some meters or kilometers away. This is a predominate mechanism of fire spread in fast moving uncontrollable fires.

stand-replacing fire: Fire that kills all or most of the overstory trees in a forest and, in doing so, initiates secondary succession or regrowth.

stormflow: The sum of channel interception, surface flow, and subsurface flow during a precipitation or snowmelt event.

subsurface flow: Also called interflow, is that part of the precipitation input that infiltrates into the soil and then flows to a stream channel in a time short enough to be part of the stormflow.

surface erosion: Caused by the actions of falling raindrops, thin films of water flowing on the soil surface, concentrated overland flow, or the erosive power of wind.

surface fire: Fire that consumes only surface fuels such as litter, low-growing plants, and dead herbaceous plants accumulated on the surface. Surface fire can ignite snags (dead standing trees), can consume shrubs and tree seedlings, and can "torch out" an occasional densely crowned mature tree. It remains a surface fire so long as its rate of spread depends on surface fuels.

surface fuels: Fuels below the aerial fuels (< 6.5 feet or 2m) and above the ground fuels.

surface runoff: Also called overland flow, this is waterflow that has not infiltrated into the mineral soil and flows off the surface to a stream channel.

USDA Forest Service Gen. Tech. Rep. RMRS-GTR-42-vol. 4. 2005

241

thermal conductivity: Expresses the quantity of heat transferred per unit length per unit time per degree of temperature gradient and is expressed in SI units as W/m/°K.

thermal cover: Cover critical to wildlife for the maintenance of body heat.

thermal energy: Energy that results from changes in the molecular activity or structure of a substance.

timelag: Measure of the rate at which a fuel approaches its equilibrium moisture content after experiencing environmental changes.

tolerance: Applies to light, soil nutrients, or other physiological requirements a species can tolerate.

total fire intensity: Rate of heat output of the fire as a whole, which is a function of the rate of area burned, fuel loading, and estimated heat yield.

total fuel: The amount of biomass that could potentially burn.

turbidity: A measure of suspended fine mineral or organic matter that reduces sunlight penetration of water, and influences photosynthesis rates and water quality.

unfavorable effects: Effects that make attainment of fire management more difficult.

urban-rural interface: The line, area, or zone where structures and other human developments meet, or intermingle with, undeveloped wildland areas.

vaporization: Is the process of adding heat to water until it changes phase from a liquid to a gas.

vegetation-replacing fire: Kills all or most of the living plants (including trees, shrubs, and herbaceous plants) on a site and, as a result, initiates secondary succession or regrowth.

vegetative resources: Plant communities of value to people and when demanded, are available through the implementation of prescribed management practices.

watershed condition: A subjective term to indicate the health (status) of a watershed in terms of its hydrologic function and soil productivity.

water quality: Refers to the physical, chemical, and biological characteristics of water in reference to a particular use.

wetlands: Areas that are saturated by surface water, groundwater, or combinations of both at a frequency and duration sufficient to support a prevalence of vegetation adapted to saturated soil conditions.

wettability: Property that can influence infiltration of the soil.

wildfire: Fire that is not meeting management objectives and, therefore, requires a suppression response.

Index

A

A-horizon 23, 55
actinomycetes 76
Africa 58, 68
Agropyron spicatum 89
Alaska 7, 111, 150, 156, 158, 162
Alnus 79
Alsea Basin 136
ammonification 79, 88
Apache-Sitgreaves National Forest
 7, 108, 115, 131, 137, 179
Appalachian Mountains 62
 Southern 62, 71
aquatic biota 135
Arctostaphylos spp. 85
Arizona
 2, 7, 56, 60, 68, 89, 109, 110, 115, 116, 127, 131,
136, 137, 138, 141, 166, 168, 179
 Apache-Sitgreaves National Forest 7
 Coronado National Forest
  Rattlesnake Fire 45, 46, 47
 Flagstaff 59
 Heber 115
 Mogollon Rim 59
 White Mountain Apache Nation 43
 White Springs Fire 43
armored crossings 195
Artemisia spp. 191
ash-bed effect 69
Aspen Fire 138
Australia 58, 62, 68, 85, 86
Australian eucalypt 68

B

B-horizon 23
bacteria 74
BAER 2, 42, 117, 131, 147, 180, 182
baseflow 102
Benton Lakes National Wildlife Refuge 165
Best Management Practices 50, 117, 132, 133
bicarbonate 128
biogeochemical cycles 165
biological crusts 76, 85
bitterbrush *See Purshia tridentata*
Bitterroot Valley 115
Black River 166
blackbrush *See Coleogyne ramosissima*
Bobcat Fire 189
Borrego Fire 138
Bradyrhizobium spp. 76
Brazil 59
Bridge Creek Fire 190
broadcast seeding 185, 189
brown trout 138
buffer capacity 62
buffer power 53

buffer strips 168
bulk density 31
bull trout 139
Burned Area Emergency Rehabilitation 2, 42, 117, 180.
  See BAER
 treatments 182
Burned Area Report 183
BURNUP Model 8

C

C-horizon 23
California
 35, 37, 63, 99, 102, 109, 116, 126, 127, 128, 129, 142,
 189, 190, 193
 San Dimas Experimental Forest 102
California chaparral 77
Canada 150, 164
Canadian Northwest Territories 9
carbon 53, 55, 162, 164
Carex spp. 163
Carex stricta 154, 162, 163
Cascade Range 109
Cascades Mountains 48
cation exchange capacity 53, 59, 62
cations 53, 62, 162
Ceanothus velutinus 89
Cercocarpus spp. 79
Cerro Grande Fire 186, 189
channel armoring 192
channel clearing 192
channel interception 102
channel stability 45
channel treatments 191, 193
 channel armoring 192
 channel clearing 192
 check dams 191
 debris basins 192
 log check dams 193
 log grade stabilizers 192
 rock cage dams 194
 rock fence check dams 191
 straw bale check dams 193
 straw bales 191
chaparral 37, 44, 88
 Arizona 89
 California 77
check dams 191
chemical changes 57
chemical characteristics 53
chemical constituents, dissolved 124
chloride 128
Clean Water Act 132
coarse woody debris 55, 56, 80, 89, 168
Coconino National Forest 107
Code of Federal Regulations 132
Coleogyne ramosissima 85
Collembola 78

Colorado Front Range   189, 190
combustion   24
   flaming   25, 157, 162
   general   24
   glowing   25
   smoldering   25, 157, 162
Composite Burn Index (CBI)   15
conceptual model   15
condensation   25, 31
conduction   25, 31
contour structures   190
contour trenches   187, 190
contour-felled logs   186
convection   25, 31
Coon Creek Fire   115
Coronado National Forest   45, 46
Costa Rica   89
crown fires   37
cryptogamic crusts   77, 85
culvert improvements   194

D

debris avalanches   43
debris basins   192
debris dam failures   115
decomposition   65, 79, 88
   rates   57
DELTA-Q model   176
denitrification   80
depth of burn   9, 14, 159, 161
Deschutes National Forest   190
Disturbed WEPP   174
Divide Fire   136
Douglas-fir   68, 89
dry ravel   42
Dude Fire   115, 136, 138, 141
duff   22, 32, 33
duff burnout model   34

E

earthworms   78
ecosystem productivity   64
ectomycorrhizae   76
endomycorrhizae   76
Entiat Fires   124
erosion   41, 183
   dry ravel   42
   gully   42
   rates   174
   rill   42
   sheet   42
eucalypt forests   69
eucalyptus   35
evapotranspiration   99
evergreen bays   159
excess water   102

F

F-layer   22, 55, 87
field capacity   100

Finland   84
fire   9
   frequency   58
   ground   157
   intensity   7, 14, 24
   retardants   129
   severity   5, 7, 13, 24, 58, 67, 161
      classification   13
      high severity burn   15
      low severity burn   15
      matrix   13, 14
      moderate severity burn   15
   smoldering   9
   surface   157
fire effects
   amphibians   140
   aquatics   135, 137
   birds   140
   first order   172
   flash floods   116
   invertebrates   140
   mammals   140
   moisture   80
   oxygen   80
   reptiles   140
   riparian   167
   species considerations   137, 139
      Bull trout   139
      Gila trout   139
      redband trout   139
fire effects models   171
fire regimes   4, 107
   mixed   4, 6
   nonfire   5
   stand replacement   4
   understory   4, 5
fire retardants   129
   lethal levels
      aquatic organisms   130
   lethal levels, aquatic   130
fireline intensity   8
fires
   Apache-Sitgreaves National Forest   115
   Aspen   138
   Bobcat   189
   Borrego   138
   Bridge Creek   190
   Cerro Grande   186, 189
   Coon Creek   115
   Divide   136
   Dude   115, 136, 138, 141
   Entiat   124
   Hayman   181, 182, 186
   Oakland Hills   193
   Picture   138
   Ponil Complex   136
   Rattlesnake   45, 46, 47
   Rodeo-Chediski   108, 115, 116, 131, 132, 137, 179, 186
   South Fork Trinity River   190
   South Fork Trinity River Fires   193
   Tillamook Burn   116

Trinity River   189
White Springs   43
Yellowstone   135, 139, 141, 142
first order fire effects   172
First Order Fire Effects Model   8, 171. *See* FOFEM
    applications   172
    description   172
Flagstaff   59
flaming   25
flaming combustion   157
flash floods   116
Florida   147, 154, 166
Florida Everglades   166
FOFEM   8, 171
forest floor   32, 56, 58
FOREST model   176
France   110
Frankia spp. *See* actinomycetes
frequency of burning   68
fuel compactness   157
Fungi   74

G

Gelisols   152, 157
Georgia   150
geotextiles   188
Gila chub   138
Gila intermedia   138
Gila River   166
Gila trout   139
glowing combustion   25
Gordonia lasianthus   159
grass fires   37
grasslands   55
Great Plains   68
Green Swamp   159
ground fires   157

H

H-layer   22, 55
Hayman Fire   181, 182, 186
Healthy Forest Restoration Act   5
heat transfer   25, 26, 31, 173
    condensation   25
    conduction   25
    convection   25
    models   32
    pathways   32
        duff   32
        forest floor   32
        litter   32
    radiation   25
    vaporization   25
heavy metals   128
Heber, Arizona   115
hillslope treatments   185, 188, 193
    broadcast seeding   185, 189
    contour structures   190
    contour trenches   187, 190
    contour-felled logs   186
    geotextiles   188

mulch   186, 189
needle cast   188
ripping   187
scarification   187
silt fences   188
slash spreading   188
straw wattles   187
temporary fencing   188
Histisols   152, 157
HR 1904. *See* Healthy Forest Restoration Act
humus   22
hydric soil
    classification   151
hydric soils
    definitions   153
hydrodynamics   153
hydrologic cycle   95, 118
    baseflow   102
    evapotranspiration   99
    infiltration   98
        hydrophobic soils   99
        interflow   98
        overland flow   98
        surface runoff   98
        wettability   99
    interception   97
        interflow   98
        overland flow   98
        snowpack   97
        stemflow   97
        streamflow   104
        subsurface flow   102
        surface runoff   101
        throughfall   97
    precipitation   95
hydroperiod   153
hydrophobic soils   99
    Quercus turbinella   99

I

Idaho   89, 138, 168
    Salmon River   49
ignition   25
Ilex glabra   159
Indonesia   32
infiltration   98
Inland Northwest   56
insects   78
interception   97
interflow   98
invertebrates   140
    response to fire   141

J

jack pine 85. *See Pinus banksiana*
Juncus   165
Juniperus spp.   30

K

Kalmia spp.   66

L

L-layer   22, 55, 87
Lake Roosevelt   131
large woody debris   22
litter   22, 32
log check dams   193
log grade stabilizers   192
longleaf pine. *See Pinus palustris*
Low-severity prescribed fire   83
Lyonia lucida   159

M

macrofauna   78, 86
  earthworms   78
  insects   78
macroinvertebrates   86, 141
macropore   31
manzanita   85. *See Arctostaphylos* spp.
mass failures   43
  debris avalanches   43
  slope creep   43
  slump-earthflows   43
mesofauna   78, 86
  Collembola   78
  mites   78
  Nematoda   78
  Rotifera   78
Michigan   158, 162, 163
  Seney National Wildlife Refuge   33
microorganisms   74, 80, 89
  low-severity prescribed fire   83
  moisture and oxygen   80
  slash burning   83
  substrate availability   81
  temperature   80
  wildfire   82
micropores   31
mineralization   79, 88
Minnesota   150, 156
Mississippi Gulf Coast   165
mites   78
model
  BURNUP   8
  conceptual   15
  DELTA-Q   176
  duff burnout   34
  First Order Fire Effects Model   8, 171
  FOREST   176
  RUSLE   173
  Universal Soil Loss Equation   183
  USLE   173
  WATSED   173
  WEPP   173, 183
Mogollon Rim   59, 131
moisture   80
Montana   85, 89, 115, 138, 165, 168
mountain mahogany. *See Cercocarpus* spp.
mulch   186, 189
mycorrhizal fungi   75, 84

N

N-fixation   89
Nantahala National Forest   62
National Climatic Data Center   200
National Fire Effects Workshop, 1978   3
National Forest Management Act   132
National Forests
  Apache-Sitgreaves National For-
    est   7, 108, 131, 137, 179
  Coconino National Forest   107
  Coronado National Forest   45, 46
  Deschutes National Forest   190
  Nantahala National Forest   62
  Pike-San Isabel National Forest   182
  San Bernardino National Forest   129
  Shasta-Trinity National Forest   189, 190, 193
  Tonto National Forest   109, 115, 138
National Interagency Fire Center   200
National Parks
  Shenandoah National Park   68
  Yellowstone National Park   128, 142
Nature Conservancy   159
needle cast   188
Nematoda   78
New Mexico   88, 116, 136, 138, 166, 189
nitrification   79, 88
Nitrobacter   79, 88
nitrogen   53, 63, 66, 119, 125, 131, 162, 163, 164
nitrogen losses   63, 64
nitrogen-fixation   79
  symbiotic   79
nitrogen-fixing bacteria   75
Nitrosomonas   79, 88
North Carolina   7, 154, 158, 159, 162, 163
nutrient
  availability   66
  cycling   65
  losses   65
nutrient cycling   78, 164
nutrients   24, 128

O

O-horizon   23
Oakland Hills Fire   193
Okefenokee Swamp   150
Oregon   49, 116, 136, 189, 190
  Oregon Cascades   48, 49
Oregon Coast Range   43
organic matter   22, 24, 32, 55, 79, 151
  accumulation   56
    forest floor   56
organic soils   151, 152
outsloping   195
overland flow   98, 101
oxygen   80
oxygen, dissolved   123

P

Pacific Northwest   64, 80, 123, 139
packing ratio   157

pathways
  channel interception 102
  subsurface flow 102
peakflow mechanisms
  debris dam failures 115
peakflows 111, 114
peatland soils 32
Persia borbonia 159
pH 53, 62, 124
Philippines 191
phosphorus 53, 64, 68, 119, 127, 131, 163, 164
Picea 7
Picture Fire 138
Pike-San Isabel National Forest 182
Pinus banksiana 9, 84
Pinus edulis 85
Pinus palustris 7
Pinus ponderosa 2, 7, 70, 88, 89, 107, 109, 110
Pinus rigida 70
Pinus serotina 159
Pinus strobus 85
Pinus sylvestris 86
pinyon pine. *See also Pinus edulis*
plant roots 87
pocosin 153, 154, 157, 159, 164
pond pine 159
ponderosa pine. *See Pinus ponderosa*
ponderosa pine forests 88
ponderosa pine stand 55
Ponil Complex 136
porosity 31
precipitation 95
preignition 25
prescribed fire
  low-severity 83
  water yield 111, 112, 113
primary standards 105
Purshia tridentata 79

Q

Quercus spp. 30
Quercus turbinella 99

R

radiation 25, 31
Rainbow Series 1, 147
raindrop splash 39
Rattlesnake Fire 45, 46, 47
redband trout 139
research needs 211
Rhizobium 76, 79
rhizosphere 75
rill formation 39
riparian
  classification 167
  definition 167
riparian area 166, 167
riparian ecosystems 149, 166
  fire effects 167
  hydrology 167
  management considerations 168

buffer strips 168
ripping 187
road treatments 193, 194
  armored crossings 195
  culvert improvements 194
  outsloping 195
  rolling dips 194
rock cage dams 194
rock fence check dams 191
Rodeo-Chediski Fire 108, 115, 116, 131, 132, 137, 179, 186
rolling dips 194
roots 87
Rotifera 78
round worms. *See Nematoda*
RUSLE 173, 174

S

Salmo trutta 138
Salmon River 49
Salt River 131
San Bernardino National Forest 129
San Dimas Experimental Forest 102
San Dimas Technology and Development Program 196
San Dimas Technology Development Center 186
savannas 55, 58, 86
scarification 187
secondary standards 105
sediment 121
  losses 47–50
  size classes 41
  suspended 121
  yield 44, 46, 123, 183
sedimentation 120
seed banks 87
Seney National Wildlife Refuge 33
Shasta-Trinity National Forest 189, 190, 193
Shenandoah National Park 68
Shenandoah Valley 58
silt fences 188
site productivity 64
slash burning 83
slash spreading 188
slope creep 43
slump-earthflows 43
smoldering 25, 157, 158, 162
smoldering combustion 157
smoldering fires 37
snow accumulation 101
snowbrush. *See Ceanothus velutinus*
snowmelt 101
snowpack 97
soil 2, 21, 29
  aggregation 31
  biology 73
  buffer capacity 62
  chemistry 53
  erosion 120, 183
  fauna 73
  flora 73
  heating 25, 160
    nitrogen losses 63, 64

heating pathways 33
  duff 33
horizons 22
  A-horizon 23, 55
  B-horizon 23
  C-horizon 23
  F-layer 22, 55, 87
  H-layer 22, 55
  L-layer 22, 55, 87
  O-horizon 23
invertebrates 86
microorganisms 24
  bacteria 74
  fungi 74
moisture 81
nutrients 162
organic 151
organic matter 59
pH 62
profile 22, 26
properties 22
structure 30
temperature profiles 34
temperatures 161
texture 29
wetland
  nutrient cycling 164
soil erosion models 171, 173
  RUSLE 173, 174
  WATSED 173, 174
  WEPP 173, 174
soil water storage 100, 108
  field capacity 100
South Africa 68, 110, 111
South Carolina 59
South Fork Trinity River Fires 190, 193
Southern Appalachian Mountains 58, 62, 71
Spain 57, 83, 189
Spartina spp. 165
springtails. *See Collembola*
spruce. *See also Picea*
stemflow 97
Stermer Ridge watersheds 131
Stipa comata 89
straw bale check dams 193
straw bales 191
straw wattles 187
streamflow discharge 104
streamflow regimes 107, 109
substrate availability 81
subsurface flow 102
sulfur 53, 64, 128
surface fires 37
surface runoff 98, 101
Sweden 86
symbiotic N-fixation 79

T

Taiwan 194
tallgrass prairie 87
temperature 80
temperature thresholds 26, 30, 54, 82

temporary fencing 188
Texas 109, 111, 121
Three Bar chaparral watersheds 109
throughfall 97
Tillamook Burn 116
"tin roof" effect 37
Tonto National Forest 109, 115, 138
total dissolved solids (TDS) 128
Trichoderma 82
Trinity River Fire 189
trout
  bull 139
  Gila 139
  redband 139
tundra 55
turbidity 121

U

U.S. Environmental Protection Agency 120, 123
U.S. Fish and Wildlife Service 151
Universal Soil Loss Equation 183. *See also* USLE
USLE 173
USLE, revised 173
Utah 191

V

vaporization 25, 31
Variable Source Area Concept 103
Verde River 166
volatilization 63

W

Washington 102, 109, 124, 127
water
  chemical characteristics 124, 126
    bicarbonate 128
    chloride 128
    nitrogen 125
    phosphorus 127
    sulfur 128
  physical characteristics 121
    sediment 121
    temperature 123
    turbidity 121, 122
Water Erosion Prediction Project 183. *See also* WEPP
water quality 105, 119
  nitrogen 119
  pH 124, 125
  phosphorus 119
  primary standards 105
  secondary standards 105
  standards 105, 119
    U.S. Environmental Protection
    Agency 120, 123, 124, 128
water repellency 37
  raindrop splash 39
  rill formation 39
  "tin roof" effect 37
water temperature 123
watershed

248

USDA Forest Service Gen. Tech. Rep. RMRS-GTR-42-vol. 4. 2005

condition   103
  rehabilitation   179
watersheds
  chemical constituents, dissolved   124
  fire effects
    flash floods   116
  sediment yield   123, 183
  Stermer Ridge   131
  Three Bar   109
  Whitespar   109
WATSED   173, 174
WEPP   173, 174, 175
wetland hydrology   **153**
wetland soil   152
  nutrients   164
    carbon   164
    nitrogen   164
    phosphorous   164
wetlands   149

classification   151, 152
  ground fire   157
  hydrology   153
  management considerations   166
  soil
    nutrient cycling   164
  surface fire   157
wettability   99
White Mountain Apache Nation   43, 131
white pine. *See Pinus strobus*
White Springs Fire   43
Whitespar Watersheds   109
Wildfire   82
Wyoming   138

Y

Yellowstone Fires   135, 139, 141, 142
Yellowstone National Park   128, 142

USDA Forest Service Gen. Tech. Rep. RMRS-GTR-42-vol. 4. 2005

249

www.ingramcontent.com/pod-product-compliance
Lightning Source LLC
Chambersburg PA
CBHW081059220326
41598CB00038B/7160